Lecture Notes in Mathematics

Edited by A. Dold and B. Eckmann

1228

Multigrid Methods II

Proceedings of the 2nd European Conference on Multigrid Methods held at Cologne, October 1–4, 1985

Edited by W. Hackbusch and U. Trottenberg

Springer-Verlag
Berlin Heidelberg New York London Paris Tokyo

Editors

Wolfgang Hackbusch
Institut für Informatik und Praktische Mathematik
Christian-Albrechts-Universität Kiel
Olshausenstr. 40–60, 2300 Kiel 1, FRG

Ulrich Trottenberg
Mathematisches Institut
Universität zu Köln
Weyertal 86–90, 5000 Köln 41, FRG

and

Gesellschaft für Mathematik und Datenverarbeitung
Institut für Methodische Grundlagen
Schloß Birlinghoven, 5205 St. Augustin 1, FRG

Mathematics Subject Classification (1980): 35JXX, 35KXX, 35LXX, 65-02, 65-06, 65F10, 65H10, 65N20, 65N30, 65NXX, 68BXX, 68C05, 68C25, 76XX, 8208

ISBN 3-540-17198-3 Springer-Verlag Berlin Heidelberg New York
ISBN 0-387-17198-3 Springer-Verlag New York Berlin Heidelberg

Printed in Germany

Printing and binding: Druckhaus Beltz, Hemsbach/Bergstr.
2146/3140-543210

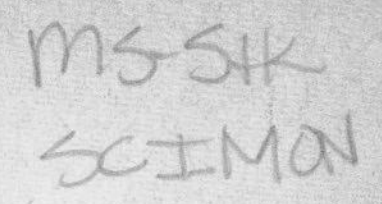

Preface:

These proceedings contain a selection of papers presented at the

2nd European Conference on Multigrid Methods

which was held at the University of Cologne on October 1–4, 1985. It was organized by:

- Gesellschaft für Mathematik und Datenverarbeitung (GMD),
  Institut für Methodische Grundlagen,
  St Augustin–Birlinghoven,
- GAMM–Fachausschuß "Effiziente numerische Verfahren für partielle Differentialgleichungen",

in cooperation with

- Mathematisches Institut der Universität zu Köln.

Whereas the First European Conference on Multigrid Methods in 1981 could be regarded as a first breakthrough of the multigrid idea to the field of application problems, this second conference reflected the systematic progress in the last four years in both theory and practice as well as the strongly grown acceptance of multigrid methods amoung users.

The papers of this volume are assumed to be of interest for a large number of scientists in academic and industrial research. They give a survey on the current multigrid development. (More specialized papers presented at the conference have been published in the GMD–Study no. 110; they are listed at the end of this volume.)

Most of the papers in this volume deal with problems arising from fluid– and aerodynamics. Special multigrid solvers are discussed for potential and Euler flows as well as for the solution of the full Navier–Stokes equations in highly complicated geometries. These problems, which are perhaps the most challenging ones for numerical mathematics of all, were also at the centre of interest at the conference.

New fields of applications considered in these proceedings are: statistical physics and magnetohydrodynamic problems. Furthermore these proceedings present papers, in which various "classical" numerical problems are treated, which are interesting in theoretical regard but which are also directly important for practical applications: local mesh refinements, error estimates, continuation techniques for nonlinear steady state problems, multigrid and conjugate gradients, indefinite problems and the treatment of singularities.

The growing importance of parallel computer architectures and – closely related to these – the need for highly efficient algorithms specially designed for this new type of computers have also introduced new aspects into multigrid development. Two of the papers in this book treat questions from this new field of multigrid research . They consider the performance of multigrid methods on special types of MIMD–architectures (bus–coupled systems and hypercubes).

We would like to express our gratitude to the Executive Board of the GMD and to the University of Cologne for all their immaterial and material support of the conference.

The practical organization was excellently carried out by Johannes Linden, Institut für Methodische Grundlagen of the GMD. He was supported by Wolfgang Joppich, several other coworkers, the secretaries and by the students of this institute. Furthermore, the Abteilung für Informationswesen of the GMD and the Institute of Mathematics of the University of Cologne provided substantial assistance to the organization of the conference. We would like to thank all persons involved.

Finally we thank all of the more than 130 participants from 12 countries and especially the lecturers for their contributions to the success of the conference.

St. Augustin, March 1986

Wolfgang Hackbusch
Ulrich Trottenberg

Contents

A MULTIGRID SOLVER FOR A STABILIZED
FINITE ELEMENT DISCRETIZATION
OF THE STOKES PROBLEM

E.M. ABDALASS, J.F. MAITRE, F. MUSY
Département de Mathématiques-Informatique-Systèmes
Ecole Centrale de Lyon B.P. 163
F 69131 ECULLY France

1. INTRODUCTION

We consider the STOKES equations in a polygonal domain $\Omega \subset \mathbb{R}^2$ with DIRICHLET boundary conditions on $\Gamma = \partial\Omega$

$$\begin{cases} -\Delta\vec{u} + \vec{\nabla}p = \vec{f} & \\ \vec{\nabla} \cdot \vec{u} = 0 & \Omega \\ \vec{u} = \vec{u}_\Gamma & \Gamma \end{cases} \tag{1}$$

assuming that the compatibility condition $\int_\Omega \vec{u}_\Gamma \cdot \vec{n}\, d\Gamma = 0$ is satisfied.

We are interested in solving finite element discretized approximations to the following variational formulation of (1)

$$\begin{cases} \text{Find } (u,p) \in X \times M \text{ such that} \\ a(u,v) + b(v,p) = (f,v), \quad \forall\, v \in X \\ b(u,q) = 0, \quad \forall\, q \in M \end{cases} \tag{2}$$

with $X = (H_0^1(\Omega))^2$, $M = L^2(\Omega)$, $f \in (L^2(\Omega))^2$

$$a(u,v) = \int_\Omega \nabla u \cdot \nabla v \;, \; b(v,q) = -\int_\Omega q \operatorname{div} v$$
$$(f,v) = \int_\Omega f \cdot v \,.$$

Let $(\mathcal{T}_h)_h$ be a family of triangular meshes of $\bar{\Omega}$ satisfying the classical conditions, and let $X_h \subset (H_0^1(\Omega))^2$ and $M_h \subset L^2(\Omega)$ be the corresponding spaces of piecewise linear functions on $\mathcal{T}_h$.

For solving problem (2), the pair (X_h, M_h) provides an instable f.e.m. scheme. BREZZI and PITKÄRANTA [4] propose two ways of stabilization at little cost :

- by adding "bubble" functions (see also [1], [5])
- by modifying the discrete equations.

We have tried the first way, and more recently the second to which we restrict ourselves in this paper. We give a multigrid solver for the STOKES equations which is very efficient, at least for the regular mesh of our first experiments.

2. STABILIZATION BY MODIFYING THE DISCRETE EQUATIONS

Let c_h be the bilinear form on $M_h \times M_h$ defined by

$$c_h(p_h, q_h) = - \sum_{T \in \mathcal{T}_h} h_T^2 \int_T \nabla p_h \cdot \nabla q_h \tag{3}$$

where h_T denotes the diameter of the triangle T.

Each solution to the problem

Find $(u_h, p_h) \in X_h \times M_h$ such that

$$\begin{aligned} a(u_h, v_h) + b(v_h, p_h) &= (f, v_h), \quad \forall v_h \in X_h \\ b(u_h, q_h) + c_h(p_h, q_h) &= 0, \quad \forall q_h \in M_h \end{aligned} \tag{4}$$

satisfies the error estimate under classical hypotheses (see [4])

$$\|u - u_h\|_{H^1(\Omega)} + \|p - p_h\|_{L^2(\Omega)/\mathbb{R}} \leq C.h\,(|u|_2 + |p|_1) \tag{5}$$

where (u,p) is a solution to problem (2) and $|.|_k$ denotes the classical SOBOLEV semi-norm.

Solving problem (4) requires the solution of a linear system of the following type

$$\begin{pmatrix} A_h & B_h^t \\ B_h & C_h \end{pmatrix} \begin{pmatrix} U \\ P \end{pmatrix} = \begin{pmatrix} b_h^1 \\ b_h^2 \end{pmatrix} \tag{6}$$

where the unknown vectors U,P and matrices A_h, B_h, C_h have the respective dimensions k_h, m_h, $k_h \times k_h$, $m_h \times k_h$, $m_h \times m_h$, with $k_h = \dim(X_h)$, $m_h = \dim(M_h)$.

For systems of type (6), PITKÄRANTA-SAARINEN [9] have introduced and studied a multigrid method. For that they have used smoothers constructed on the "squared system", as HACKBUSCH [6] and VERFÜRTH [11] have done in the case $C_h = 0$. Here we propose a smoother constructed on the system (6) itself, as we have already done in [8]. This smoother seems to be more adapted to this type of problem.

3. THE MULTIGRID SOLVER

One step of the smoother, applied to the system (6), transforms (U°, P°) into $(\bar{U}, \bar{P})$ as follows :

- $\bar{U}$ results from one sweep of the pointwise GAUSS-SEIDEL iteration on $A_h\, U = b_h^1 - B_h^t\, P^\circ$, starting from $U = U^\circ$

- introducing the blocks corresponding to the partition of the pressure nodes into interior and boundary ones, the matricial equation $- C_h\, P = B_h\, U - b_h^2$ is rewritten

$$- C_{11}\, P_i = B_1\, U + C_{12}\, P_b - b_1^2$$
$$- C_{22}\, P_b = B_2\, U + C_{12}^t\, P_i - b_2^2;$$

starting from $P^\circ = (P_i^\circ\ , P_b^\circ)$, $\bar{P} = (\bar{P}_i\ , \bar{P}_b)$ is obtained from

$$(I + \rho\ (D_1 - E_1))\ \bar{P}_i = (I + \rho\ F_1)\ P_i^\circ + \rho\ h^{-2}(B_1\bar{U} + C_{12}\ P_b^\circ - b_1^2)$$
$$\bar{P}_b = P_b^n, \text{ with, for } r = 0 \text{ to } n - 1 :$$
$$(I + \rho\ (D_2 - E_2))\ P_b^{r+1} = (I + \rho\ F_2)\ P_b^r + \rho\ h^{-2}\ (B_2\ \bar{U} + C_{12}^t\ \bar{P}_i - b_2^2)$$

where, for $j = 1,2$, $- h^{-2}\ C_{jj} = D_j - E_j - F_j$ according to the classical decomposition into the diagonal, lower triangular and upper triangular parts.

Notice that in this smoother two parameters appear, the real $\rho > 0$ and the integer n.

In the sequel, we give some numerical results which show the efficiency of the multigrid method using such a smoother. This multigrid scheme has been contructed as usually, the matrix at each level h being exactly that of system (6). In our examples, the meshes $\mathcal{T}_h$ are nested so that the finite element spaces X_h and M_h satisfie the inclusions

$$X_H \subset X_h\ ,\ M_H \subset M_h \quad \text{for } H > h,$$

and the transfer operators are naturally constructed from the canonical injection (see for example [7]).

4. NUMERICAL RESULTS

All the numerical experiments reported here correspond to the solution of the STOKES equations in the square $\Omega =]0,1[^2$ with solution $\vec{u} = (u_1\ , u_2)$, p given by

$$u_1(x,y) = - u_2(x,y) = \sin 3\ (x + y)$$
$$p\ (x,y) = - 6 \cos 3\ (x + y).$$

The "regular" meshes T_h are constructed by decomposing Ω in the classical way into equal triangles of edgesizes $h,h,h\sqrt{2}$.

The interior nodes are numbered in lexicographic order.

The convergence rate of the multigrid schemes is estimated for i steps by

$$\sigma(i) = \left(\frac{e(i)}{e(o)}\right)^{1/i}$$

where $e^2(i) = ||U^i - u_h||_o^2 + h^2||P^i - p_h||_o^2$, $||\cdot||_o$ denoting the discrete L^2-norm.

All the MG results given here correspond to a four grids method with h = 1/4 , 1/8, 1/16, 1/32 ; (ν_1, ν_2) means that ν_1 presmoothing and ν_2 postsmoothing are performed.

Table 1 list, both for the V- and W-cycle, the values of σ (20) corresponding to some values of the parameter ρ, the parameter n being fixed and equal to 3.

ρ	0.1	0.5	1	10	100
(0,1)-V cycle	0.759	0.624	0.609	0.650	0.656
(1,1)-W cycle	0.562	0.339	0.332	0.334	0.335

Table 1. The convergence rate as a function of ρ

We remark that the convergence rate is not very sensitive to the choice of ρ. In the sequel we always take ρ = 1.

The following results show that n = 2 or 3 relaxation steps on the boundary pressure nodes improve the convergence rate (at a very little cost). It agrees with what is recommended by BRANDT [3].

n	1	2	3	4
(0,1)-V cycle	0.652	0.612	0.608	0.608
(1,1)-W cycle	0.390	0.332	0.332	0.332

Table 2. The convergence rate as a function of n

The first two tables suggest to choose (ρ,n) = (1,3) and we shall take this choice in all the following tests.

Table 3 gives some values of $\sigma(10)$ corresponding to various smoothing strategies (ν_1,ν_2).

(ν_1,ν_2)	(1,0)	(0,1)	(2,0)	(0,2)	(1,1)
V cycle	0.548	0.543	0.422	0.420	0.416
W cycle	0.339	0.343	0.232	0.221	0.224

Table 3. The convergence rate as a function of (ν_1,ν_2)

Finally we give some Full Multigrid results for various values of the finest mesh-size, the coarsest level being fixed at H = 1/4. We denote the FMG solution by $(\tilde{u}_h, \tilde{p}_h)$ and recall that (u_h, p_h), (u,p) are the exact solutions of problems (4), (2) respectively. We indicate the CPU time on a VAX 780 computer. The number of MG calls at each level is 2 for V-cycle and 1 for W-cycle.

	h	$\|\|u-u_h\|\|_o$	$\|\|u-\tilde{u}_h\|\|_o$	$\|\|p-p_h\|\|_o$	$\|\|p-\tilde{p}_h\|\|_o$	CPU time in sec.
(0,1) V-cycle	1/32	$1.77\ 10^{-2}$	$3.82\ 10^{-2}$	$2.42\ 10^{-1}$	$4.75\ 10^{-1}$	14
	1/64	$4.59\ 10^{-3}$	$1.05\ 10^{-2}$	$7.37\ 10^{-2}$	$1.62\ 10^{-1}$	28
	1/128	$1.15\ 10^{-3}$	$2.55\ 10^{-3}$	$2.3\ 10^{-2}$	$4.88\ 10^{-2}$	82
(1,1) W-cycle	1/32	$1.77\ 10^{-2}$	$3.08\ 10^{-2}$	$2.42\ 10^{-1}$	$3.9\ 1^{-1}$	17
	1/64	$4.59\ 10^{-3}$	$5.86\ 10^{-3}$	$7.37\ 10^{-2}$	$9.8\ 10^{-2}$	36
	1/128	$1.15\ 10^{-3}$	$1.24\ 10^{-3}$	$2.3\ 10^{-2}$	$2.64\ 10^{-2}$	100

Table 4. Error of the F.M.G. solution. Comparison with the discretization error

5. CONCLUDING REMARKS

The results reported here show that the smoother seems to be well adapted to the solution of system (6). This good convergence rate agrees with the Fourier analysis. In fact the smoothing factor (see for exemple [10] for the definition) for $\rho = 1$ is very closed to 0.5, value of the smoothing factor of the GAUSS-SEIDEL smoother on the Laplacian.

We must notice that the solution of (2) by means of the "bubble" elements leads also to systems of type (6), but with a different C_h . It implies modifications for the smoother. The efficiency of our multigrid solvers is of the same order for both types of systems.

All these first experiments have been limited to a particular domain with a regular mesh. In the next step we shall experiment our solvers on a general 2D mesh. For this purpose we are developing a more general program following the techniques of BANK [2].

The ultimate goal is to solve 3D problems for which the "mini-elements" must be particularly competitive.

REFERENCES :

[1] ARNOLD, D.N., F. BREZZI and M. FORTIN, A stable finite element for the Stokes equations, Preprint, University of Pavia, 1983.

[2] BANK, R., PLTMG Users'Guide. Technical Report, Department of Mathematics, University of California, San Diego, 1982.

[3] BRANDT, A., Multigrid Techniques : 1984 Guide with Applications to Fluid Dynamics. G.M.D. - Studien Nr. 85, St. Augustin, 1984.

[4] BREZZI, F. and J. PITKÄRANTA, On the stabilization of finite element approximations of the Stokes equations. In : W. HACKBUSCH (ed.), Efficient Solutions of Elliptic Systems, Proceedings, Kiel, Jan. 1984. Notes on Numerical Fluid Mechanics, vol. 10, pp. 11-19, Vieweg, 1984.

[5] CROUZEIX, M. and P.A. RAVIART, Conforming and non-conforming finite element methods for solving the stationary Stokes equations. R.A.I.R.O. Anal. Numer. 7 R-3, pp. 33-76, 1977.

[6] HACKBUSCH, W., Analysis and multigrid solutions of mixed finite element and mixed finite difference equations. Preprint, Institut für Angemandte Mathematik, Ruhr-Universität Bachum, 1980.

[7] MAITRE, J.F. and F. MUSY, Multigrid Methods : convergence theory in a variational framework. S.I.A.M. Journal on Numerical Analysis, vol. 21, n° 4, pp. 657-671, 1984.

[8] MAITRE, J.F., F. MUSY and P. NIGON, A fast solver for the Stokes equations using multigrid with a Uzawa smoother. In : D. BRAESS, W. HACKBUSCH, U. TROTTENBERG (eds.), Advances in Multi-Grid Methods, Proceedings, Oberwolfach, Dec. 1984. Notes on Numerical Fluid Mechanics, vol. 11, pp. 77-83, Vieweg, 1985.

[9] PITKÄRANTA, J. and T. SAARINEN, A multigrid version of a simple finite element method for the Stokes problem. Report - MAT - A217, Helsinki University of Technology, 1984.

[10] STÜBEN, K. and U. TROTTENBERG, Multigrid Methods : Fundamental Algorithms, Model Problem Analysis and Applications. In : W. HACKBUSCH and U. TROTTENBERG (eds.), Multigrid Methods, Proceedings, Köln-Porz, Nov. 1981. Lecture Notes in Math. 960, pp. 1-176, Springer-Verlag, 1982.

[11] VERFÜRTH, R., A multilevel algorithm for mixed problems. S.I.A.M. Journal on Numerical Analysis, vol. 21, n° 2, pp. 264-271, 1984.

A-POSTERIORI ERROR ESTIMATES, ADAPTIVE LOCAL MESH REFINEMENT AND MULTIGRID ITERATION

Randolph E. Bank[+]
Department of Mathematics
University of California at San Diego
La Jolla, California 92093

1. INTRODUCTION

Multi-level iteration and adaptive local mesh refinement are separately very effective techniques for solving partial differential equations. Since local mesh refinement procedures produce a hierarchy of (nearly) nested meshes, and multi-level iteration requires a sequence of meshes, it is natural to consider ways of combining the two. In the realm of a general purpose elliptic equation solver, the two procedures enhance each other's effectiveness such that the overall algorithm is more robust and powerful than the sum of its parts.

I have studied several aspects of adaptive local mesh refinement used in conjunction with multi-level iteration, often as joint work. In this manuscript, both procedures are discussed, with particular emphasis placed on the algorithmic aspects of combining the two. Many details and proofs are omitted, but these can be found elsewhere. The ideas and algorithms described here have been developed, implemented and tested in the software package PLTMG [8]. This package uses adaptive local mesh refinement, multi-level iteration, and piecewise linear triangular finite elements to solve a fairly general class of parameter dependent nonlinear elliptic pdes in general regions of the plane.

To keep the present discussion simple, here we restrict attention to the model linear elliptic pde.

$$\begin{aligned} Lu = -\nabla(a\nabla u) + bu &= f \qquad \text{in } \Omega\ , \\ a\frac{\partial u}{\partial n} &= g \qquad \text{on } \partial\,\Omega \end{aligned} \tag{1.1}$$

[+] This work was supported in part by the Office of Naval Research under Contract N00014-82-K-0197.

where $a > 0$, $b > 0$ in $\overline{\Omega}$. Sufficient (piecewise) smoothness of all coefficients and the boundary is assumed. The weak form of (1.1) is: find $u \in H^1(\Omega)$ such that

$$a(u,v) = (f,v) + \langle g,v \rangle \tag{1.2}$$

for all $v \in H^1$, where

$$a(u,v) = \iint_\Omega a \nabla u \cdot \nabla v + b u v \, dx$$

$$(f,v) = \iint_\Omega f v \, dx$$

$$\langle g,v \rangle = \int_{\partial\Omega} g v \, dx$$

and $H^1(\Omega)$ denotes the usual Sobolev space. We denote the evergy norm of $v \in H^1(\Omega)$ by $|||v|||^2 = a(v,v)$.

Let τ_h be a triangulationof Ω and M_h the space of continuous piecewise polynomials of degree k associated with τ_h. Then the finite element approximation $u_h \in M_h$ satisfies

$$a(u_h, v) = (f,v) + \langle g,v \rangle \tag{1.3}$$

for all $v \in M_h$.

Fundamental to the adaptive local mesh refinement procedure is the computation of a posteriori estimates for the error $e = u - u_h$. Much of the pioneering work in this area has been done by Babuska, Rheinboldt and their coworkers (see [1-7], for example). In addition to their use in adaptive refinement, such a posteriori estimates are also used to provide the user of a finite element program with an estimate of the global quality of the computed solution. These dual objectives require different, and sometimes conflicting, properties of the a posteriori error estimate.

As a guide to global errors for the user, it is important that the a posterior error estimate $\bar{e}$ satisfies inequalities of the form

$$C_1 \|e\| \leqslant \|\bar{e}\| \leqslant C_2 \|e\| \tag{1.4}$$

for some norm $\|\cdot\|$ and positive constants C_1 and C_2 of order one. Usually the norm is the H^1 or energy norm, where bounds like (1.4) have been proved for several schemes, including the one proposed here. This scheme computes a (discontinuous) function $\bar{e}$, as opposed to computing the scalar $\|\bar{e}\|$ directly, so it is possible to provide the user with error estimates in several norms (even if (1.4) cannot be proved for all).

Local error indicators based on $\bar{e}$ are used for a sequence of discrete choices (to refine or not to refine a given element). Since $\|e\|$ is very "flat" with respect to changes in the mesh near an optimum [5] , the ability to distinguish "large" from "small" errors is all that is required. Having $\bar{e}$ satisfy a local version of (1.4) or even (1.4) itself is not particularly useful in this context. Indeed, given the continuous and global dependence of the solution on the data and boundary conditions, it is unlikely that a local version of (1.4) could be shown for any computable error estimate which used only local computations.

In somewhat the same vein, since the implicit assumption of adaptive local mesh refinement procedures is that the "best" elements to refine are those with the largest error indicators, the choice of norm is very important. The L^2 norm is a poor choice, since large L^2 errors in an element can result from a relatively distant singularity (the well known pollution effect), and refining the elements near the singularity may have a greater impact on reducing the error in the element than refining the element itself. Energy or H^1 norms are more strongly correlated to such singularities, and practical experience has shown they are most appropriate for elliptic equations.

Finally, $\bar{e}$ should be inexpensive to compute. Schemes which involve the assembly and solution of a global set of equations are likely to be prohibitively costly. Our scheme uses only local computations based on solving a Neumann problem in each element of the mesh. For piecewise linear elements (as in PLTMG), the cost is comparable to assembling a global stiffness matrix and right hand side by the usual elementwise procedure. For polynomials of degree k , the size of the local system to be solved grows linearly in k , while the order of the element stiffness matrix used for computing u_h grows as k^2 . Thus the error estimate becomes comparatively more cost effective with increasing k .

The remainder of this paper is organized as follows: In Section 2, we describe the computation of the a posteriori error estimate $\bar{e}$. In Section 3, we describe the local mesh refinement procedure used in conjunction with multi-level iteration. Finally, in Section 4 we present a numerical illustration.

2. AN A-POSTERIORI ERROR ESTIMATE

In this section we briefly describe an error estimate for the model equation (1.1). A more complete discussion may be found in Bank and Weiser [12].

Let $v \in H^1(t)$ for some triangle $t \in \tau_h$. Then by integration by parts, we obtain

$$a(e,v)_t = (Le,v)_t + \left\langle a\frac{\partial e}{\partial n}, v \right\rangle_{\partial t} \tag{2.1}$$

where for some region $r \subset \Omega$

$$a(v,w)_r = \iint_r a\nabla u \nabla v + buv\,dx$$

$$(v,w)_r = \iint_r vw\,dx$$

and for some 1-dimensional manifold $\partial r \subset \Omega$,

$$\langle v,w \rangle_{\partial r} = \int_{\partial r} vw\,dx$$

Let B denote the set of triangle edges coincident with part of $\partial\Omega$ and E be the set of edges lying in the interior of Ω . For interior edges, we define the jump and average (in the sense of traces) of a function $v \in H^1(t_i)$, $i = 1,2$, by $[v] = v_1 - v_2$, $\{v\} = (v_1 + v_2)/2$, respectively, where v_i , $i = 1,2$ are the limit values of v for the two triangles sharing the edge. For purposes of computing the jump, the "normal" direction for an interior edge is arbitrary but fixed.

Let $H_p = \prod_t H^1(t)$. If (2.1) is summed over elements, we obtain, after some manipulation,

$$a(e,v) = (R,v) + \langle S,v \rangle_B + \langle J, \{v\} \rangle_E + \langle \{a\frac{\partial e}{\partial n}\}, [v] \rangle_E \tag{2.2}$$

for $v \in H_p$, where

$$a(e,v) \equiv \sum_t a(e,v)_t$$

$$(R,v) = \sum_t (R,v)_t, \qquad R = f - Lu_h$$

$$\langle S,v \rangle_B = \sum_{e\in B} \langle S,v \rangle_e, \qquad S = g - a\frac{\partial u_h}{\partial n}$$

$$\langle J,\{v\} \rangle_B = \sum_{e\in B} \langle J,\{v\} \rangle_e, \qquad J = \left[a\frac{\partial u_h}{\partial n} \right]$$

$$\langle \{a\frac{\partial e}{\partial n}\},[v] \rangle_E = \sum_{e\in E} \langle \{a\frac{\partial e}{\partial n}\}, [v] \rangle_e .$$

For the special case $v \in H^1(\Omega)$, $[v] = 0$, $\{v\} = v$ and (2.2) becomes

$$a(e,v) = (R,v) + \langle S,v \rangle_B + \langle J,v \rangle_E , \tag{2.3}$$

while if $v \in M_h$, we obtain the standard orthogonality relation

$$a(e,v) = 0 \tag{2.4}$$

Proceeding formally for the moment, (2.2) suggests we seek an approximation $\hat{e} \in H_p$, where $\hat{e}$ is defined for $t \in \tau_h$ as the solution of

$$a(\hat{e},r)_t = (R,v)_t + \langle S,v \rangle_{/B\cap\partial t} + \tfrac{1}{2}\langle J,v \rangle_{/E\cap\partial t} \tag{2.5}$$

for all $v \in H^1(t)$. If we sum (2.5) over t , we obtain, for $v \in H_p$.

$$a(\hat{e},v) = (R,v) + \langle S,v \rangle_{/B} + \langle J,\{v\} \rangle_{/E} \tag{2.6}$$

Comparing (2.6) with (2.3) we see that for v $H^1(\Omega)$,

$$a(e,v) = a(\hat{e},v) , \tag{2.7}$$

so that the true error e is the elliptic projection of $\hat{e}$ onto $H^1(\Omega)$. Thus we obtain, (taking $v = e$ in (2.7)),

$$|||e||| \leqslant |||\hat{e}||| \tag{2.8}$$

The residual R , boundary residual S , and jumps J can all be evaluated once the approximate solution u_h is known. However, there remain several unresolved issues regarding this scheme.

First, in the limit $h_t \to 0$ (h_t = diameter of t) with the geometry of t held fixed, we have $a(v,1)_t \to 0$, indicating that the local problems (2.5) converge to singular Neumann problems. We do not expect the right hand side of (2.5) to be consistent data in this limiting case, so the problems may not be well posed (this also occurs if $b \equiv 0$ in some element). Second, we must numerically approximate $\hat{e}$ to obtain a computable error estimate. Third, in solving the first two problems, we would like to maintain a bound like (2.8), as well as obtain a lower bound relation as in (1.4).

Let $P_k(t)$ denote the polynomials of degree at most k on t , and let I be the local interpolation operator corresponding to the restriction of M_h to t . For the case of C° Lagrangian elements under consideration, $I(u) \in P_k(t)$ interpolates u as the $(k+1)(k+2)/2$ points in t with barycentric coordinates $(i/k,\ j/k,\ 1-(i+j/k))$, $0 \leqslant i \leqslant k$, $0 \leqslant j \leqslant k-i$.

Let $k' > k$ and let $P_I(t) = \{v \in P_{k'}(t) \text{ such that } I(v) = 0\}$ and set $P_d = \prod_t P_I(t) \subset H_d$. In words, P_d is a space of discontinuous piecewise polynomials of degree at most k' which are zero at all the degrees of freedom of the original approximation space M_h . Our computable approximation $\bar{e} \in P_d$ is defined by

$$a(\bar{e},v) = (R,v) + \langle S,[v] \rangle_{/B} + \langle J,\{v\} \rangle_{/E} \tag{2.9}$$

for all $v \in P_d$. Notice that (2.9) may be decomposed into local problems for $\bar{e}$ defined on each element t ; find $\bar{e} \in P_I(t)$ such that

$$a(\bar{e},v)_t = (R,v)_t + \langle S,v \rangle_{/B\cap\partial t} + \tfrac{1}{2}\langle J,v \rangle_{/E\cap\partial t} \tag{2.10}$$

for all $v \in P_I(t)$. Since $I(1) = 1$, the constant function is not in $P_I(t)$ and (2.10) remains well posed in the limit $h_t \to 0$. One can prove in the linear case considered here [12,14] as well as in the mildly nonlinear case [11] that the computed error estimate $\bar{e}$ satisfies bounds of the form (1.4). For many problems we have tried $|||\bar{e}||| \to |||e|||$ as $N \to \infty$, but we cannot yet prove such a result.

In practice taking $k' = k + 1$ seems adequate. In the case of standard C° Lagrange triangular elements, one must assemble and solve a linear system of order $k+2$ for each element in order to compute $\bar{e}$ (this holds also for the nonlinear case [11]). This compares with the assembly of the element stiffness matrices of order $(k+2)(k+1)/2$ which are required for the original computation of u_h. Thus the cost of computing $\bar{e}$ is of the same order (for small k) or far less (for larger k) than the cost of assembling the original set of equations for u_h.

Finally, we remark that splitting the jumps J equally between the two elements sharing an edge is somewhat arbitrary. The bound (1.4) holds for more general splitting strategies. In PLTMG, $\bar{e}$ is a discontinuous piecewise quadratie polymonial which is zero at all triangle vertices. The jumps are partitioned between the two elements sharing an edge in proportion to the relative sizes of the gradients of u_h in the two elements (with the element with the larger gradient getting the larger fraction). This has an insignificant effect on $\bar{e}$ as a global error estimate but does improve its performance in terms of a local error indicator.

3. A LOCAL MESH REFINEMENT PROCEDURE

Our local mesh refinement scheme is motivated by the work of Babuska and Rheinboldt [3,5,7]. For triangular elements, it is based on the "regular" refinement of an element t into 4 "son" elements of similar geometry by pairwise connecting the midpoints of the edges of t.

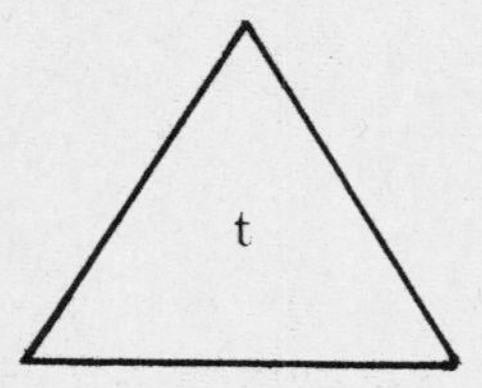

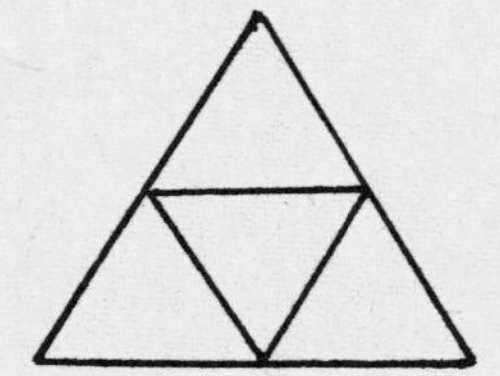

Figure 3.1. Regular refinement of t

For each element $t \in \tau_h$ we compute a local error indicator θ_t . In our case $\theta_t \equiv \|\bar{e}\|_{H^1(t)}$ where $\bar{e}$ is computed as in Section 2. As with Babuska and Rheinboldt we compute a threshold value $\theta_{max}\sigma$ where

$$\theta_{max} = \max_t \theta_t,$$

and $\sigma \in (0,1)$ is an error reduction factor which is (initially) computed as follows: if $\theta_f \equiv \theta_{max}$ for the previous adaptive step and θ_s is the largest error in element for the four sons of f , then $\sigma = \theta_s / \theta_f$. We then require the refinement of all elements with

$$\theta_t \geqslant \theta_{max} \cdot \sigma \tag{3.1}$$

The overall effect of (3.1) is to produce meshes in which all elements have roughly equal error indicators, and these meshes are quasi optimal in the sense of Babuska and Rheinboldt. Unlike Babuska and Rheinboldt, however, we also require the refinement of an element for other reasons. In particular, we require all refined meshes to satisfy the 1-irregular and 2-neighbor rules.

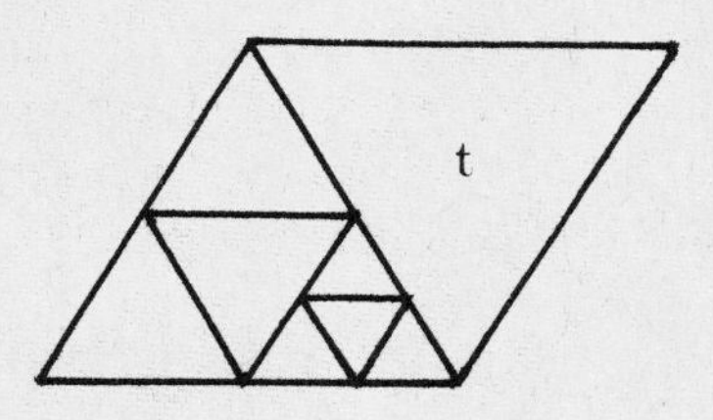

Figure 3.2a. Refinement of t required by 1-irregular rule.

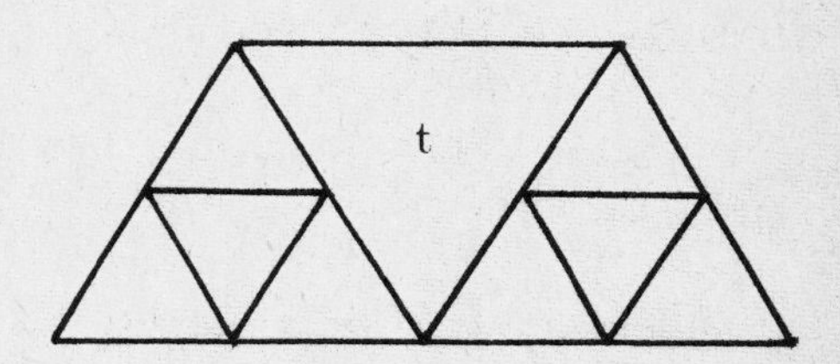

Figure 3.2b. Refinement of t required by 2-neighbor rule.

Irregular vertices occur on triangle edges at interfaces between refined and unrefined elements. The 1-irregular rule requires that any element with 2 or more irregular vertices on one of its edges must be refined (see Figure 3.2a). The 2-neighbor rule requires any element with irregular vertices on two or more edges to be refined (Fig. 3.2b).

As one might expect, in practice relatively few elements are refined on the basis of the 1-irregular and 2-neighbor rules. However, their importance lies in the fact that they guarantee the transition between more refined and less refined regions will be smooth, they allow the use of relatively simple and efficient data structures to represent the refinement process, and they guarantee that the linear system to be solved will be sparse and easy to assemble (see [10,14] for a complete discussion of these points).

In any event, at the conclusion of any refinement step, the interface between refined and unrefined elements must be as in Figure 3.3a. As a final "cleanup" step we temporarily refine elements with irregular vertices into 2 "green" elements as illustrated in Figure 3.3b. This allows the mesh to become a triangulation.

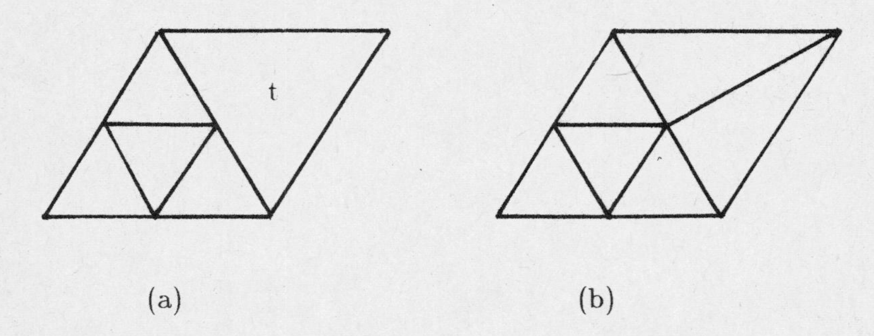

Figure 3.3. Temporarily adding a "green" edge.

Before proceeding to the next refinement step, green edges are removed, allowing the affected elements to (possibly) be refined in the normal way. Thus there can be no deterioration in the geometrical quality of the elements as a result of the adaptive refinement process. (It does mean, however, that the triangulations may not be strictly nested.)

In combining local adaptive mesh refinement with multi grid, several interesting issues arise. Our overall strategy is to embed the local mesh refinement scheme within a nested iteration (or full multi grid) process. That is, we wish to compute a sequence of meshes $\tau_1, \tau_2, \cdots \tau_k$ and corresponding finite element spaces $M_1, M_2, \cdots M_k$ of increasing dimension. We then sequentially solve for approximations $u_j \in M_j$ using u_{j-1} as the initial guess on level j and employing a j-level iteration. The triangulations can be computed adaptively because u_{j-1} is computed before τ_j is required.

For the multigrid method to be optimally efficient, we would like the dimensions N_j of M_j to increase geometrically. For 2-dimensional problems $N_j \sim 4N_{j-1}$ is typical.

On the other hand, our local mesh refinement procedure is designed to selectively refine the mesh, and in general it should not be expected that this procedure by itself will always produce the required geometric increase in problem size. Thus we are lead to consider modifications of the basic procedure allowing an element to be refined more than once between multi grid solutions.

Such a procedure is sketched below:

<u>Algorithm Refine</u>

$$\text{input} \ \ \tau_1, M_1, \ k = 1, \ N_0 = N_1$$

(1) Compute $u_k \in M_k$ using a fixed number r of k-level iterations.

(2) Set $j = 1$; if $N_k << 4N_{k-1}$ set $\ell = k$; else set $\ell = k+1$ and $N_\ell = N_k$.

(3) While $j \leqslant 5$ and $N_\ell << 4N_{\ell-1}$ do:

- (a) compute or update $\bar{e}$
- (b) if $j = 1$ and $|||\bar{e}|||$ is small enough, stop
- (c) compute σ , θ_{max} , and create or refine τ_ℓ using (3.5)
- (d) interpolate (as necessary) u_k onto τ_ℓ
- (e) $j \leftarrow j+1$

(4) Set $k = \ell$ and go to (1).

In Algorithm Refine, the adaptive loop is executed a maximum of 5 times (5 was empirically chosen). At each step we compute or update $\bar{e}$, refine the mesh and interpolate the current solution onto the newly created mesh points. As j increases, we expect the quality of the error indicators based on $\bar{e}$ to deteriorate since the solution is not recomputed but merely interpolated onto the refined mesh.

If $j > 1$ or $\ell = k$, the value of σ given in (3.1) may be increased if it appears that too many elements would be refined ($N_\ell >> 4N_{\ell-1}$. The value of σ is <u>never</u> decreased in order to increase the number of elements to be refined, as this could very easily result in nonoptimal meshes.

In PLTMG, the solution u_k is piecewise linear. However, in defining u_k at newly created mesh points, a local quadratic interpolation scheme is used. The obvious choice of linear interpolation leads to many triangle edges with zero jumps, causing the quality of $\bar{e}$ to deteriorate much more rapidly than with quadratic interpolation.

Most of the time the loop (3) terminates with $N_\ell \approx 4N_{\ell-1}$. Very strong singularities can cause termination because $j > 5$. When this happens we solve (via multigrid iteration) the problem of the (partially) completed level τ_k and then re-enter the adaptive loop with $\ell = k$ on line (2). This intermediate solution is used to stop the deterioration of the local error indicators with increasing j , and thus preserve the integrity of the adaptive refinement process.

4. A NUMERICAL EXAMPLE

In this section we present some results from a numerical experiment.

$$-\Delta u = 0 \text{ in } \Omega \tag{4.1}$$

in the region Ω illustrated in Figure 4.1. As for the boundary conditions, on the "top" of the crack we set $u = 0$, while on the "bottom" we set $\partial u/\partial n = 0$. The combination of an interior angle of 2π and the change in boundary conditions causes a singularity with dominant term $r^{1/4}\sin(\theta/4)$ to propagate from the crack tip. Dirichlet boundary conditions were set on the circle such that $u = r^{1/4}\sin(\theta/4)$. was the exact solution of the problem. A contour map of the solution is shown in Figure 4.2.

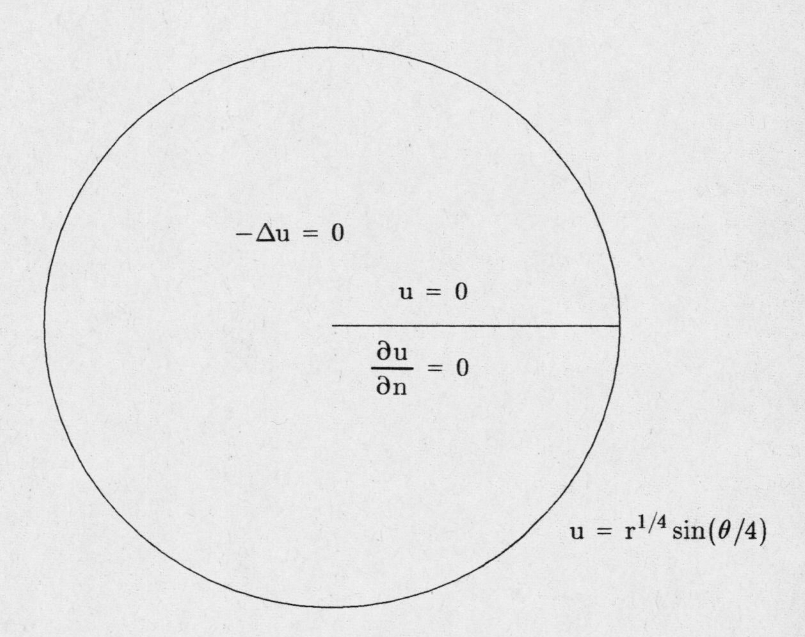

Figure 4.1 – Ω for test problem

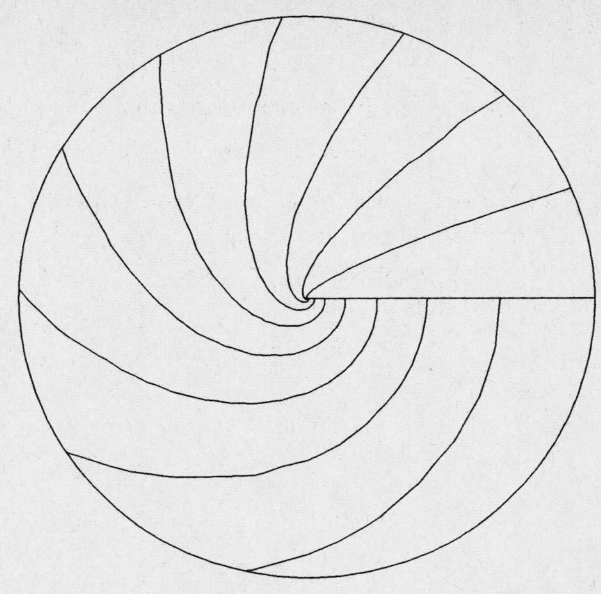

Figure 4.2 – Contour map of $u = r^{1/4} \sin(\theta/4)$

This problem was solved using the refinement procedure outlined in Section 3 and the a posteriori error estimates of Section 2. Beginning with a uniform mesh of eight elements and ten vertices, three refined meshes were constructed. These meshes are illustrated in Figures 4.3-4.6. Due to the severity of the singularity, the last two meshes required intermediate solutions as described in Section 3.

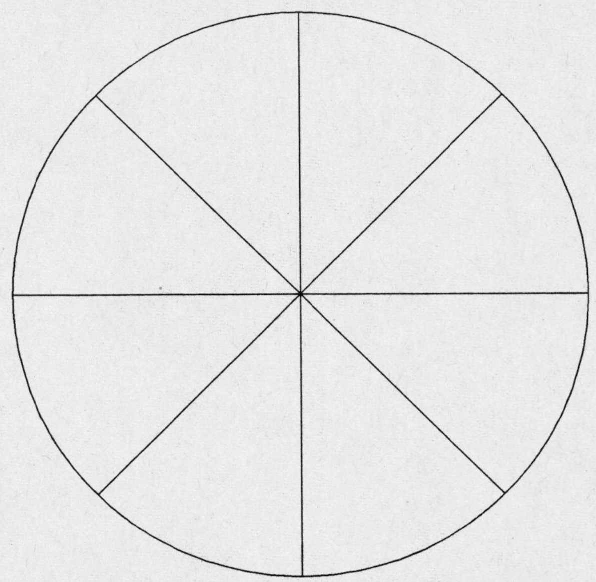

Figure 4.3 – Triangulation τ_1 ; N = 10

Figure 4.4 – Triangulation τ_2 ; N = 41

Figure 4.5 – Triangulation τ_3 ; N = 165

Figure 4.6 – Triangulation τ_4 ; N = 666

In Table 4.1 we compare the a posteriori error estimate with the true error for the case of H^1 norm. For the a posteriori error estimate the estimated number of digits is given by

$$\text{estimated digits} = -\log_{10}(\|\bar{e}\|_1 / \|u_h\|_1) , \tag{4.2}$$

while for the exact error,

$$\text{exact digits} = -\log_{10}(\|e\|_1 / \|u\|_1) \tag{4.3}$$

Table 4.1

Level	N	Estimated digits	Exact digits
1	10	0.303	0.124
2	41	0.509	0.316
3	165	0.793	0.691
4	666	1.105	1.145

From the data it appears the effectivity ratio of Babuska and Rheinboldt, given by $\|\tilde{e}\|/\|e\|$ is converging to 1. Fitting the (exact error) data by least squares to a function of the form $CN^{-q/2}$, we obtain $q \cong 1.13$, $C \cong 3.26$. For piecewise linear approximation on an optimum sequence of grids the value of q is 1. For uniformly refined meshes, the value of q for this problem is (effectively) 1/4 due to the singularity. The least squares fit suggests that the adaptive refinement procedure is working quite well for this problem, and that the sequence of generated meshes are not far from optimal in terms of approximation of u.

One worry here with respect to multigrid is that such highly nonuniform meshes may retard the convergence of the multigrid iteration. To some extent, this fear is well founded, but there are several mitigating circumstances (see [16]). The smoothing iteration used here was symmetric Gauss-Seidel accelerated by a minimum residual algorithm (orthomin [13]). The effectiveness of such acceleration schemes in the multigrid context was shown in [9]. For the case of highly nonuniform meshes, acceleration is particularly useful, since the adaptivity inherent in the acceleration process can cope with the somewhat distorted spectrum of eigenvalues produced by nonuniform refinement, and thus offset to some extent the shortcomings of the basic smoother.

Let $\tilde{e}(m,r)$ be the difference between the exact u_h and the approximation of u_h computed using a multigrid iteration. Here m is the number of smoothing iterations and r is the number of cycles (W-cycles in this case). In our algorithm one alternates m smoothing iterations with coarse grid corrections in a symmetrical pattern. The case r=2 for a 3-level scheme is illustrated symbolically in Figure 4.7.

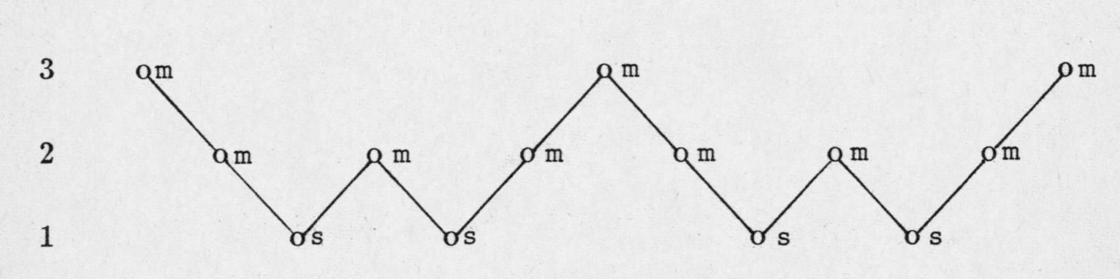

r = 2 cycles
s = solve directly (sparse Gaussian elimination)
m = m smoothing iterations (SGS with orthomin)

Figure 4.7 – 3-level multigrid algorithm

In current terminology, we use $m/2$ pre-smoothing and $m/2$ post-smoothing steps except for presmoothing the first cycle and post-smoothing the last cycle where m iterations are used.

Let

$$\text{digits}\,(m,r) = -\log_{10}(\|\bar{e}(m,r)\|_1 \,/\, \|u_h\|_1) \tag{4.4}$$

denote the number of correct digits in the computed solution relative to the exact finite element solution. (This should not be confused with comparing the computed solution with the exact solution u of (1.1).) In Table 4.2, we tabulate the results for $1 \leqslant m \leqslant 2$, $1 \leqslant r \leqslant 2$, for the 4-level iteration using the nonuniform meshes in Figures 4.3-4.6.

Table 4.2

Correct digits in H^1 norm

	r = 1	r = 2
m = 1	2.87	3.00
m = 2	3.05	3.17

In producing Table 4.2, the initial guess was zero (nested iteration or full multigrid was not used). The results show that for $m = r = 1$, the accuracy is already well below discretization error (see Table 4.1). On the other hand, the other cases suggest that the asymptotic covergence rate may not be as good as one commonly expects in multigrid.

We remark that the work of Yserentant [15] on hierarchical basis methods presents an alternative procedure for dealing with highly nonuniform meshes, and one which we believe merits serious consideration.

REFERENCES

1. Babuska, I. and A. Miller, *A posteriori error estimates and adaptive techniques for the finite element method.* Technical Report BN-968, Institute for Physical Sciences and Technology, University of Maryland, 1981.
2. Babuska, I. and W. C. Rheinboldt, A posteriori error analysis of finite element solutions or one-dimensional problems. *SIAM Journal on Numerical Analysis* 18:565-589, 1981.

3. Babuska, I. and W. C. Rheinboldt, A posteriori error estimates for the finite element method. *International Journal for Numerical Methods in Engineering* 12:1597-1615, 1978.
4. Babuska, I. and W. C. Rheinboldt, Analysis of optimal finite-element meshes in R^1. *Mathematics of Computation* 33:435-463, 1979.
5. Babuska, I. and W. C. Rheinboldt, Error estimates for adaptive finite element computations. *SIAM Journal on Numerical Analysis* 15:736-754, 1978.
6. Babuska, I. and W. C. Rheinboldt, On the reliability and optimality of the finite element method. *Computer and Structures* 10:87-94, 1979.
7. Babuska, I. and W. C. Rheinboldt, Reliable error estimation and mesh adaptation for the finite element method. In *Computational Methods in Nonlinear Mechanics*, 67-108, North-Holland, New York, 1980.
8. Bank, R. E., *PLTMG User's Guide, Edition 4.0.* Technical Report, Department of Mathematica, University of California at San Diego, La Jolla, California, March, 1985.
9. Bank, R. E. and C. Douglas, Sharp estimates for multigrid rates of convergence with general smoothing and acceleration. *SIAM Journal of Numerical Analysis* 22:617-633 (1985).
10. Bank, R. E., A. H. Sherman, and A. Weiser, Refinement algorithms and data structures for regular local mesh refinement. In *Scientific Computing*, R. Stepleman et al., Eds., IMACS/North-Holland, New York, 3-17, 1983.
11. Bank, R. E. Analysis of a local a-posteriori error estimate for elliptic equations. In *Accuracy Estimates and Adaptive Refinement in Finite Element Computations*, Babuska et al., Eds., John Wiley, 1985.
12. Bank, R. E. and A. Weiser, Some a posteriori error estimates for elliptic partial differential equations. *Mathematics of Computation*, 44:283-301, 1985.
13. Elman, H. C., *Iterative Methods for Large Sparse Nonsymmetric Systems of Linear Equations.* Technical Report 229, Computer Science Department, Yale University, 1982.
14. Weiser, A., *Local-mesh, local-order, adaptive finite element methods with a posteriori error estimators for elliptic partial differential equations.* Technical Report 213, Computer Science Department, Yale University, 1981.
15. Yserentant, Y., *On the Multi-level Splitting of Finite Element Spaces.* Technical Report 21, University of Aachen, 1983.
16. Yserentant, Y., *The Convergence of Multi-level Methods for Solving Finite Element Equations in the Presence of Singularities.* Technical Report 14, University ofAachen, 1982.

CONTINUATION AND MULTI-GRID FOR NONLINEAR ELLIPTIC SYSTEMS

R. E. Bank
Department of Mathematics
UC San Diego
La Jolla, CA 92093

H. D. Mittelmann
Department of Mathematics
Arizona State University
Tempe, AZ 85287

Abstract. Recently the authors have developed and successfully applied a continuation technique for the numerical solution of parameter-dependent nonlinear elliptic boundary value problems. The method was integrated into an existing multi-grid package based on an adaptive finite element discretization. An extension to nonlinear systems of differential equations is considered. Since one important field of application is the VLSI device simulation we discuss this problem and present preliminary numerical results for a MOSFET device.

1. Introduction

In the following we consider the parameter-dependent nonlinear problem

$$G(u,\alpha) = 0, \tag{1.1}$$

where $G : X^m \times R^p \to X^m$, X a suitable function space and α a vector of real parameters. (1.1) will typically represent a parameter-dependent nonlinear elliptic system of dimension m and order two. An important example are the VLSI device simulation equations (3.1) for which m is equal to three.

The solution manifold $G^{-1}(0)$ of (1.1) may have a very complex structure. In practical applications it is frequently desirable and sufficient to compute certain cross-sections of it, i.e. all parameters α are kept fixed at certain values and for simplicity

---

The work of the first author was supported by the Office of Naval Research under Grant N00014-82-K-0197. The second author was supported by the Air Force Office of Scientific Research under Grant AFOSR-84-0315.

suppressed except one which we will call λ. A simple graphical representation of this cross-section may be obtained if a functional of the solution is depicted versus λ. It should be noted, however, that usually in these applications such functionals are in fact important pieces of information that are of considerable interest, again we refer to the VLSI problem in 3.

We assume that a vector

$$(1.2) \qquad R(u(\lambda)),\ R : X^m \to R^k$$

of differentiable functionals is given and that most of the important information about the problem can be obtained from the $k \times p$ diagrams

$$(1.3) \qquad R_i = R_i(u(\alpha_j)),\ i = 1,\ldots,k,\ j = 1,\ldots,p$$

and the corresponding solutions.

The solutions $u(\lambda)$, $\lambda = \alpha_j$, in (1.3) and from them the solution curves $R_i(u(\lambda))$ usually have to be computed in a continuation fashion. Starting from a known solution $u(\lambda_0)$ another solution is computed by utilizing a predictor-corrector scheme. Frequently such a starting solution is not known and has to be determined itself from an approximate solution. Along the solution curves singular points have to be expected and our goal is to provide methods to locate these points, to overcome turning points and to switch branches at (simple) bifurcation points.

In the following section we will outline a continuation algorithm that has been successfully used for the scalar case ($m = 1$ in (1.1)). It provides an efficient and robust method and has been implemented for a general class of second order elliptic boundary value problems discretized by finite elements in [1]. The continuation method is applied on a coarse triangulation of the given domain and fine grid approximations at arbitrarily specified points are obtained by multi-grid.

The equations of VLSI device simulation will be considered in Section 3. It will be shown how naturally several physical parameters $\lambda = \alpha_j$ and functionals $R_i(u)$ arise. An important special case is the latchup phenomenon recently solved in [3,4] for the time-dependent system. Since, however, the main interest is in steady-state solutions, we propose to solve a continuation problem of the above type for which turning points are to be expected. The last section contains first numerical results.

2. The Continuation Method

In the following we assume that a choice of the parameter $\lambda = \alpha_j$ in (1.1), (1.2) and also of the functional R_i in (1.3) has been made and simply denote (1.1) by

$$(2.1) \qquad G(u,\lambda) = 0, \quad G : X^m \times R \to X^m,$$

and R_i by r.

The parametrization of the solution u of (2.1) by λ and thus of r by λ need not be possible in general. Let s denote the arclength along a solution curve $(u,\lambda) = (u(s),\lambda(s))$ of (2.1). Then under sufficient smoothness and regularity assumptions this latter dependence is differentiable and thus ($G_u^0 = G_u(u_0,\lambda_0)$)

$$(2.2a) \qquad G_u^0 \dot{u}_0 + G_\lambda^0 \dot{\lambda}_0 = 0,$$

$$(2.2b) \qquad \|\dot{u}_0\|^2 + \dot{\lambda}_0^2 = 1$$

in a solution $(u_0,\lambda_0) = (u(s_0),\lambda(s_0))$ of (2.1). Here, subscripts stand for partial derivatives and a dot for differentiation with respect to s.

In singular points (u_0,λ_0) of (2.1) the linear operator G_u^0 in (2.2a) is singular and, for example, a solution by Newton's method breaks down. This singularity may be removed or reduced by augmenting (2.1) by a (normalizing) equation, a well-known example of which is

$$(2.3) \qquad N_\sigma(u,\lambda,\sigma) \equiv \Theta\, \dot{u}_0(u-u_0) + (2-\Theta)\dot{\lambda}_0(\lambda-\lambda_0) - \delta\sigma = 0$$

which approximates (2.2b). σ is called the pseudo-arclength parameter. Θ satsifying $0 \le \Theta \le 2$ may be used as a weighting parameter.

Instead of (2.3) we propose to use

$$(2.4) \qquad N_r(u,\lambda,\sigma) \equiv \Theta\, \dot{r}_0(r-r_0) + (2-\Theta)\dot{\lambda}_0(\lambda-\lambda_0) - \delta\sigma = 0,$$

where $r_0 = r(u_0)$, $\dot{r}_0 = (\frac{d}{d\sigma} r(u))_{\sigma=\sigma_0}$ and r denotes the above functional. From (2.4) we see that the three values $\Theta = 0,1,2$ are special in the sense that for $\Theta = 0(2)$ the augmenting equation characterizes points with a fixed $\lambda(r)$-value, while for $\Theta = 1$ the points lie on a hyperplane orthogonal to the solution arc in (r_0, λ_0).

This suggests to use target values in the two (physically relevant) variables r and λ. Such a strategy seems appropriate as long as starting from a point on the solution curve another point corresponding to a suitably picked target value can be computed in a small number of (corrector) iterations. This was achieved for the scalar case (m=1) in (2.1) and the choice

$$r(u) = \|u\| \tag{2.5}$$

in [1,10,11]. The following class of nonlinear boundary value problems in the plane

$$\begin{aligned} -\nabla \cdot a(x,y,u,\nabla u,\alpha) + f(x,y,u,\nabla u,\alpha) &= 0 \text{ in } \Omega, \\ u = g_1(x,y,\alpha) &\quad \text{on} \quad \partial\Omega_1, \\ a \cdot n = g_2(x,y,u,\alpha) &\quad \text{on} \quad \partial\Omega_2 = \partial\Omega \setminus \partial\Omega_1, \end{aligned} \tag{2.6}$$

was considered. Here, Ω is a connected domain in R^2 with boundary $\partial\Omega_1$, n is the unit normal vector on $\partial\Omega$, $a = (a_1, a_2)^T$ and a_1, a_2, f_1, g_1, g_2 are given scalar functions. In [1] the program PLTMG, an implementation of an adaptive finite element multi-grid method combined with the continuation technique is described in detail. In addition to theoretical results on the continuation method [10,11] contain numerical results for several problems from the applications including problems with turning points, symmetry-breaking and other bifurcation points as well as parameter-switching for cases depending on several ($p > 1$ in (1.1)) parameters. Other problems to which PLTMG has been applied successfully are Steklov eigenvalue problems for which the Dirichlet condition in (2.6) is parameter-dependent.

The predictor step determines predicted values

$$\begin{pmatrix} u_p \\ \lambda_p \end{pmatrix} = \begin{pmatrix} (1+\gamma)u_0 + \beta\dot{u}_0 \\ \lambda_0 + \alpha\dot{\lambda}_0 \end{pmatrix} \tag{2.7}$$

such that α, β, γ satisfy the three equations

$$N(u_p, \lambda_p, \delta\sigma) = 0,$$
$$(2.8) \qquad (u_0, G(u_p, \lambda_p)) = 0,$$
$$(\dot{u}_0, G(u_p, \lambda_p)) = 0,$$

in a least-squares sense. Here it is assumed that X is a Hilbert space and (.,.) denotes its inner product.

In the corrector iteration the augmented system

$$(2.9) \qquad F(y) = \begin{pmatrix} G(u,\lambda) \\ N(u,\lambda,\sigma) \end{pmatrix} = 0 \quad , \; y^T = (u^T, \lambda)$$

is solved by Newton's method starting at (u_p, λ_p). The restriction of the problem (2.1) to the (r,λ)-plane in case $N = N_r$ is used in (2.8) introduces an additional singularity in exceptional points. It was shown, however, that for the proper choice of the continuation parameter and N the Jacobian F_y is regular except in bifurcation points and has minimum deficiency there (cf. [11]).

Preliminary results for continuation in the case of systems ($m > 1$ in (2.1)) will be reported in the last section. The choice of a norm $r = \|u\|$ seems less appropriate for $u \in X^m$ and the components of u representing very different quantities.

3. VLSI Device Simulation

There is not the space here to outline the derivation and significance of the coupled 3 × 3 - system describing the behavior of carriers in semiconductors under the influence of external fields, instead we refer to [3,4] and in particular to [6]. Under several assumptions the following dimensionless (normalized) form of the equations is obtained

$$g_1(u,v,w) \equiv -\Delta u + e^{u-v} - e^{w-u} - k_1 = 0,$$
$$(3.1) \qquad g_2(u,v,w) \equiv \nabla \mu_n e^{u-v} \nabla v + k_2 = 0,$$
$$g_3(u,v,w) \equiv -\nabla \mu_p e^{w-u} \nabla w + k_2 = 0.$$

Here u denotes voltage, v and w the quasi-Fermi variables related to the electron and hole carrier densities n,p through

$$n = e^{u-v} \quad , \quad p = e^{w-u}.$$

μ_n, μ_p denote the mobility coefficients, k_1 the doping profile and k_2 the recombination term.

To complete the definition of the problem the domain, the above coefficients and right-hand sides and boundary conditions for the unknowns u,v,w have to be prescribed. Typically the set of parameters α (cf. (1.1)) considered for (3.1) are the Dirichlet boundary values applied at the contacts of the device, while the functionals R(u) in (1.2) are the contact currents, i.e. certain contour integrals involving u, v and w. We refer again to the above mentioned papers for more details and numerical results. In general the desired output of the device simulation stage is a set of I-V curves where V denotes the applied contact voltage and I one of the resulting contact currents. These curves are needed as input for the final stage of circuit simulation (cf. [4]).

In the following we will use the coupled system of VLSI device simulation equations as an example to outline of which form the nonlinear problem (1.1) will in general be that is to be solved by the continuation method.

We first derive the weak form of the equations (3.1). Let $H^1(\Omega)$, $H^1_0(\Omega)$ denote the usual Sobolev spaces while $H^1_s(\Omega)$, s = u,v,w, denotes the affine spaces whose elements satisfy the Dirichlet boundary conditions prescribed for the respective functions. Let $H = H^1_u \times H^1_v \times H^1_w$, $H_0 = (H^1_0)^3$. A weak form of (3.1) is: find $(u,v,w) \varepsilon H$, such that for all $(\phi,\psi,\chi) \varepsilon H_0$

$$\int_\Omega \nabla u \nabla \phi + (e^{u-v} - e^{w-u} - k_1)\phi \, dxdy = 0,$$

$$(3.2) \qquad - \int_\Omega \mu_n e^u \nabla(e^{-v}) \nabla\psi + k_2 \, \psi \, dxdy = 0,$$

$$\int_\Omega \mu_p e^{-u} \nabla(e^w) \nabla\chi + k_2 \chi \, dxdy = 0.$$

These equations are now discretized by piecewise linear finite elements on a triangulation T of Ω. Let $t \varepsilon T$ denote a triangle with vertices p_i, $1 \leq i \leq 3$, interior angles Θ_i at p_i and let the length of the edge opposite p_i be denoted by ℓ_i and its midpoint by m_i. We define the dot products d_i, $1 \leq i \leq 3$, by

$$(3.3) \qquad d_i = \ell_j \ell_k \cos \Theta_i,$$

where the triple (i,j,k) is any cyclic permutation of (1,2,3). Let ϕ_i, $1 \le i \le 3$, finally denote the usual piecewise linear basis functions satisfying $\phi_i(p_j) = \delta_{ij}$. The box method discretization (cf. [2,3]) is based on Gauss' Divergence Theorem applied to a "box" around each of the vertices p of the triangulation. This polygonal box has as its edges the perpendicular bisectors of all triangle edges having p as one of its endpoints.

As is standard in finite elements the equations (3.2) and the Jacobian matrix of these equations needed for Newton's method are assembled elementwise. As in [2] we give the results of the box method for one of the terms only (correcting a few misprints). The essential term in the diagonal part of the Jacobian corresponding to the second equation of (3.2) (the "v-equation") is

$$(3.4) \qquad \int_t e^u \nabla(e^{-v}\phi_i)\nabla\phi_j \, dx\,dy, \qquad 1 \le i,j \le 3.$$

The corresponding 3×3 element matrix is

$$\frac{1}{4|t|}\begin{bmatrix} d_3e^{u(m_3)-v(p_1)} + d_2e^{u(m_2)-v(p_1)} & -d_3e^{u(m_3)-v(p_2)} & -d_2e^{u(m_2)-v(p_3)} \\ -d_3e^{u(m_3)-v(p_1)} & d_3e^{u(m_3)-v(p_2)} + d_1e^{u(m_1)-v(p_2)} & -d_1e^{u(m_1)-v(p_3)} \\ -d_2e^{u(m_2)-v(p_1)} & -d_1e^{u(m_1)-v(p_2)} & d_2e^{u(m_2)-v(p_3)} + d_1e^{u(m_1)-v(p_3)} \end{bmatrix}$$

where $|t|$ is the area of $t \in T$. Using analogous discretizations a 9×9 element stiffness matrix containing 9 such blocks is obtained. Let in some ordering p_i, $i = 1,\dots,N$, denote all vertices of T and denote the vector of unknowns $z = (u_1, v_1, w_1, \dots, u_N, v_N, w_N)$, where $u_i = u(p_i)$ etc. Element matrices and right-hand sides are assembled according to this ordering of equations and nodal values to yield the system

$$(3.5) \qquad G(z,\alpha) = 0, \qquad G : R^{3N} \times R^P \to R^{3N}$$

and the Jacobian (w.r.t.z) G_z. As mentioned above α consists of the Dirichlet boundary values applied at the contacts.

Often the system of equations (3.1) does not have a unique solution. For instance, for a given set of applied voltages, some devices may exist in several physical states. A typical behaviour, called latchup or snapback in the semiconductor literature is illustrated in Figure 3.1.

Figure 3.1 I-V curve in case of latchup

The latchup phenomenon may occur when several devices which are designed to be logically independent are physically located close to one another within the chip, creating the possibility of so-called parasitic devices, i.e. unwanted interactions between the actual devices, under certain operating conditions. See [3,4] for an example. The study of the latchup is thus a very important aspect of the device simulation problem.

Within the continuation framework, the problem is that of continuing beyond and of computing turning points, cf. Fig. 3.1. The algorithmic issues related to the numerical computation of turning points are now reasonably well understood, and many algorithms have been proposed for this purpose. For latchup problems, the curvature near the turning points appears to be quite large in many cases, and we anticipate that the traditional approach of pseudo-arclength continuation coupled with an Euler predictor (cf. [8]) may require many small steps in such regions. Our present belief is that more general N of the form (2.4) coupled with a several-parameter predictor will prove to be a more efficient and robust combination. See example 2 in [10] for the application of the continuation method utilizing (2.4), (2.7) as implemented in PLTMG to a problem with a nearly angular turning point. To compute turning points the program utilizes a bisection/secant method on $\dot{\lambda}(\sigma) = 0$.

4. Numerical Results

The latchup problem described in the previous section will be solved by a suitable generalization of the continuation method of 2.

These results will then be presented in a forthcoming paper. Here we present continuation results for a MOSFET device. We assume that the system (3.1) has to be solved on a region schematically given in Figure 4.1.

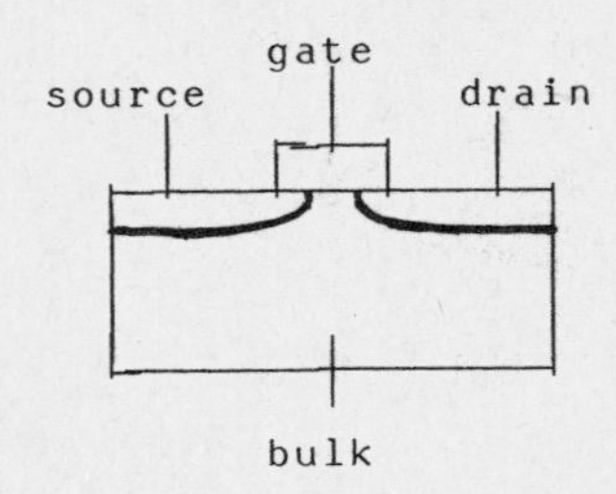

Fig. 4.1 Schematic diagram of a MOSFET

Such a device was considered in detail in [6] to which we refer. We specify now which coefficients and boundary conditions will be used for (3.1). The doping profile is $k_1 = 10^{10}$ in source and drain, while $k_1 = -10^5$ in the bulk, k_2 is zero and $\mu_n = \mu_p = 1$. Homogeneous Neumann boundary conditions for u, v and w are prescribed on the sides of the device, for u on the sides of the gate (oxide) and for v, w on the interface between oxide and silicon. v and w are not defined in the oxide region. Homogeneous Dirichlet conditions for v and w are prescribed on the bottom surface and the top of the source region, while u is -5 log 10 on the bottom, 10 log 10 at the source, u = 40, physically equivalent to 1 volt, on the top of the oxide. In the oxide region only $-\frac{1}{3}\Delta u = 0$ has to be solved. u has to be continuous across the silicon-oxide interface and the jump condition $\frac{1}{3}\nabla u.n\big|_{oxide} = \nabla u.n\big|_{silicon}$ has to be satisfied. The λ-dependent boundary condition is prescribed at the top of the drain as

$$(4.1)\qquad u = 40\lambda + 10 \log 10,$$

$$v = w = 40\lambda.$$

So λ corresponds to applied volts.

The triangulation serving as the coarse grid for a multi-grid method is shown in Figure 4.2. This triangulation was generated from a very coarse initial subdivision of the domain by the subroutine TRIGEN of the program PLTMG [1]. A generalization of this program, written by the authors, to 3 $\times$ 3 systems was used to derive a box method finite element discretization of (3.1) as outlined in 3. The

resulting discrete problem (3.5) had dimension 3N = 1401.

In the following we present the results of λ-continuation, i.e. using $\Theta = 0$ in (2.3). Starting from the solution u = v = w = 0 except for the fixed Dirichlet values first the solution for $\lambda = 0$ is computed. Then λ-steps of .25 are taken up to $\lambda = 3$.

λ	it_s	k_s	it_d	k_d	$\kappa(G_z)$
0	21	0	28	0	2.12E14
.25	5	6	8	4	
.5	13*	7	18*	0	1.43E15
.75	7	5	6	3	
1	20	3	12	2	6.35E17
1.25	14	3	17	1	
1.5	15	0	15	0	
1.75	6*	0	19	0	
2	6*	0	4	0	1.47E22
2.25	3	0	9	0	
2.5	4	3	10	0	
2.75	8	0	5	0	
3	5*	9	7	0	3.95E26

Table 4.1 Continuation results for MOSFET, λ drain voltage

Table 4.1 lists the number of the required corrector iterations using single respectively double precision VS-FORTRAN on an IBM 3081 and for some λ the condition numbers with respect to the spectral norm of the Jacobian G_z, the matrix of the linear system in the damped Newton corrector. An asterisk at an iteration number denotes that this step had to be subdivided further by PLTMG. The stopping criterion is adapted automatically by the program to the available machine precision. The values k_s, k_d denote the number of times the steplength in the damping process had to be cut back, k = 0 corresponds to the steplength 1.

We note the high number of corrector iterations needed for most of the continuation steps to λ about 2. Taking the size of the condition numbers into account it seems surprising that the single precision iterations converge at all. It is, however, not necessary that the Newton system is solved with a small relative error, it is sufficient that a descent direction is found. We see that considerably more damping is needed in single precision. The number of corrector steps for increasing values of λ up to (and exceeding) 3 is going down because the changes in the solution are smaller, the predictor yields already starting guesses with small residuals and a further reduction by only one or two orders of magnitude is required.

The approximate solutions are depicted in Figures 4.3-4.9. They were produced on a VAX 780 with approximately the same precision as single precision on the IBM. It was, however, only possible to continue beyond λ = .83 after relaxing the stopping criterion. This shows how close to failure these computations are. Figures 4.3-4.6 are surface plots as viewed from the SW-corner of the domain. Figures 4.7-4.9 are contour plots in which one contour line is drawn for each tenth of the total range of the depicted function. Although we do not present the results of multi-grid computations here we note that typically a MG run using additionally one or two adaptive refinements of the triangulation in Figure 4.2 will yield a sufficiently accurate solution for practical purposes.

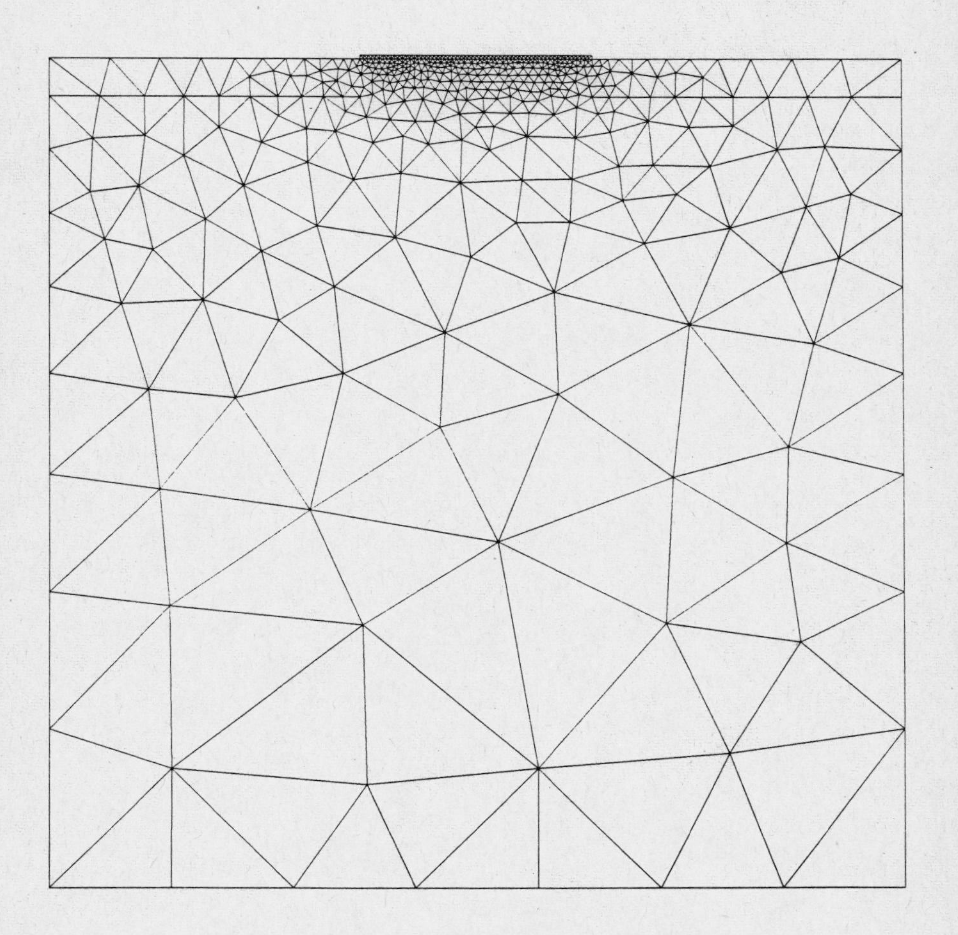

Fig. 4.2 Triangulation for MOSFET (467 vertices)

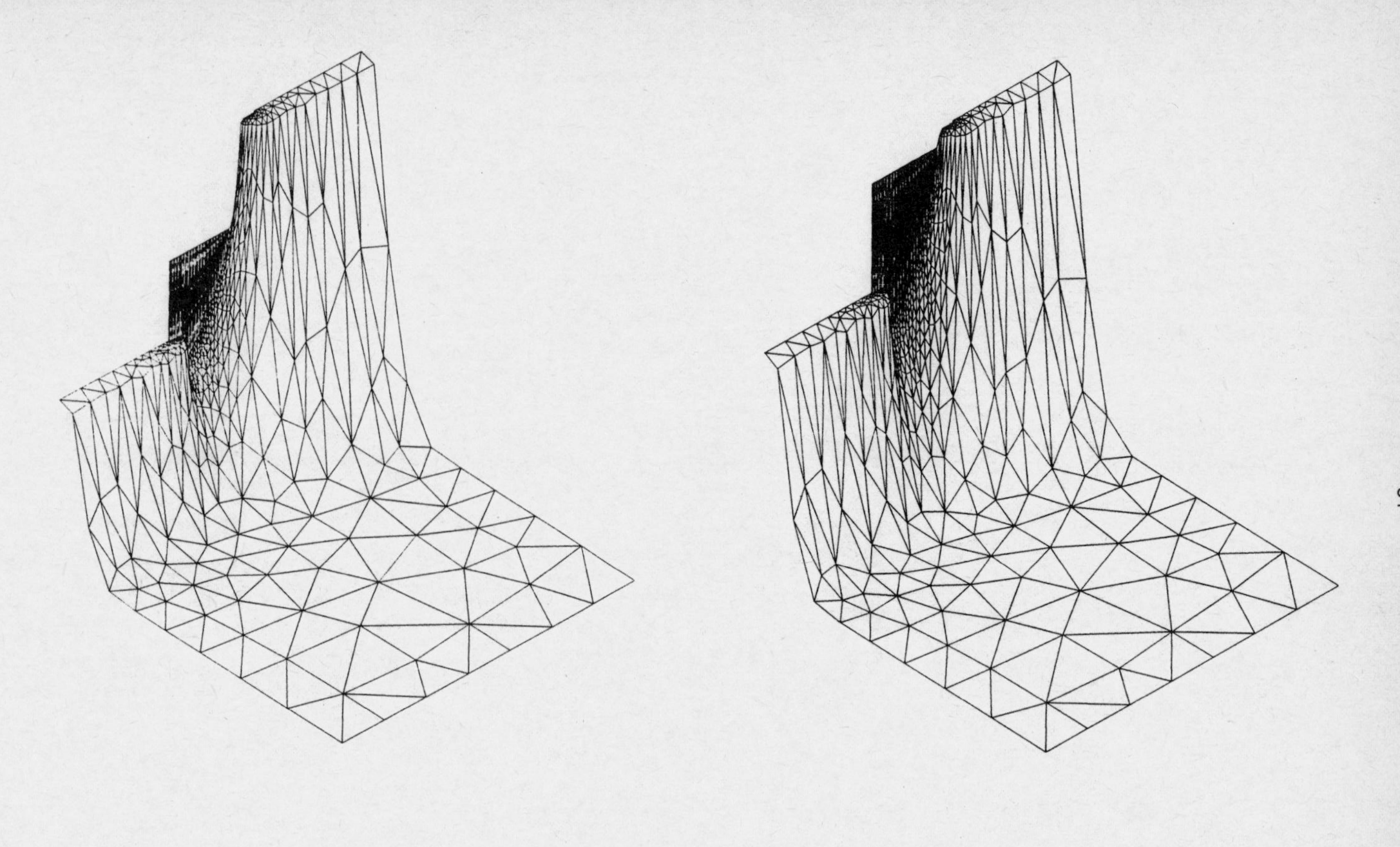

Fig. 4.3 u for 1 volt

Fig.4.4 u for .5 volt

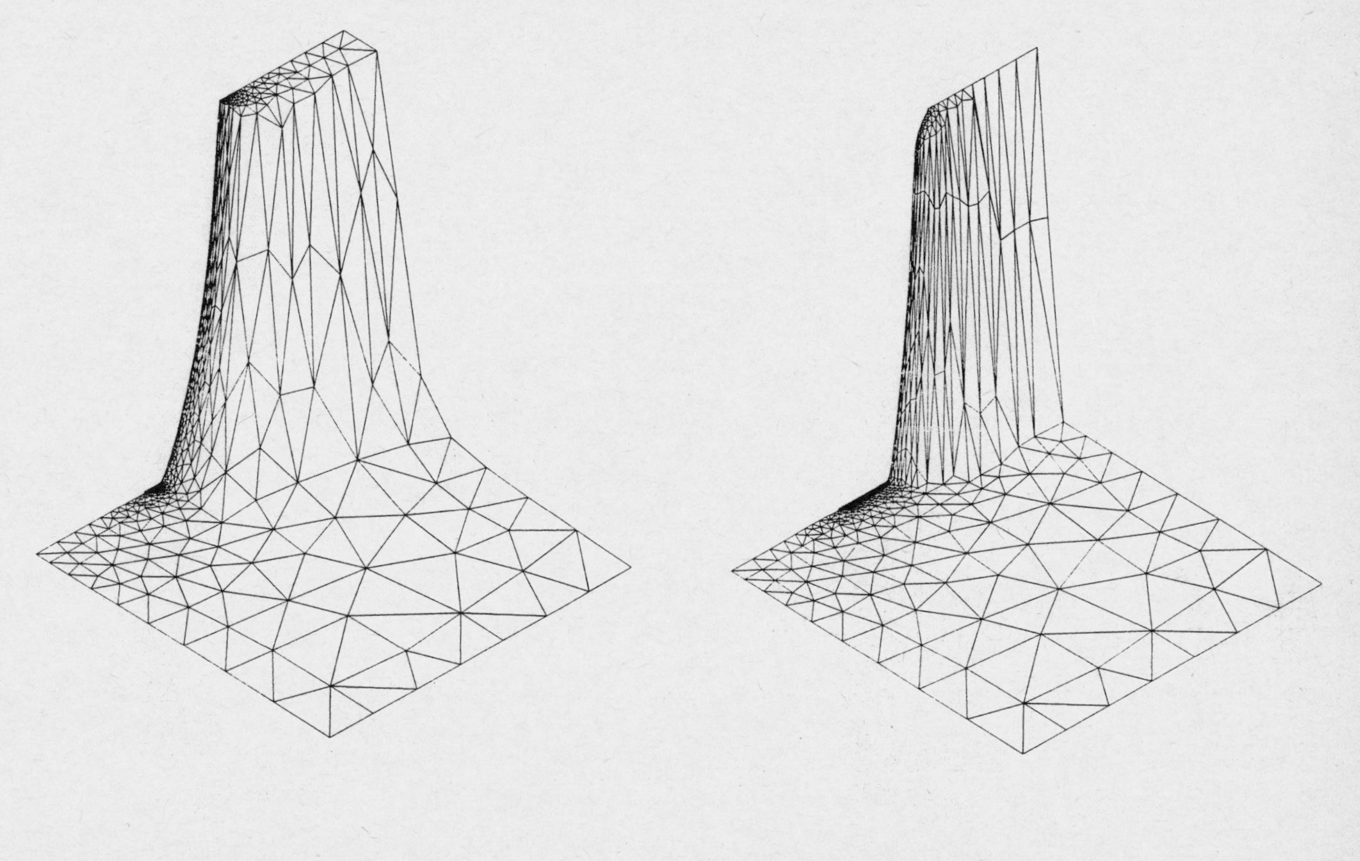

Fig. 4.5 v for 1 volt

Fig. 4.6 w for 1 volt

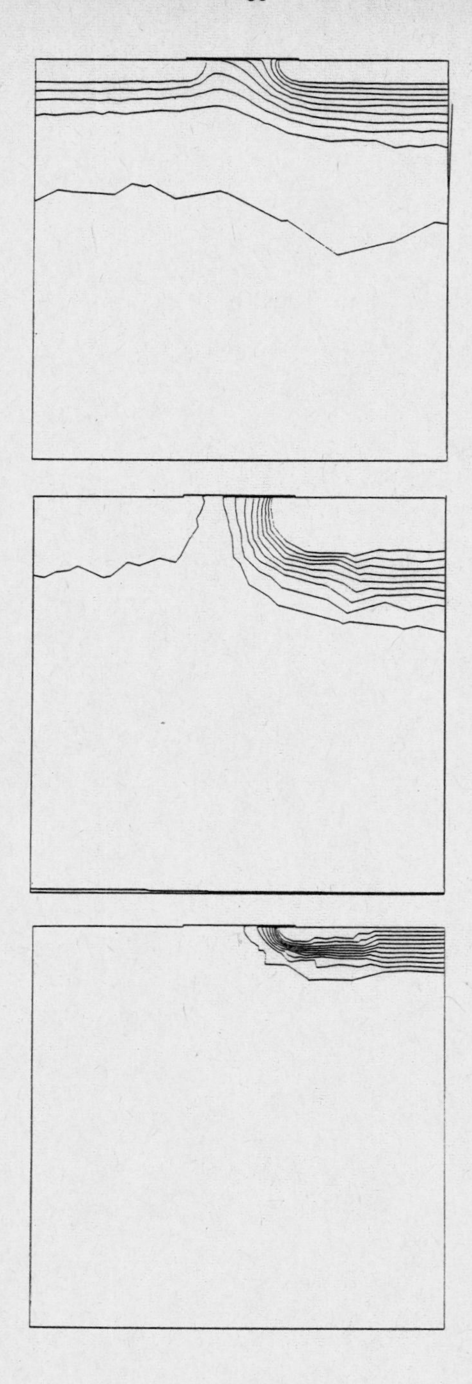

Figures 4.7-9   Contour plots of u (top), v and w for 1 volt

References

[1] R.E. Bank, PLTMG User's Guide, Edition 4.0, Tech. Report, Dept. Math., University of California, San Diego (1985).

[2] R.E. Bank, D.J. Rose and W. Fichtner, Numerical Methods for Semiconductor Device Simulation, SIAM J. Sci. Stat. Comp. 4, 416-435 (1983) and IEEE Trans. Electron Devices, Vol. ED-30, 1031-1041 (1983).

[3] R.E. Bank, W.M. Coughran, Jr., W. Fichtner, D.J. Rose and R.K. Smith, Computational Aspects of Semiconductor Device Simulation, Numerical Analysis Manuscript 85-3, AT&T Bell Laboratories, Murray Hill, NJ 07974.

[4] R.E. Bank, W.M. Coughran, Jr., W. Fichter, E.H. Grosse, D.J. Rose and R.K. Smith, Transient Simulation of Silicon Devices and Circuits, Numerical Analysis Manuscript 85-8, AT&T Bell Laboratories, Murray Hill, NJ 07974.

[5] R.E. Bank and T.F. Chan, PLTMGC: A Multi-grid Continuation Program for Parametrized Nonlinear Elliptic Systems, SIAM J. Sci. Stat. Comp. 3, 173-194 (1982).

[6] W. Fichtner, D.J. Rose and R.E. Bank, Semiconductor Device Simulation, SIAM J. Sci. Stat. Comp. 4, 391-415 (1983) and IEEE Trans. Electron Devices, Vol. ED-30, 1018-1030 (1983).

[7] W. Hackbusch, Multi-grid Solution of Continuation Problems, in "Iterative Solution of Non-linear Systems (R. Ansorge, T. Meis, W. Törnig, eds.), Lecture Notes in Mathematics 953, Springer-Verlag, Berlin (1982).

[8] P.A. Markowich, C.A. Ringhofer and A. Steindl, Computation of Current-Voltage Characteristics using Arclength Continuation, IMA J. Appl. Math. 33, 175-187 (1984).

[9] H.D. Mittelmann, Multi-grid methods for simple Bifurcation Problems, in "Multi-grid Methods (W. Hackbusch, U. Trottenberg, eds.), Lecture Notes in Mathematics 960, Springer-Verlag, Berlin (1982).

[10] H.D. Mittelmann, Multi-level Continuation Techniques for Nonlinear Boundary Value Problems with Parameter Dependence, Tech. Report No. 85, Dept. Math., Arizona State University, Tempe, AZ 85287 (1985) (to appear in Appl. Math. Comp.).

[11] H.D. Mittelmann, A Pseudo-Arclength Continuation Method for Nonlinear Eigenvalue Problems (submitted to SIAM J. Numer. Anal.) (1985).

[12] H.D. Mittelmann, Continuation near Symmetry-Breaking Bifurcation Points, in "Numerical Methods for Bifurcation Problems" (T. Küpper, H.D. Mittelmann, H. Weber, eds.), ISNM 70, Birkhäuser-Verlag, Basel (1984).

[13] W.C. Rheinboldt and J.V. Burkhardt, A locally parametrized Continuation Process, ACM Trans. Math. Software 9, 215-235 (1983).

MAGNETOHYDRODYNAMIC EQUILIBRIUM CALCULATIONS USING MULTIGRID

B.J. BRAAMS

F.O.M. Instituut voor Plasmafysica
Postbus 1207, 3430 BE Nieuwegein, The Netherlands

attached to Max-Planck-Institut für Plasmaphysik
D-8046 Garching bei München, Fed. Rep. Germany

ABSTRACT

The multigrid method has been applied to the solution of the two-dimensional elliptic equation that governs axisymmetric ideal magnetohydrodynamic equilibrium. The possibility of applying multigrid to the computation of axisymmetric equilibria in the 'inverse coordinates' formulation and to three-dimensional equilibrium and evolution calculations is investigated.

INTRODUCTION

In the magnetic confinement approach to controlled thermonuclear fusion, the hot plasma is prevented from escaping by the use of a strong magnetic field. The $\mathbf{v} \times \mathbf{B}$ force (velocity times magnetic field) inhibits motion of electrons and ions perpendicular to the magnetic field, and the plasma particles are constrained in lowest order to follow helical orbits 'tied' to a magnetic field line. In order to confine the plasma also with regard to its motion along the field, configurations are employed in which the magnetic field lines are enclosed in a toroidal region. The two main contenders in magnetic confinement research are the tokamak, which is an axisymmetric device, and the stellarator, which does not have a continuous symmetry. In either device, the magnetic field is generated by a combination of external currents and currents flowing in the plasma, making the configuration of plasma and magnetic field a nonlinear system.

We are concerned here with the application of multigrid methods to the computation of magnetic confinement configurations. In this context the plasma may be considered as an ideal magnetohydrodynamic (MHD) fluid, and the static equilibrium of plasma and field is described by the system of equations,

$$\nabla \cdot \mathbf{B} = 0, \tag{1^a}$$

$$(\nabla \times \mathbf{B}) \times \mathbf{B} = \nabla p, \tag{1^b}$$

where $\mathbf{B}$ is the magnetic field and p is the kinetic pressure. (Rationalized units in which $\mu_0 = 1$ are employed). Despite its simple appearance the system (1) is extremely difficult to solve numerically for general three-dimensional configurations, and a substantial fraction of the production time of the Cray-1 computer at Garching is spent on this problem. Typical calculations require $\sim 10^4$ iterations and take between 2 and 4 hours of CPU-time in order to compute a single equilibrium to acceptable accuracy. To find efficient methods for the computation of three-dimensional solutions to Eq. (1) is thus a major challenge for computational plasma physics. In the axisymmetric case the system (1) may be reduced to a single elliptic partial differential equation in the plane, which can be solved efficiently by a variety of methods.

Specifically, we consider in this paper the application of multigrid to three classes of problems in computational MHD: (1) Computation of axisymmetric equilibria on an Eulerian grid. (2) Axisymmetric equilibria in the inverse coordinates formulation. (3) Three-dimensional equilibrium and evolution calculations. Although fast numerical methods for these three problems are of considerable interest, no previous investigation into the use of multigrid for computations in magnetic confinement theory seems to exist.

Problem (1) requires the solution of an almost linear, uniformly elliptic equation (a nonlinearity occurs only in the right hand side), and application of multigrid is straightforward. Our code achieves full multigrid efficiency, and is several times faster than codes based on direct rapid elliptic solvers.

In regard to problem (2) we discuss the relationship between the inverse coordinates approach on one hand and grid generation through elliptic systems on the other. This leads to a formulation of the inverse coordinates equilibrium problem as a quasilinear elliptic system, which is suitable for multigrid treatment. (A code has not been written).

The discussion of problem (3) is also of an analytical nature. Local mode analysis is employed in order to develop a relaxation procedure, rather than for a posteriori validation only. The best local relaxation scheme will still be slowly converging for two classes of disturbances, viz. the slow magnetosonic and the shear Alfvén modes, in both cases with the wavevector nearly transverse to the magnetic field. These are the lowest frequency modes in the MHD spectrum. A satisfactory treatment of these troublesome modes requires distributive line relaxation along the magnetic field and semi-coarsening on magnetic surfaces; this emphasizes the need for an adaptive, field-tied grid for 3-D MHD calculations. The proposed simple relaxation scheme can be suitable for implicit ideal MHD evolution calculations on the slowest timescale.

The reader is assumed to be familiar with multigrid methods, as presented in particular in Refs. [**1**] and [**2**]. Useful reviews of the relevant MHD theory may be found in Refs. [**3**]–[**6**].

1. A Multigrid Code for Axisymmetric Equilibrium

The Grad-Schlüter-Shafranov equation. A very substantial simplification in the system (1) is afforded by the assumption of axisymmetry. Under this assumption the equation $\nabla \cdot \mathbf{B} = 0$ may be solved by choosing the representation,

$$\mathbf{B} = F\nabla\phi + \nabla\psi \times \nabla\phi,$$

where ϕ is the ignorable angle in the cylindrical (r, ϕ, z) coordinate system, and F and ψ are axisymmetric scalar functions. A similar representation is found for the current,

$$\nabla \times \mathbf{B} = -\Delta^*\psi\nabla\phi + \nabla F \times \nabla\phi,$$

where the operator Δ^* is given by

$$\Delta^*\psi = r\frac{\partial}{\partial r}\left(r^{-1}\frac{\partial\psi}{\partial r}\right) + \frac{\partial^2\psi}{\partial z^2}. \tag{2}$$

From the force balance equation (1^b) one may next derive $\nabla\psi\times\nabla p=\mathbf{0}$, $\nabla F\times\nabla p=\mathbf{0}$, and $\nabla\psi\times\nabla F=\mathbf{0}$. It is taken to follow that ψ, F, and p are functionally related, and one writes $F=F(\psi)$ and $p=p(\psi)$. (These must be understood as local relations in case a surface of constant ψ has disconnected parts). Finally, consideration of force balance along $\nabla\psi$ leads to the Grad-Schlüter-Shafranov equation [**7**]–[**9**],

$$\Delta^*\psi=-r^2\frac{dp}{d\psi}-F\frac{dF}{d\psi}. \tag{3}$$

This equation is the basis for the study of axisymmetric ideal MHD equilibrium. For given profiles $p(\psi)$ and $F(\psi)$ it is an almost linear elliptic p.d.e., the nonlinearity occurring only in the right hand side. The various methods (not including multigrid) that have been used in the past to solve Eq. (3) have been reviewed in Refs. [**10**] and [**11**].

Discretization scheme. Conventional second order accurate discretization methods for the equilibrium equation (3), written as $\Delta^*\psi=f(\mathbf{r},\psi)$, are of the five point molecule form,

$$\begin{pmatrix} & * & \\ * & * & * \\ & * & \end{pmatrix}\psi=f.$$

All previous finite difference codes for the solution of Eq. (3) employ such a second order method. Better methods are available for smooth f, in particular a fourth order accurate 'compact' discretization of the shape,

$$\begin{pmatrix} * & * & * \\ * & * & * \\ * & * & * \end{pmatrix}\psi=\begin{pmatrix} & * & \\ * & * & * \\ & * & \end{pmatrix}f.$$

Specifically: Consider a uniform rectangular mesh with spacing $\delta r=h$ and $\delta z=k$. Consider the natural splitting, $\Delta^*=\mathcal{L}_r+\mathcal{L}_z$, and define the second order accurate difference approximations $\mathcal{L}_r^h$ and $\mathcal{L}_z^k$ by,

$$[\mathcal{L}_r^h\psi](r,z)=\frac{r}{h^2}\Big(\frac{\psi(r+h,z)-\psi(r,z)}{r+h/2}-\frac{\psi(r,z)-\psi(r-h,z)}{r-h/2}\Big),$$

$$[\mathcal{L}_z^k\psi](r,z)=\frac{1}{k^2}(\psi(r,z+k)-2\psi(r,z)+\psi(r,z-k)).$$

Then a fourth order discretization of $\Delta^*\psi=f$ is obtained from the identity,

$$\begin{aligned}&\Big[\mathcal{L}_r^h+\mathcal{L}_z^k+\frac{1}{12}(h^2+k^2)\mathcal{L}_r^h\mathcal{L}_z^k+O(h^4+k^4)\Big]\psi\\&=\Big(1+\frac{1}{12}h^2\mathcal{L}_r^h+\frac{1}{12}k^2\mathcal{L}_z^k\Big)f.\end{aligned} \tag{4}$$

A special treatment on the plasma–vacuum interface, on which f or its first order derivatives may be discontinuous, is required in order to gain full advantage of the higher accuracy obtained for the interior equations. (Such a treatment has not been implemented in our code).

Performance of multigrid. A demonstration code has been written, which employs multigrid relaxation to solve Eq. (3) on a rectangular region subject to Dirichlet boundary conditions. The mesh at level i has size $(2^i + 1) \times (2^i + 1)$, where $1 \leq i$; the coarsest grid therefore contains only one single interior point. The cycling algorithm is of the full multigrid, full approximation storage variety, and employs adaptive switching. Full-weighting transfer is used for both the solution and the residuals, and bi-cubic interpolation is used for the corrections. Red-black point relaxation is employed on all grids except on the coarsest, where a nonlinear root-finding algorithm is employed. The special treatment on the coarsest grid is necessary because the equation generally admits multiple solutions; the algorithm that is employed on the coarsest grid is designed to find the interesting solution.

For an example calculation we assumed a right hand side in Eq. (3) of the form,

$$f(\mathbf{r}, \psi) = \begin{cases} -r^2 g(\psi) - c, & \psi > 0 \\ 0, & \psi \leq 0 \end{cases}$$

in which g is a second-degree polynomial and c is a constant. The contour defined by $\psi = 0$ is the free boundary of the plasma. A calculation using 7 levels (the finest grid has size 129×129) required 120 msec on the Cray-1 to solve Eq. (3) to the level of the discretization error. The total number of passes over each of the levels 1–7 was 10, 19, 14, 8, 6, 4, and 2 respectively. The computing time was divided about evenly between evaluation of the r.h.s., evaluation of the residuals, bi-cubic interpolation, and all other chores.

Thus, full multigrid efficiency for the solution of the equation (3) has been achieved. For a linear problem this code is about 2–3 times slower than the well-optimized fast Buneman solver used at Garching [10], but for the more relevant nonlinear problems the codes based on a direct elliptic solver require Picard iteration, and multigrid relaxation is easily the fastest procedure available.

2. Axisymmetric Equilibrium in Inverse Coordinates

For stability and transport calculations related to axisymmetric configurations it is required to have an explicit representation of the magnetic surfaces of the equilibrium (the contours of constant ψ), which are in this case assumed to form a nested set that converges to a single 'magnetic axis'. Such a representation may be found by a numerical mapping after having computed the equilibrium on a fixed spatial grid. In recent years, however, a class of procedures has become popular in which the equilibrium is computed in a formulation that immediately gives the spatial coordinates r and z as functions of ψ and η, where η is some angular variable. This 'inverse variables' method has been employed in Refs. [**12**]–[**15**]. There is considerable freedom in the choice of the angular variable η. It is defined via orthogonality in Ref. [**12**], via a specification of the Jacobian in Refs. [**13**] and [**14**], and in Ref. [**15**] it is suggested to choose η such that contours of constant η are straight rays.

The definition of the angular variable via orthogonality or via a Jacobian constraint leads to a system of differential–algebraic equations that is not easy to solve numerically. On the other hand, to choose the contours of constant η to be straight rays is rather restrictive. A more general formulation for defining the angular variable is suggested by an analogy with the method of grid generation through elliptic equations [**15**], [**16**], [**17**]. This leads to a formulation of the inverse equilibrium problem in which the coordinate η is defined as the solution of an elliptic equation, resulting in a quasilinear elliptic system of equations for $r(\psi, z)$ and $z(\psi, z)$. This formulation is discussed below, as it is the most suitable for multigrid treatment.

Grid generation through elliptic equations. We first consider by way of example the problem of constructing a boundary-fitted curvilinear coordinate system (ξ^1, ξ^2) to cover the pseudo-rectangular region $G \in \mathbf{R}^2$. The Cartesian coordinates on $\mathbf{R}^2$ are (x_1, x_2). G is to be mapped to the unit square in the (ξ^1, ξ^2) plane, and the points A, B, C, and D on the boundary ∂G are to be mapped to $(0,0)$, $(0,1)$, $(1,1)$, and $(1,0)$. Using the method of grid generation through elliptic equations, the curvilinear coordinates ξ^i are defined by Poisson equations,

$$\Delta \xi^i = F^i, \tag{5}$$

subject to Dirichlet conditions on ∂G: $\xi^1 = 0$ on AB, $\xi^1 = 1$ on CD, etc. In the simplest case one chooses $F^i = 0$, but a nonzero right hand side in Eq. (5) may be used to obtain more control over the resulting mesh.

The expression for the Laplacian on the curvilinear coordinates ξ^i is,

$$\Delta u = \frac{1}{\sqrt{g}} \frac{\partial}{\partial \xi^i} \left(\sqrt{g}\, g^{ij} \frac{\partial u}{\partial \xi^j} \right). \tag{6}$$

(Summation over repeated indices is understood). Substituting ξ^k for u in the above equation and considering Eq. (5) one finds,

$$\frac{1}{\sqrt{g}} \frac{\partial}{\partial \xi^i} (\sqrt{g}\, g^{ij}) = F^j$$

and therefore,

$$\Delta u = g^{ij} \frac{\partial^2 u}{\partial \xi^i \partial \xi^j} + F^j \frac{\partial u}{\partial \xi^j} \tag{7}$$

To obtain numerically the transformation $\vec{x} \to \vec{\xi}$ it is convenient to solve in inverse coordinates, and obtain $\vec{x}$ as a function of $\vec{\xi}$ rather than $\vec{\xi}$ as a function of $\vec{x}$. (The whole point of the grid generation is that differential equations may be more easily solved on the transformed region). The components x_i are obtained by solving

$$\Delta x_i = 0, \quad i = 1, 2 \tag{8}$$

on the unit square in the (ξ^1, ξ^2) plane, again subject to Dirichlet boundary conditions. With Δ given by Eq. (7) this is a quasilinear elliptic system.

Application to axisymmetric equilibrium. The analogy between the above model problem of grid generation and the inverse coordinates approach to axisymmetric MHD equilibrium is quite obvious; one only has to replace Δ by Δ^* and change from a rectangular to a polar geometry, with the singularity at the magnetic axis. In a curvilinear coordinate system (ξ, η) the operator Δ^* has the representation,

$$\begin{aligned} \Delta^* u = |\nabla \xi|^2 \frac{\partial^2 u}{\partial \xi^2} + 2(\nabla \xi \cdot \nabla \eta) \frac{\partial^2 u}{\partial \xi \partial \eta} + |\nabla \eta|^2 \frac{\partial^2 u}{\partial \eta^2} \\ + \Delta^* \xi \frac{\partial u}{\partial \xi} + \Delta^* \eta \frac{\partial u}{\partial \eta}, \end{aligned} \tag{9}$$

which corresponds to Eq. (7). For the transformed coordinate ξ we may choose $\xi = \psi$, or any function of ψ alone, so that $\Delta^* \xi$ follows from the equilibrium equation

(3), and may be assumed known. The natural choice of the differential equation for η is $\Delta^*\eta = 0$. From Eq. (2) it is seen that

$$\begin{cases} \Delta^* r = -r^{-1} \\ \Delta^* z = 0 \end{cases} \tag{10}$$

Considering r and z as functions of ξ and η, Eq. (10) naturally remains valid, but the operator Δ^* is given by Eq. (9) instead of by Eq. (2). This quasi-linear elliptic system of equations (10) governs the equilibrium in inverse coordinates. The boundary conditions require periodicity in η, specified r and z on the plasma boundary, and an appropriate expansion near the magnetic axis.

The system (10) appears suitable for multigrid treatment, although it is more complicated than the equilibrium equation in the form (3). The standard second order discretization has a symmetric nine-point stencil, and an incomplete Cholesky decomposition should probably be used for relaxation. (Alternating direction line relaxation is an alternative, but point relaxation must not be employed on a polar grid). Higher order discretizations would be of interest, in particular a spectral method in the angular coordinate.

3. Prospects for Three-Dimensional Calculations

In this Section an analytical study of the use of multigrid for the difficult and (at present) very expensive area of 3-D MHD computations is initiated. The system of equations (1) is equivalent to the system that governs steady, inviscid, incompressible flow, as can be seen by making the substitutions,

$$\mathbf{B} \to \mathbf{v}, \qquad p \to -\frac{p}{\rho} - \frac{v^2}{2}.$$

Progress in solving the corresponding hydrodynamic equations is therefore of immediate interest for magnetic confinement studies.

The Chodura and Schlüter approach. In attempting to develop a multigrid relaxation procedure for 3-D equilibrium I have found it convenient to take as the starting point the approach of Chodura and Schlüter [**18**], who employ a finite difference discretization on a fixed spatial mesh of the system (1) in primitive variables.

The solution procedure employed by Chodura and Schlüter is designed to find a constrained minimum of the potential energy, W, which is given by

$$W = \int_T \left(\frac{B^2}{2} + \frac{p}{\gamma - 1} \right) dV, \tag{11}$$

where T is the toroidal computational region, and γ is the adiabatic index. Minimization of W is performed through displacements of the form,

$$\begin{cases} \delta \mathbf{B} = \nabla \times (\boldsymbol{\xi} \times \mathbf{B}) \\ \delta \rho = -\nabla \cdot (\rho \boldsymbol{\xi}) \end{cases} \tag{12}$$

subject to $\boldsymbol{\xi} = 0$ on the boundary ∂T. The relation $p = \rho^\gamma$ is assumed (ρ corresponds to the mass density of the MHD fluid). Through these displacements, an arbitrary initial plasma and field configuration is transformed under the constraints of mass and flux conservation into a minimum energy state.

In leading order the change in energy due to the displacements (12) is given by

$$\delta W = -\int_T (\mathbf{F} \cdot \boldsymbol{\xi})\, dV, \tag{13}$$

where $\mathbf{F} = (\nabla \times \mathbf{B}) \times \mathbf{B} - \nabla p$. Therefore, a state of minimum energy under the displacements (12) satisfies $\mathbf{F} = 0$, and is a solution to the force balance equation (1^b). The equation (1^a) remains satisfied if it is satisfied initially. Eq. (13) also shows a possible route to the energy minimum, viz. to choose at each iteration $\boldsymbol{\xi} = \alpha \mathbf{F}$, where $\alpha > 0$ and α is sufficiently small to ensure stability, but this steepest descent algorithm is prohibitively slow. Chodura and Schlüter have employed both conjugate gradient acceleration and a second-order Richardson scheme, with good results, but some 10^3–10^4 iterations are still required for practical calculations.

A multigrid relaxation procedure. We now attempt to derive a relaxation procedure for the system (1) that effectively reduces the short wavelength error components, and that may therefore be suitable in the context of multigrid. A finite difference discretization of Eq. (1) on a staggered mesh is envisaged, but it turns out that the analysis of relaxation procedures can largely be carried out without reference to the discretized system of equations.

To satisfy flux conservation, $\nabla \cdot \mathbf{B} = 0$, is of course easy. At each relaxation sweep $\mathbf{B}$ may be updated by a distributive scheme of the form $\mathbf{B} \leftarrow \mathbf{B} + \delta \mathbf{B}$, where

$\delta\mathbf{B} = \nabla\chi$. To satisfy exactly $\nabla \cdot \mathbf{B} = 0$ after the iteration sweep one would have to find χ as the solution to a Poisson equation, but here it suffices to approximate χ locally by any kind of relaxation prescription that is suitable for Poisson equations. Notice that the replacement $\mathbf{B} \leftarrow \mathbf{B} + \nabla\chi$ does not affect the force balance equation.

For the second equation, $\mathbf{F} = \mathbf{0}$, the work of Chodura and Schlüter suggests a relaxation scheme based on the coupled replacements $\mathbf{B} \leftarrow \mathbf{B} + \delta\mathbf{B}$ and $p \leftarrow p + \delta p$, where $\delta p = \gamma\,(p/\rho)\delta\rho$, and where $\delta\mathbf{B}$ and $\delta\rho$ are given by Eq. (12). These replacements do not affect $\nabla \cdot \mathbf{B} = 0$. The question is how to choose the displacement vector $\boldsymbol{\xi}$ in Eq. (12) as a function of the current residual $\mathbf{F}$.

To answer this question one must consider the principal terms of the change $\delta\mathbf{F}$ under the displacements (12), viz. those terms in which $\boldsymbol{\xi}$ is twice differentiated:

$$\begin{aligned} \delta\mathbf{F} \simeq (\mathbf{B}\cdot\nabla)(\mathbf{B}\cdot\nabla)\boldsymbol{\xi} + (B^2 + \gamma p)\nabla(\nabla\cdot\boldsymbol{\xi}) \\ - \mathbf{B}(\mathbf{B}\cdot\nabla)(\nabla\cdot\boldsymbol{\xi}) - (\mathbf{B}\cdot\nabla)\nabla(\mathbf{B}\cdot\boldsymbol{\xi}) \end{aligned} \tag{14}$$

(The operator ∇ acts on $\boldsymbol{\xi}$ only). A desirable relaxation scheme should have $\delta\mathbf{F} \simeq -\mathbf{F}$, at least for the short wavelength components. Fourier analysis transforms $\delta\mathbf{F}$ into $\delta\tilde{\mathbf{F}}$ and $\boldsymbol{\xi}$ into $\tilde{\boldsymbol{\xi}}$, related by $\delta\tilde{\mathbf{F}} \simeq \boldsymbol{A}\cdot\tilde{\boldsymbol{\xi}}$, where

$$\boldsymbol{A} = -B^2(k_{\parallel}^2\boldsymbol{I} + (1+\beta)\mathbf{k}\mathbf{k} - k_{\parallel}(\mathbf{b}\mathbf{k} + \mathbf{k}\mathbf{b})) \tag{15}$$

where $\mathbf{k}$ is the wavevector, $\beta = \gamma p/B^2$, $\mathbf{b} = \mathbf{B}/B$, and $k_{\parallel} = \mathbf{k}\cdot\mathbf{b}$. ($\beta$ is a small parameter for magnetic confinement).

The operator $\boldsymbol{A}$ may be inverted:

$$\boldsymbol{A}^{-1} = -(B^2 k_{\parallel}^2)^{-1}(\boldsymbol{I} + \beta^{-1}\mathbf{b}\mathbf{b} - \mathbf{k}\mathbf{k}/k^2), \tag{16}$$

and a desirable relaxation scheme should approximate $\tilde{\boldsymbol{\xi}} = -\boldsymbol{A}^{-1}\tilde{\mathbf{F}}$ for short wavelength components. Of course $\boldsymbol{A}^{-1}$ contains $\mathbf{k}$, and is therefore not a local linear operator. Various things can be tried, for instance to drop the term $\mathbf{k}\mathbf{k}/k^2$ and to replace $k_{\parallel}^2$ by $2\omega^{-1}(h_x^{-2} + h_y^{-2} + h_z^{-2})$, where ω is a constant of order unity, and h_x, h_y, and h_z are the local mesh spacings. Via this prescription one obtains a relaxation procedure based on

$$\boldsymbol{\xi} = \boldsymbol{R}\cdot\mathbf{F}, \tag{17}$$

where $\boldsymbol{R}$ is the operator,

$$\boldsymbol{R} = \frac{\omega}{2}B^{-2}(h_x^{-2} + h_y^{-2} + h_z^{-2})^{-1}(\boldsymbol{I} + \beta^{-1}\mathbf{b}\mathbf{b}). \tag{18}$$

The large coefficient on $\mathbf{b}\mathbf{b}$ in this relaxation prescription is worthy of note.

Further analysis of the proposed procedure. The relaxation scheme (17) must now be analyzed in order to see whether all short wavelength error modes are effectively reduced. Continuing with the linear analysis, and still considering principal terms only, one finds,

$$\begin{aligned}\delta\tilde{\mathbf{F}} &= \boldsymbol{A}\cdot\boldsymbol{R}\cdot\tilde{\mathbf{F}}\\ &= -\frac{\omega}{2}(h_x^{-2}+h_y^{-2}+h_z^{-2})^{-1}(k_\parallel^2\boldsymbol{I}+(1+\beta)\mathbf{kk}-k_\parallel\mathbf{bk})\cdot\tilde{\mathbf{F}}.\end{aligned}$$

It may be seen that the scheme is not satisfactory, as those modes for which

$$\mathbf{k}\perp\mathbf{B}\quad\text{and}\quad\mathbf{k}\perp\tilde{\mathbf{F}}$$

(approximately) are not well eliminated. (In addition there may be problems related to the occurrence of different values of the grid spacing, but those difficulties are easily taken care of by line relaxation). The troublesome modes are the slow magnetosonic mode, for which $\mathbf{F}$ and $\mathbf{B}$ are nearly parallel, and the shear Alfvén mode, for which $\mathbf{B}$, $\mathbf{F}$, and $\mathbf{k}$ form an orthogonal triad. These are the lowest frequency modes ($\nu \to 0$) in the MHD spectrum.

The reason that these modes are not well eliminated can also be understood on physical grounds. The perturbation related to the slow magnetosonic mode concerns the pressure only, and is characterized by a long wavelength along the magnetic field and a short wavelength across the field. As the restoring force for this perturbation acts along field lines, the relaxation procedure only becomes effective when the mesh spacing corresponds to the length scale along the field, but on such a mesh (assuming equal coarsening in all directions) the perturbation will be invisible due to the rapid variation across the field. Similarly, the restoring force for the shear Alfvén mode is located in a plane in which the mode has a long wavelength, but perpendicular to this plane there is a rapid variation.

Both the form of the operator $\boldsymbol{A}^{-1}$ in Eq. (16) and the physical picture outlined above point the way to a remedy. One needs line relaxation along the magnetic field (which allows to retain $k_\parallel$ in going from $\boldsymbol{A}^{-1}$ to $\boldsymbol{R}$) to eliminate effectively the slow magnetosonic mode, and either plane relaxation or (more likely) semi-coarsening within flux surfaces to deal with the shear Alfvén mode. As the magnetic field configuration is unknown a priori this requires an adaptive grid, approximately tied to the field. Development of adaptive grid methods for 3-D MHD is also important for reasons of numerical accuracy, but no satisfactory algorithm exists at present. Nevertheless, multigrid in conjunction with adaptive grid methods seems the most promising area of investigation towards efficient 3-D MHD equilibrium computations.

For time dependent three-dimensional calculations the scheme derived above may be more promising, as it would allow to follow accurately the evolution on the longest ideal MHD timescale, while eliminating efficiently the faster modes.

CONCLUSIONS

One objective in writing this paper has been to point out to both plasma physicists and multigrid experts that certain problems in computational MHD are of shared interest.

The axisymmetric equilibrium problem lends itself to a straightforward application of the multigrid procedure, and this has resulted in a code that is ~ 3 times faster than a code which uses a well optimized Buneman solver and Picard iteration. The main interest in very fast 2-D equilibrium calculations is for real-time data interpretation and control of an experiment, on a timescale of ~ 10 msec or less. Considering that in monitoring an experiment one is solving a chain of similar problems, and that a grid of modest size will suffice, our study has demonstrated at least the near-term feasibility of this application.

The problem of computing axisymmetric equilibrium in the inverse coordinates formulation is a more challenging (although hardly speculative) application of multigrid, for which furthermore the relative gain over competing methods would be much larger, as rapid direct solvers are not available. Not all previous formulations of the inverse equilibrium problem are well suited for multigrid treatment, but the analogy with grid generation through elliptic equations shows the correct approach. In particular, any code for elliptic grid generation that can handle a polar geometry should almost immediately be applicable to the inverse coordinates MHD equilibrium problem.

The really difficult and expensive areas of work in computational MHD are the stability eigenvalue problem for axisymmetric equilibria (which has not been addressed in this paper), and the three-dimensional equilibrium and evolution problems. An impression of the complexity of the 2-D stability problem can be gained by noticing that it has required nearly a decade of work and the advent of the Cray-1 computer before the main result from the existing stability codes was obtained, viz. the Troyon scaling law [**19**]. For three-dimensional equilibrium and evolution problems a multigrid approach has been initiated here, but a fully satisfactory procedure has not yet been obtained. The main outstanding problem for these 3-D computations is to develop adaptive methods, in which the grid is adjusted to the (unknown) magnetic configuration.

ACKNOWLEDGEMENTS

I am grateful to Drs. L.M. Degtyarev and V.V. Drozdov for comments on an earlier laboratory preprint of this paper, which have led to a corrected discussion of the literature on the inverse variables method.

This work was performed as part of the research program of the association agreement of Euratom and the "Stichting voor Fundamenteel Onderzoek der Materie" (FOM) with financial support from the "Nederlandse Organisatie voor Zuiver-Wetenschappelijk Onderzoek" (ZWO) and Euratom. The author's stay at Garching was supported through a Euratom mobility agreement.

REFERENCES

1. A. Brandt, *Multi-Level Adaptive Solutions to Boundary Value Problems*, Math. Comp. **31** (1977), 333–390.
2. A. Brandt, *Guide to Multigrid Development*, in "Multigrid Methods", Proceedings of the Conference held at Kōln–Porz, Nov. 1981 (W. Hackbusch, U. Trottenberg, Eds.), Lecture Notes in Mathematics, Springer, Berlin, 1982.
3. G. Bateman, "MHD Instabilities", The MIT Press, Cambridge, Mass., 1978.
4. J.P. Freidberg, *Ideal Magnetohydrodynamic Theory of Magnetic Fusion Systems*, Rev. Mod. Phys. **54** (1982), 801–902.
5. B.B. Kadomtsev and V.D. Shafranov, *Magnetic Plasma Confinement*, Sov. Phys. Usp. **26** (1983), 207–227; Usp. Fiz. Nauk (USSR) **139** (1983), 399–434.
6. R.M. Kulsrud, *MHD Description of Plasma*, in "Handbook of Plasma Physics", Vol. 1: Basic Plasma Physics I, (A.A. Galeev and R.N. Sudan, Eds.), Elsevier, Amsterdam, 1984, pp. 115–145.
7. V.D. Shafranov, *On Magnetohydrodynamical Equilibrium Configurations*, Sov. Phys. JETP **6** (1958), 545–554; J. Exper. Theor. Phys. **33** (1957), 710–722.

8. R. Lüst and A. Schlüter, *Axialsymmetrische Magnetohydrodynamische Gleichgewichtskonfigurationen*, Z. Naturforsch. **12a** (1957), 850–854.
9. H. Grad and H. Rubin, *Hydromagnetic Equilibria and Force-Free Fields*, in "Proceedings of the Second United Nations International Conference on the Peaceful Uses of Atomic Energy", Geneva, 1958.
10. K. Lackner, *Computation of Ideal MHD Equilibria*, Comput. Phys. Commun. **12** (1976), 33–44.
11. L.E. Zakharov and V.D. Shafranov, *Equilibrium of Current-Carrying Plasmas in Toroidal Systems*, in "Reviews of Plasma Physics", vol. 11, Energoisdat, Moscow, 1982 (Russian, translation not yet available).
12. P.N. Vabishchevich, L.M. Degtyarev and A.P. Favorskii, *Variable-Inversion Method in MHD-Equilibrium Problems*, Sov. J. Plasma Phys. **4** (1978), 554–556, Fiz. Plazmy **4** (1978), 995–1000.
13. J. DeLucia, S.C. Jardin and A.M.M. Todd, *An Iterative Metric Method for Solving the Inverse Tokamak Equilibrium Problem*, J. Comput. Phys. **37** (1980), 183–204.
14. H.R. Hicks, R.A. Dory and J.A. Holmes, *Inverse Plasma Equilibria*, Comput. Phys. Reports **1** (1984), 373–388.
15. L.M. Degtyarev and V.V. Drozdov, *An Inverse Variable Technique in the MHD-Equilibrium Problem*, Comput. Phys. Reports **2** (1985), 341–387.
16. J.F. Thompson (Ed.), "Numerical Grid Generation", North Holland, New York, 1982.
17. J.F. Thompson, Z.U.A. Warsi and C.W. Mastin, *Boundary-Fitted Coordinate Systems for Numerical Solution of Partial Differential Equations—A Review*, J. Comput. Phys. **47** (1982), 1–108.
18. R. Chodura and A. Schlüter, *A 3D Code for MHD Equilibrium and Stability*, J. Comput. Phys. **41** (1981), 68–88.
19. F. Troyon, R. Gruber, H. Saurenmann, S. Semenzato and S. Succi, *MHD-Limits to Plasma Confinement*, 11th European Conference on Controlled Fusion and Plasma Physics, Aachen, 1983, Plasma Phys. Contr. Fusion **26** (1984), 209–215.

On the Combination of the Multigrid Method and Conjugate Gradients

Dietrich Braess
Fakultät für Mathematik
Ruhr-Universität, D-4630 Bochum, F.R. Germany

Multigrid algorithms and conjugate gradient methods with appropriate preconditioning are both efficient tools for the solution of equations which arise from the discretization of partial differential equations. Sometimes it is favourable to combine both methods. We will discuss two typical examples which elucidate different reasons for the combination of both methods. 1. When solving elasticity problems with multigrid methods, conjugate gradients are useful for avoiding the locking effect. 2. When the biharmonic equation or plate bending problems are treated by using conjugate gradients, the fast Poisson solvers provide good preconditioners. The analysis of both problems leads to different mathematical problems.

1. Introduction

The finite element discretization of elliptic equations leads to large systems of equations. Efficient tools for solving them are multigrid methods and conjugate gradient methods. The combination of both methods was already treated by Bank and Douglas [2] who investigated conjugate gradients in the smoothing step of multigrid algorithms. Braess and Peisker [4] considered the solution of the biharmonic equation by conjugate gradients with a multigrid procedure for the Poisson equation used for preconditioning. This has the advantage that a fast and well tuned Poisson solver [11] can be used, and it is not necessary to establish an individual multigrid code for the biharmonic problem.

Moreover, Kettler [9] proposed to accelerate multigrid iterations by using it as a preconditioner for a conjugate gradient method. Recently, G. Brand [5] reported on a multigrid algorithm for solving an elasticity problem, which only became an efficient algorithm after applying this kind of acceleration.

We emphasize that the mathematical analysis of each case is different from the other ones. Nevertheless, we will focus our attention to those arguments which are not restricted to the special cases under consideration.

Principally, there are two different ways of a combination of both methods:

1. Conjugate gradient methods are used in the smoothing step of a multigrid iteration. This was recommended in cases where an unknown direction in the domain is distinguished and point relaxation yields poor smoothing rates. Another reason for using this modification of the smoothing step seems also to be importnat. In some situations, the finest grid is almost the coarsest grid which is still reasonable for physical reasons, i.e., the approximation property of the coarse-grid-correction may be unsatisfactory. The use of conjugate gradients will not only provide the smoothing but at the same time improve the coarse-grid correction.

2. Multigrid algorithms are used as preconditioners for conjugate gradient algorithms. If the error reduction in one multigrid cycle is larger than, say 1/3, the convergence is improved in this way. Moreover, if the multigrid method is only used for preconditioning, one may use it with a (modified and) simpler problem as long as the original and the modified problem are spectral equivalent. In this way it becomes possible to apply fast Poisson solvers and other well tuned multigrid codes to a broader class of problems.

2. A Multigrid Algorithm for the 2-dimensional Linear Elasticity Problem

In elasticity theory one encounters systems of partial differential equations of second (or fourth) order. If in particular the displacements are small and independent of the third variable, one is led to the linear system for the membrane [6,10]:

$$\left.\begin{aligned} u_{xx} + \tfrac{1}{2}(1-\nu)u_{yy} + \tfrac{1}{2}(1+\nu)v_{xy} &= f \\ \tfrac{1}{2}(1-\nu)v_{xx} + v_{yy} + \tfrac{1}{2}(1+\nu)u_{xy} &= g \end{aligned}\right\} \quad (x,y)\in\Omega. \tag{2.1}$$

Dirichlet boundary conditions are found on that part of the boundary, where the displacements u and v are prescribed. On the other part of the boundary natural boundary conditions are encountered.

The Poisson coefficient ν depends on the material $(0<\nu<1/2)$. Typical values are $\nu\approx 1/3$.

The differential equations (2.1) are the Euler equations for the minimization of the energy:

$\Pi(u,v) = a(u,v) - \int(fu-gv)\,dxdy$ where

$$a(u,v) = \int \begin{pmatrix} \varepsilon_{11} \\ \varepsilon_{22} \\ \varepsilon_{12} \end{pmatrix} \begin{pmatrix} 1 & \nu & 0 \\ \nu & 1 & 0 \\ 0 & 0 & 2(1-\nu) \end{pmatrix} \begin{pmatrix} \varepsilon_{11} \\ \varepsilon_{22} \\ \varepsilon_{12} \end{pmatrix} dxdy \qquad (2.2)$$

and the strains are the symmetric derivatives

$$\varepsilon_{11} = u_x, \quad \varepsilon_{22} = v_y, \quad \varepsilon_{12} = \frac{1}{2}(u_y+v_x).$$

Recently, G. Brand [5] investigated a multigrid algorithm for a finite element discretization. Specifically, the algorithm referred to piecewise linear functions in regular triangular grids (see the figure below) and to 2-block Gauss-Seidel relaxation for the smoothing.

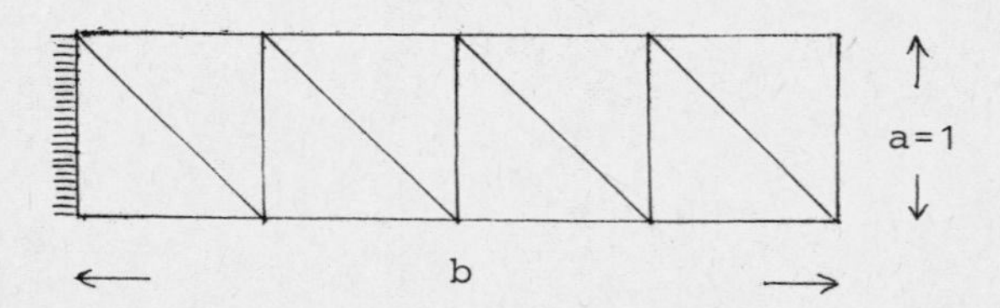

Fig. Coarsest grid ($h_{max}=1$) in the test of a multigrid algorithm for the membrane problem (Dirichlet b.c. on the left hand side).

The first results reported by Brand [5] were as follows.

1. Multigrid algorithms with point relaxation lead to error reduction factors

$$\eta \leq 0.6.$$

In particular, in most cases $\eta \leq 0.22$ is observed, while in the case $b=4$, $h_{max}=1$, $h_{min}=1/8$ the factor seems to be not (substantially) smaller than 0.6.

2. If a multigrid cycle is used as a preconditioner in a conjugate gradient iteration, then an

error reduction to $\leq 10^{-5}$ in 7 steps

is obtained.

Obviously, the pure multigrid method cannot be considered as satisfactory, but the version with conjugate gradient is. Before we will analyse the behaviour of the pure multigrid method, we want to understand the improvement which results from the use of conjugate gradients.

Given a linear equation Ax=b with a positive definite matrix, assume that the iteration

$$x_{\nu+1} = x_\nu + C^{-1}(b-A_{x_\nu})$$

is convergent with C being also positive definite. If the error reduction by ν steps of the iteration is given by

$$\eta^\nu,$$

then ν steps of the pcg-method (preconditioned congugate gradient method) with preconditioning by the matrix C will yield a factor

$$(T_\nu(\tfrac{1}{\eta}))^{-1} \leq 2\ (\frac{\eta}{1+\sqrt{1-\eta^2}})^\nu .$$

Here, T_ν is the ν-th Chebyshev polynomial [10]. Usually, the pcg-method is even more favourable due to the clustering of eigenvalues (and if the bounds of the eigenvalues are not symmetrical to the origin).The formulas yield a reduction better than 10^{-3} for η=0.6 and ν=7. They also show that 3 steps would be sufficient for a nested iteration in the multigrid framework [8].

3. The Influence of the Locking Effect on Multigrid Methods.

In this section two other combinations of multigrid methods and the use of conjugate gradients will be discussed. For this purpose we want to understand better the behaviour of the pure multigrid algorithm.

The poor convergence of the multigrid iteration was first understood to have its origin in the terms with the mixed derivatives. This argument seems plausible since point relaxation is used in the smoothing step.

However, a glance on the results in Table 1 shows that there should be other reasons for the behaviour of the algorithm. The table contains some results from [5] . Convergence is good for b/a=1. It is more important that the result for the two-grid method in the critical case (b/a=4) is also good, though the 4-grid version leads to slow convergence here.

Therefore, a poor smoothing rate cannot be the reason for the unsatisfactory behaviour. We obtain a better understanding of the situation by the explanation that the solution on the coarsest grid with h=1 does not provide a good approximation of the long range effects on the next grid with h=1/2. The poor coarse-grid-correction is inherited by the fine grids.

Here we encounter a phenomenon which is well known to engineers working with finite elements and which is called *locking*.

Table 1. Some convergence results [5] for the **W**-cycle of a multigrid cycle for the elasticity problem.

b	h_{max}	h_{min}	η
1	1	1/8	0.2
	1	1/16	0.2
4	1	1/8	0.6
	1	1/2	> 0.3
	1/4	1/8	0.22
8	1	1/8	0.4

It is known that the elasticity problem is treated more appropriately as a beam than as a membrane, if the ratio b/a is very large. Specifically, in this case the displacements are obtained by a differential equation of fourth order. Likewise, if there are only a small number of elements in the vertical direction, then the discretization with membrane elements does not provide good results. This is denoted as locking effect. The locking effect seems to be less severe for quadrilateral elements than for triangular ones. For this reason, quadrilateral elements are frequently used in continuum mechanics.

The discussion shows that the coarse-grid-correction may be not good in multigrid methods for elasticity problems, in particular on the coarser grids. The efficiency of the multigrid algorithm is now improved, if the smoothing steps combine relaxation with conjugate gradient methods. Specifically, instead of performing μ relaxation steps, we perform μ steps of a pcg-iteration with relaxation as preconditioning [2].

The improvement by this version for the combination of multigrid method and cg-method is easily understood. The *modified smoothing step* does not only damp the highly oscillating terms as a classical smoothing procedure, but it can also reduce a smooth term in the error which has survived in the coarse-grid correction. In principle, the cg-method does not know a preference for short or long waves.

We note that the insertion of such a cg-procedure is not costly, since the part of the code which has to be added to the multigrid code is almost problem independent. On the other hand, a modification of a code which would employ corrections of membrane elements for the reduction of the locking effect, would depend on the problem and also on the level in the hierarchy of grids.

There is a third version for a combination. It uses the cg-method for an outer iteration as Brand's version does. However, it goes one step farther and applies Poisson solvers for the preconditioning. Korn's inequality [6] which is the crucial point in the proof of the ellipticity of (2.1), provides the nontrivial part of the estimate

$$c_1 \int_\Omega \{(\nabla u)^2+(\nabla v)^2\}dxdy \leq a(u,v) \leq c_2\int_\Omega\{(\nabla u)^2+(\nabla v)^2\}\, dxdy . \qquad (3.1)$$

Here, the positive constants c_1 and c_2 depend only on Ω and on the kind of the boundary conditions. The inequalities imply that the elasticity problem is spectral equivalent to two Poisson problems and that the condition number $\kappa = c_2/c_1$ is independent of h.

Therefore, when using a multigrid method as a fast Poisson solver, then we also get a pcg-iteration with a convergence rate which is independent of h. Since there are very efficient Poisson solvers [12], this variant is preferable to the variant in the last section whenever the constants in (3.1) satisfy $c_2/c_1<20$.

4. A Mixed Method for the Biharmonic Equation

Multigrid methods are even more important for elliptic problems of fourth order. The condition numbers of the matrices of the arising equations grow faster than for second order problems. Therefore, pure pcg-methods are less efficient than in the case of second order problems (see the comments in the appendix).

We regard the biharmonic equation, but the same arguments apply to plate bending problems:

$$\Delta^2 u = f \qquad \text{in } \Omega\subset\mathbb{R}^2,$$

$$u= \frac{\partial u}{\partial n} = 0 \qquad \text{on } \partial\Omega .$$

To formulate a mixed method, let $S_h \subset H^1(\Omega)$ be a finite element space and set $S_h^o = S_h \cap H_o^1(\Omega)$.

Determine a pair $(u,w) \in S_h \times S_h^o$ such that

$$(w,\psi) + (\nabla u, \nabla \psi) = 0, \quad \text{for } \psi \in S_h,$$

$$(\nabla w, \nabla \varphi) \qquad = -(f,\varphi) \quad \text{for } \varphi \in S_h^o.$$

Here $(.,.)$ refers to the usual inner product in $L_2(\Omega)$. After choosing bases for the finite element spaces S_h and S_h^o we obtain linear equations in matrix vector form. These will be written without changing symbols:

$$\begin{pmatrix} M & B^T \\ B & 0 \end{pmatrix} \begin{pmatrix} w \\ u \end{pmatrix} = \begin{pmatrix} 0 \\ -f \end{pmatrix} \tag{4.1}$$

The matrix M is nonsingular and the vector w may be eliminated

$$BM^{-1}B^T w = f. \tag{4.2}$$

Equations of the form (4.1) are typical for mixed finite element methods. The reduction to the form (4.2) is frquently performed, although this reduction may be concealed. The matrix $BM^{-1}B^T$ is positive definite, and the application of the cg-method to (4.2) is equivalent to the application of a *conjugate Uzawa algorithm*.

Nevertheless, the features of (4.2) heavily depend on the actual problem, specifically on the order of the differential operators of which B and M are discretizations. For this reason, there are several variants of the algorithms of Uzawa type. The main differences are in the choice of the preconditioning and in the accuracy when M is inverted.

Here, B corresponds to a differential operator of second order. Therefore,

$$\varkappa(BM^{-1}B^T) = \mathcal{O}(h^{-4}). \tag{4.3}$$

(On the other hand, the analogous condition number is bounded independently of h if the mixed method for the Stokes problem is considered [13].)

Recently, Gustafsson [7] investigated a preconditioning procedure for the system (4.2) which he obtained in a two-step approximation of $BM^{-1}B^T$.

1. Step. The boundary terms in the matrices M and B are ignored. The resulting system corresponds to a finite element solution of an elliptic problem with modified boundary conditions:

$$\Delta u = w, \qquad u=0 \quad \text{on } \partial\Omega$$

$$\Delta w = f, \qquad w=f_2 \quad \text{on } \partial\Omega$$

or in matrix-vector notation:

$$B_o M_o^{-1} B_o u = f.$$

Since $\kappa(M_o)$ is bounded, this approximation may in turn be replaced by

$$B_o \cdot B_o u = f. \tag{4.4}$$

Here, as in [7] , the matrix B_o corresponds to the discretization of the Poisson equation.

2. Step. Replace B_o by a good preconditioner of B_o. Specifically, following [7] , let LL^T be the result of the (modified) incomplete Choleski decomposition of B_o, and replace (4.4) by

$$(LL^T)^2 u=f. \tag{4.5}$$

An alternative for the second step is the use of a multigrid iteration [4] . To be specific, one or two steps of a multigrid iteration for solving the Poisson equation are used as a preconditioner for this equation.

[We note that the approximate equations are only solved to obtain *preconditioned gradients*.]

The analysis of the approximation of the matrix in (4.2) for a pcg-method differs slightly from the study of the same approximation for a classical iteration procedure. In particular, the implications of the fact that a square of a matrix is found in (4.4), will be investigated.

The first step of the procedure above implies that the condition number cannot be better than

$$\text{const} \cdot h^{-1}.$$

This follows from

$$\|B_o u\|^2 \leq \|B^T u\|^2 \leq ch^{-1} \|B_o u\|^2 . \tag{4.6}$$

The inequalities, which are sharp in contrast to Lemma 3.1 in [7] , were derived in [4] from à priori estimates for discrete harmonic functions. The growth as h^{-1} arises from the fact that the original Neumann boundary condition has been approximated by a boundary condition for the Laplacian, i.e., the difference between the order of the boundary operators is one. Consequently, the number of iteration steps in a pcg-algorithm which usually is proportional to $\kappa^{1/2}$, increases as

$$h^{-1/2} \sim N^{1/4}$$

if N denotes the number of unknowns.

For more details see [4] . Here, we are more interested in the analysis of the second step, since here the multigrid iteration is involved.

5. Estimates of Spectral Condition Numbers

An iterative procedure like a multigrid algorithm is mostly evaluated by its error reduction factor. Here we are interested in the spectral condition number, if the procedure is used for preconditioning. We will investigate the connection between the two features. In particular, influence of the squaring of matrices will be studied.

Given a linear equation $Ax=b$, assume that we choose the matrix C for the preconditioning. This means that the new direction $x^{k+1}-x^k$ is evaluated from $C^{-1}(Ax^k-b)$ and x^k-x^{k-1} by the well known formulas for the cg-method [1,10,11] . The convergence **rate** of the pcg-iteration depends on the spectral condition number

$$\kappa(C^{-1}A) = \frac{\lambda_{max}(C^{-1}A)}{\lambda_{min}(C^{-1}A)} .$$

Usually, the following property is used for calculating κ. If α and β are optimal bounds such that

$$\alpha C \leq A \leq \beta C, \tag{5.1a}$$

then

$$\kappa(C^{-1}A) = \frac{\beta}{\alpha} . \tag{5.1b}$$

Here and in the sequel $A\leq B$ means that $B-A$ is positive semidefinite.

In view of (5.1) there arises the question whether $A\leq B$ does imply $A^2\leq B^2$?

If A and B commute, the answer is affirmative, but the following example shows that generally the implication is not true. Let $a>2$ and

$$A = \begin{pmatrix} 1 & a \\ a & 2a^2 \end{pmatrix} \qquad B = \begin{pmatrix} 2 & 0 \\ 0 & 3a^2 \end{pmatrix}$$

$$A^2 = \begin{pmatrix} 1+a^2 & a+2a^3 \\ a+2a^3 & a^2+4a^4 \end{pmatrix} \qquad B^2 = \begin{pmatrix} 4 & 0 \\ 0 & 9a^4 \end{pmatrix} .$$

Obviously, we have $A\leq B$ and $A^2 \nleq B^2$. On the other hand, if A and B are positive definite matrices, then $A^2\leq B^2$ implies that $A\leq B$. This follows from Heinz' inequality.

The basic idea of the following estimates is a simple observation. If $\|Ax\| \leq \alpha \; \|Bx\|$ holds with respect to the Euclidean norm for some $\alpha > 0$, then $x^T A^T Ax \leq \alpha^2 x^T B^T Bx$ holds which may be rewritten as $A^T A \leq \alpha^2 B^T B$. In the particular case when A and B are symmetric, the matrices A^2 and B^2 are compared.

Let K_m be the linear operator which assign to each $y \in \mathbb{R}^M$ an approximation $x_m = K_m y$ to the solution of the equation $B_o x = y$. Specifically, x_m may be the result of m steps of a convergent iteration with starting value zero. Then there is a sequence (η_m) with $\eta_m \to 0$ such that

$$\|K_m y - B_o^{-1} y\| \leq \eta_m \|B_o^{-1} y\| \; . \tag{5.2}$$

This inequality means that the relative error in the Euclidean norm is small. The adjoint operator K_m^T provides an approximative solution such that the residual is small

$$\|B_o K_m^T y - y\| \leq \eta_m \|y\| . \tag{5.3}$$

The factors η_m are the same in both equations. Indeed, we deduce from (5.2) that

$$\|K_m B_o - Id\| = \sup_{x \neq 0} \frac{\|K_m B_o x - x\|}{\|x\|}$$

$$= \sup_{y \neq 0} \frac{\|K_m y - B_o^{-1} y\|}{\|B_o^{-1} y\|} \leq \eta .$$

Since a matrix and its adjoint matrix have the same norm, we obtain $\|B_o K_m^T - Id\| \leq \eta$ which in turn yields (5.3).

On the other hand, it follows from (5.2) and the triangle inequality that

$$(1-\eta_m) \|x\| \leq \|K_m B_o x\| \leq (1+\eta_m) \|x\| .$$

Hence, $(1-\eta_m)^2 Id \leq B_o K_m^T K_m B_o \leq (1+\eta_m)^2 Id$ and

$$\kappa(K_m^T K_m B_o^2) \leq \left(\frac{1+\eta_m}{1-\eta_m}\right)^2 . \tag{5.4}$$

Now we are in a position to understand how a good preconditioner for the solution of $B_o^2 x = f$ is obtained. Let K_m be as above with $\eta_m < 1$ and assume that the "preconditioned gradient" is computed as follows

$$f \longmapsto w = K_m f \longmapsto u = K_m^T w. \tag{5.5}$$

This means that in the first step an approximate solver is used which yields a small error in the Euclidean norm, while in the second step the (adjoint) solver is used which yields a small residue (=small error in the H^2-norm). For this case, the condition number is given by (5.4).

Finally, we turn to the consequences for the choice of appropriate multigrid procedures. Usually, the fine-to-coarse-grid restriction is the adjoint of the coarse-to-fine-grid interpolation in multigrid algorithms for finite element equations (see [8]). Furthermore, let Jacobi or checkered Gauss-Seidel relaxation be used as smoothing procedures. Denote the number of presmoothing and postsmoothing iterations by ν_1 and ν_2, respectively. Writing $K_m(\nu_1,\nu_2)$, we have the symmetry relation $K_m^T(\nu_1,\nu_2) = K_m(\nu_2,\nu_1)$.

From the general convergence theory we know that η_m is small if ν_1 is sufficiently large [8]. This means that the multigrid method which is used in the first step in (5.5) should contain sufficiently many presmoothing operations while the multigrid iteration for the second step should contain sufficiently many postsmoothing operations.

This is in contrast to the solution of the original Poisson equation (when no square is involved). Then the (symmetrical) multigrid iteration with the same number of presmoothing and postsmoothing steps would be the appropriate choice when used as a preconditioner.

Appendix

When comparing classical Gaussian elimination, preconditioned conjugate gradient methods and multigrid methods, we know that multigrid methods are the most efficient methods as $N\to\infty$. Here, N is the number of unknown values. In the design of algorithms for real life problems one would like to know where the cross-over-points are located.

Clearly, there is no general answer. Nevertheless, we have made the experience that a thumb rule can be given for one kind of problems, namely scalar elliptic equations of second order in two dimensions. Then the computing effort is

$\mathcal{O}(N^2)$ for Gaussian elimination (not nested dissection),

$\mathcal{O}(N^{5/4})$ for pcg-methods with preconditioning by ILU-decomposition,

$\mathcal{O}(N)$ for multigrid methods.

According to our experience with examples with nonregular meshes, the cross-over between Gaussian elimination and pcg-methods is for N between 100 and 300.

The cross-over between pcg-methods and multigrid methods is for N between 500 and 2000. The latter point is not very sharp, since a gain of, say 20% in computer time in one algorithm will change the cross-over point by a factor of two.

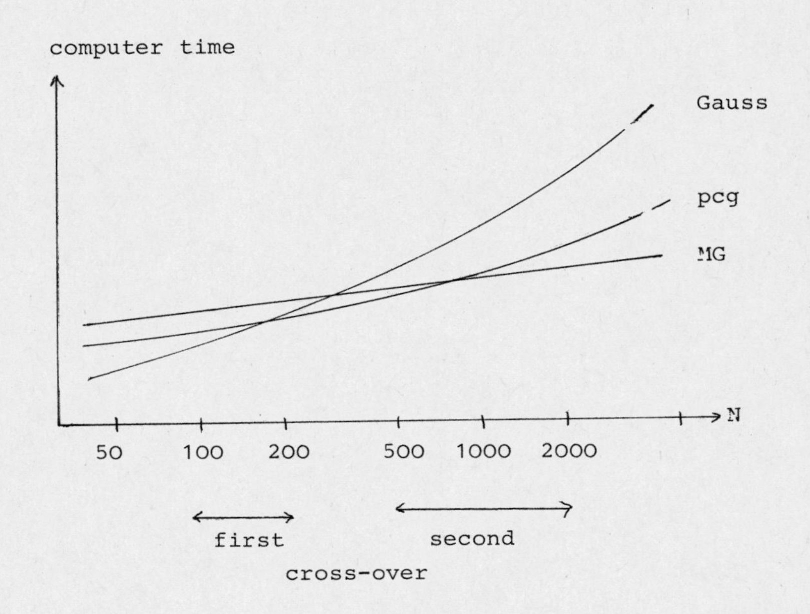

It was interesting to see that the pcg-methods are superior to Gaussian elimination already for fairly small N. This is supported by other results given in the literature. Note that N=89 was given as the critical point for a model problem in [1] , while a larger value namely $N \approx 300$ corresponds to the results in [10].

For 3-dimensional problems (of second order) the domain where the pcg-methods are the favorite, is probably substantially larger and covers the region $10^2 \leq N \leq 10^4$ (cf.[3]). If, on the other hand, problems of fourth order in two dimensions are considered, the multigrid methods are expected to the best already for small N.

References

1. O. Axelsson and V.A. Barker. Finite Element Solution of Boundary Value Problems. Theory and Computation. Academic Press 1984

2. R.E. Bank and C.C. Douglas. Sharp estimates for multi-grid rates of convergence with general smoothing and acceleration. SIAM J. Numer.Anal. 22, 617-633 (1985)

3. A. Behle and P.A. Forsyth,Jr. Multigrid solution of three dimensional problems with discontinuous coefficients. Appl.Math.Comp. 13, 229-240 (1983)

4. D. Braess and P. Peisker. On the numerical solution of the biharmonic equation and the role of squaring matrices for preconditioning (submitted).

5. G. Brand. Multigrid methods in finite element analysis of plane elastic structures (in prep., augmented by private communications).

6. G. Duvaut and J.L. Lions. Inequalities in Mechanics and Physics. Springer, Berlin-Heidelberg-New York 1976

7. I. Gustafsson. A preconditioned iterative method for the solution of the biharmonic problem. IMA J.Numer.Anal. 4, 55-67 (1984)

8. W. Hackbusch. Multi-Grid Methods and Applications. Springer, Berlin-Heidelberg-New York-Tokyo 1985

9. R. Kettler. Analysis and comparison of relaxation schemes in robust multigrid and preconditioned conjugate gradient methods. In "Multigrid Methods"(W. Hackbusch and U. Trottenberg, eds.) Springer, Berlin-Heidelberg-New York 1982

10. H.R. Schwarz. Methode der Finiten Elemente. Teubner 1984

11. J. Stoer. Solution of large linear systems of equations by conjugate gradient type methods. In "Mathematical Programming: The State of the Art", pp.540-565. Springer, Berlin-New York 1983

12. R. Stüben and U. Trottenberg. Multigrid methods: Fundamental algorithms, model problem analysis and applications. In "Multigrid Methods" (W. Hackbusch and U. Trottenberg, eds.) Springer, Berlin-Heidelberg-New York 1982

13. R. Verfürth. A combined conjugate gradient-multigrid algorithm for the numerical solution of the Stokes problem. IMA J.Numer. Anal. 4, 441-455 (1984)

Multi-Level Approaches to Discrete-State and Stochastic Problems

Achi Brandt* and Dorit Ron*

Department of Applied Mathematics
The Weizmann Institute of Science
Rehovot, Israel

and

Daniel J. Amit**

Racah Institute of Physics
The Hebrew University
Jerusalem, Israel

Abstract

Fast multi-level techniques are developed for large-scale problems whose variables may assume only discrete values (such as spins with only "up" and "down" states), and/or where the relations between variables is probabilistic. Motivation and examples are taken from statistical mechanics and field theory. Detailed procedures are developed for the fast global minimization of discrete-state functionals, or other functionals with many local minima, using new principles of multi-level interactions. Tests with Ising spin models are reported. Of special interest to physicists are the Ising model in a random field and spin glasses, which are known to lead to difficulties in conventional Monte-Carlo algorithms.

Content

---

* Research supported by the Air Force Aeronautical Laboratories, Air Force Systems Command, United States Air force, under Grant AFOSR 84-0070.

** Research supported in part by the Fund of Basic Research of the Israel Academy of Science and Humanities.

1. Introduction

The present effort has been motivated by the recognition that while Monte-Carlo studies of the statistical properties of systems in statistical mechanics and in field theory have become a central methodological tool, on par with analytical and experimental investigations, there is an increasing number of systems for which this tool is too slow. In fact, for some problems of central interest in both fields, present day computers seem inadequate to obtain significant results in a reasonable amount of time. To mention some such systems one recalls the computation of masses in quantum chromodynamics [22]; the phase structure of spin-glasses or magnets in random magnetic fields [1]; or problems of combinatorial optimization, of which the travelling salesman is a simple representation [9].

Most of the current effort to deal with this difficulty is concentrated in the development of new hardware, more atune to the specific Monte-Carlo task [24]. Only modest attention has been devoted to the improvement of the algorithm. In fact, in the thirty years since the invention of the technique [23], almost all improvements have been concerned with the speeding up of the basic algorithm, improvements which have increased the specialization to particular problems. This line of development has culminated with the microcanonical technique [7].

Another line of development has been the shift from the original idea [23] of generating a sequence of configurations, with the correct probability distribution – a Fokker-Planck approach [29], to a Langevin approach, in which one solves stochastic time dependent differential equations for the various physical quantities [11], [26].

The closest, in spirit, to the class of algorithms to be presented here, has been the Monte-Carlo renormalization group approach [19], [30]. Yet, in this approach configurations are still generated in the original Monte-Carlo process, even though physical quantities are computed in each configuration on different scales (to produce renormalized coupling constants [19] or renormalized expectation values [30]).

All the techniques mentioned above suffer from two basic deficiencies: slow transitions and slow balancing of deviations (see Sec. 2.3 below). By a proper modification of the statistical technique (cf. Sec. 2.3) the latter deficiency can be reduced to the first, and the origin of both is traced to the purely local nature of the classical algorithms.

This situation strongly suggests the need to introduce multi-level processes. Such processes (multigrid methods) have in the past resolved quite similar difficulties in the solution of partial differential equations, by combining local processing on different scales, employing interactions between several grids of widely varying meshsizes (see [10] and references therein).

One cannot, however, readily use classical multigrid to solve the problems of statistical mechanics and field theory. These problems differ from the differential problems in two basic aspects:

(A) Discrete state. In differential problems each variable (the value of the unknown solution at each site) is a real number (or a real vector), hence the entire solution can be regarded as a linear combination of components of varying scale (e.g., varying wavelengths), each of which is mainly treated by processing on that scale (i.e., on a grid with a comparable meshsize), more or less independently of the processing on other levels. In statistical physics, by contrast, and in many other important fields such as combinatorial optimization and pattern recognition, the variables are often either discrete or at least have discrete components. The simplest examples are Ising spin problems, in which each variable can only assume the values 1 and -1, or optimization problems with various dichotomous decisions. In such problems the solution is no longer a combination of more or less independent scales. As a result the inter-scale interactions are more intricate, and a good deal more interesting.

Indeed, the technique of inter-scale interactions for discrete-state problems is the main finding of the present study from the point of view of multigrid methodology. This technique can also provide a better treatment for highly nonlinear *continuous*-state problems, such as minimization problems plagued with too many local minima (see Sec. 4).

(B) Statistical nature. The purpose of calculations in differential equations is to find an approximate solution for one particular set of data (forcing terms, boundary conditions, etc.). In statistical physics the data are random, and/or the solution is stochastic (governed by probabilistic relations), hence the aim is not only to obtain (probable) solutions, but mainly to calculate their statistical properties. Consequently, in addition to the traditional role of multigrid in accelerating numerical processes toward equilibrium (here meaning toward highly probable configurations), it should here also provide for fast statistics. At low temperatures (low stochasticity) this means providing for fast transition between approximate ground states. At higher temperatures this also means large-scale balancing of deviations, to allow fast reliable convergence of long-range correlations (necessary for example for renormalization calculations).

The easiest approach is first to deal with these statistical aspects in separation from the discrete-state feature, by studying a statistical model with continuous variables, such as the XY model, or the Heisenberg model, etc. In such models the discrete (or highly nonlinear) nature of the problem appears only at much coarser levels (e.g., at the scale of probable vortices or instantons), so that on finer levels multigrid processes similar to the classical ones can be used.

The present paper deals mainly with the discrete-state aspect, and chiefly at low temperatures, where it is clearly separated from statistical elements.

2. Model Problem and Classical Solution Techniques

2.1 The physical problem

We will discuss the algorithms in the context of nearest neighbor Ising models, with uniform or random interactions and possible site dependent external magnetic fields. The discussion will further be restricted to two dimensional lattices, for which the new algorithms have been tried. All these restrictions are not essential. The logic, as well as the effectiveness of the algorithms should not be much affected if the number of local states becomes greater than two; if the interactions involve more neighbors than just the nearest ones; or if the number of lattice dimensions becomes greater than two.

The framework is defined by an $L \times L$ lattice, whose sites are labelled by an index i. At each site i there is an *Ising spin*, or briefly a "*spin*", i.e., a variable S_i which can assume one of the two values: $+1$ and -1. Thus, the total number of spins is $N = L^2$, and the lattice has 2^N possible *configurations* (arrangements of spin values).

With each configuration $C = \{S_i\}$ we associate the "*energy*"

$$E(C) = -\sum_{<i,j>} J_{ij} S_i S_j - \sum_i h_i S_i \tag{2.1}$$

where $< i,j >$ denotes a "*bond*", i.e., a pair of nearest neighbor sites, so the first summation is over all such bonds, while the second sums over all the lattice sites. J_{ij} and h_i are given parameters, representing, respectively, the couplings and the external magnetic fields.

In principle, the energy (2.1) should be supplemented by boundary conditions. Instead, for convenience, periodic geometry is often used, namely one sets the lattice on a torus. Thus, if the coordinates of site i are denoted (x_i, y_i), where x_i and y_i are integers and $1 \leq x_i \leq L$ and $1 \leq y_i \leq L$, then i and j are called nearest neighbors iff there exist integers x_{ij} and y_{ij} such that $|x_i - x_j + x_{ij}L| + |y_i - y_j + y_{ij}L| = 1$.

At zero temperature, this physical system settles to a "*ground state*", that is, a configuration with the lowest possible energy. The computational task is to find such a ground state, or, if possible, all of them. More practically, what is usually given is some statistical properties of J_{ij} and h_j, and the goal of computations is to find statistical properties of the ground states.

At higher temperatures random fluctuations enter, and configurations other than ground states can be obtained, although the larger their energy the less probable they are. The probability $P(C)$ of obtaining a configuration C is given by the *Gibbs* (or *Boltzmann*) *distribution*

$$P(C) = e^{-\beta E(C)} / Z(\beta) \tag{2.2}$$

where $1/\beta$ is the temperature (in suitable units), and $Z(\beta)$ is a normalization factor (the partition function) derived from the condition $\sum_C P(C) = 1$. Here the computational task is to calculate statistical properties of the system, such as the average magnetization

$$\bar{M} = \sum_C M(C) \cdot P(C)$$

where $M(\{S_i\}) = N^{-1} \sum_i S_i$. It is desired to calculate such averages as functions of temperature, with special interest in temperatures close to 0 (large β) and close to certain critical values (phase transitions).

Note the relation between the two computational tasks: at sufficiently low temperature (sufficiently high β), only ground states are probable; so to obtain the desired statistics, the ground states should effectively be found.

In most problems N is large. One is typically interested in N being anywhere between several hundreds to many many thousands. It is therefore impractical to gather the desired statistics, or locate the ground states, by scanning all the 2^N configurations. Instead, Monte-Carlo methods are used.

2.2 Classical Monte-Carlo method

The Monte-Carlo method is a computational process which generates a sequence of configurations such that ultimately each configuration appears a number of times proportional to its physical probability (2.2). On some finite section of this sequence, the statistical calculations are done. The problem is of course to generate a feasible small section which is sufficiently representative in terms of measured statistics.

The classical Monte-Carlo generates the next configuration in the sequence from the previous one, by treating a single spin at a time. The spins are taken in some order; lexicographic, red-black and random orders are common. Each spin S_i, in its turn, is either flipped or left unchanged, according to one of the following two rules:

(1) *Heat bath rule*: Let C be the previous configuration and C' the one obtained from it by changing S_i to $-S_i$. The decision whether to choose C or C' as the next configuration is taken randomly (by some random number generator), with the "*transition probability*":

$$P(C \to C') = \frac{P(C')}{P(C) + P(C')} = \frac{1}{1 + e^{\beta \cdot \delta E_i}} \tag{2.3}$$

where

$$\delta E_i = E(C') - E(C) = 2h_iS_i + 2\sum_j J_{ij}S_iS_j, \tag{2.4}$$

j running in the summation over the four nearest neighbors of i. Note, thus, that this transition probability is an easily calculable local quantity.

(2) *Metropolis rule* [23]: If $E(C') < E(C)$ then flipping is performed; energy is lowered whenever possible. On the other hand, if $E(C') \geq E(C)$ then flipping is decided randomly again, with the probability (2.3) assigned to it.

Any one of those rules is widely used. It is easy to show for each of them that, the only stationary distribution (i.e., the only distribution which stays unchanged by every Monte-Carlo step) is (2.2). Since any configuration can be reached from any other with positive probability, the elementary theory of Markov chains implies that ultimately (i.e., in a sufficiently long sequence) each configuration C will appear a number of times proportional to $P(C)$. Moreover, that theory actually implies that whatever configuration one starts with, the probability $P_n(C)$ of obtaining C at the n-th step tend to $P(C)$ as $n \to \infty$. When n is sufficiently large so that $P_n(C) \approx P(C)$, to the approximation needed in the desired statistics, we say that the system has reached *equilibrium*.

Since $P_n(C) \to P(C)$ for any large n, it is not really necessary to sample all the configurations in the sequence. Since each step treats another spin, it is indeed equivalent to sample (take into account) only the configurations obtained at the end of a full sweep (visiting once all sites).

2.3 Monte-Carlo slowness

To produce reasonable approximations to the desired statistics, the Monte-Carlo sequence of configurations usually needs to be very long. There are two intrinsic reasons for this slowness:

(A) Slow transition. This is a low temperature difficulty, which to an important degree also affects intermediate (e.g., critical) temperatures. At such temperatures, the important (i.e., more probable) configurations are locally stable; that is, in the space of configurations they lie near the bottom of extended "*attraction basins*". Each Monte-Carlo step leads just to a neighboring configuration, with higher probability of moving downhill (toward the bottom of the basin) than uphill. Hence, most probably, it will require a great number of steps to escape from a given attraction basin and visit a neighboring one. Moreover, all those neighboring attraction basins are likely themselves to be near the bottom of a larger-scale attraction basin, and so on. The process will therefore take a very long time both to equilibrate and to effectively sample the space, in order to produce reliable averages.

More specifically, for example in case of uniform ($J_{ij} \equiv 1$) Ising models not dominated by external fields (e.g., $|h_i| \ll 1$), at low temperatures the statistically important configurations are made of large blocks of mostly aligned spins. These blocks can only slowly be changed by single spin flips. At very low temperatures such flips cannot change such blocks at all – the familiar experience of "walls".

Near critical temperatures changes are still probable, but they only slowly affect the large-scale blocks, exactly those blocks whose statistics is important for studying critical phenomena. The equilibration time is proportional to ξ^2, where ξ, the correlation length, tends to infinity as the temperature approaches the critical point – the familiar "critical slowing down".

This difficulty is even more acute in disordered frustrated systems, such as spin-glasses ($J_{ij} = \pm 1$ in some random manner) and magnets in quenched random fields. A spin-glass has very large number of essentially degenerate low-lying states (this number diverges as $N \to \infty$), and crossing the barriers between them by single-spin flips can take anywhere from short to infinitely long times. This is the origin of such phenomena as the remenant magnetization and irreversible cooling [12], [13], [20], [21], [27], [28].

(B) Slow balancing of deviations. Even when equilibrium has been reached, accumulation of precise statistics is slowed down by the presence of random deviations, whose effect on the desired averages decreases only as $n^{-1/2}$, where n is the number of sampled configurations. This slowness appears at any temperature, including very high ones. For example, even for uniform ($J_{ij} \equiv 1$) and homogeneous ($h_i \equiv 0$) Ising models at very high temperatures ($\beta \ll 1$), to obtain $\bar{M} = 0$ to an accuracy of 10^{-6} would require some 10^{12} Monte-Carlo steps.

The high temperature aspect of this trouble is easy to overcome, by changing the order through which averages are calculated. For example, the average magnetization can be rewritten in the form

$$\bar{M} = N^{-1} \sum_i \sum_{C^{(i)}} P(C^{(i)})[P_{i+}(C^{(i)}) - P_{i-}(C^{(i)})]$$

where $C^{(i)}$ are spin configurations on the lattice excluding site i, and $P_{i+}(C^{(i)})$ is the conditional probability of getting $+1$ at site i given $C^{(i)}$, and $P_{i-} = 1 - P_{i+}$. Since P_{i+} is readily calculable, and since each $C^{(i)}$ appears in the Monte-Carlo process a number of times proportional to $P(C^{(i)})$, $\bar{M}$ can be calculated by averaging, over the sampled configurations C, the quantity $\sum_i \bar{S}_i(C)/N$, instead of $\sum_i S_i(C)/N$, where $\bar{S}_i(C) = 2P_{i+}(C^{(i)}) - 1$ and $C^{(i)}$ here is the configuration C without its spin at i. It is easy to show that the standard deviation in averages calculated in this way is $O(\beta)$ times the standard deviation in the original Monte-Carlo averages. By excluding more sites in the definition of $C^{(i)}$, this factor can be reduced to $O(\beta^m)$, for any prescribed integer m, but the work of computing the conditional probabilities grows exponentially in m.

Thus, for high enough temperatures, the statistical deviations can instantly be eliminated. At lower temperatures the above device will no longer work: averaging each spin value in its given neighborhood eliminates deviations related to single spin fluctuations, but does little to eliminate fluctuations of *blocks* of spins. To eliminate those, blocks of spins should be weighted against their chance of being simultaneously flipped.

The question is which blocks to use. There are certainly too many to be all included. Only the more highly probable block flipping should therefore be considered. Those, however, yield configurations which lie at the bottom of different attraction basins, thus requiring long transition computations.

So, in a sense, the second problem (slow balancing of deviations) can be reduced to the first one (slow transition). Since this is basically a low temperature (high β) problem, we have decided to confront it directly by first studying the difficult case of zero temperature ($\beta \to \infty$). The next sections describe this study, returning to the case of finite β in the conclusion.

3. Approaches to the Minimization Problem

At zero temperature, the problem is to quickly find ground states, i.e., configurations C with minimal energy $E(C)$. If there is more than a single ground state, the problem is also to have fast transition from one to another, so as to be able to calculate any desired statistics.

Special cases of the minimization problem have been solved by showing their equivalence to certain graph-theoretic problems for which polynomial-time algorithms are known. The Ising spin-glass problem ($J_{ij} = \pm 1$ with equal probability; $h_i = 0$) in *two dimensions* has been shown to be equivalent to the Chinese Postman problem [2], [5], [6], [8], while the ferromagnetic Ising model ($J_{ij} = 1$) with arbitrary (e.g., randomly generated) magnetic field in *any* dimension has been reduced to the Min-cutset problem in graph theory [3], [4]. In both cases the resulting algorithms produce ground states in $O(N^3)$ computer operations.

Although this complexity is not intractable, and allowed the production of many model results, it is still too large to be easily applied, especially to three and four dimensional problems. More importantly, this graph-theoretic approach is limited to very particular model problems; it does not provide the general computational tool needed in statistical mechanics and field theory. By contrast, the multi-level stochastic algorithm developed below is very general, and its typical complexity is $O(N^{3/2})$. Indeed, the main significance of the graph-theoretic approach is that its model results can be used in testing the more general algorithms (cf. Sec. 6.1).

3.1 Point-by-point minimization

A natural computational process that immediately suggests itself when one wants to minimize a function of many variables is the *point-by-point minimization*, or *point-by-point relaxation*, which is actually the zero temperature limit of the Monte-Carlo process described above. The process involves scanning the variables one by one in some prescribed order. Each variable in its turn is changed to the value which brings about the greatest reduction in the energy. For the spin systems discussed here, this means scanning the spins, flipping (changing the sign of) each one in its turn, if and only if that flip reduces the energy.

For *continuous*-state (continuous S_i) problems, this point-by-point relaxation converges very slowly. (Classical multigrid methods were developed to overcome exactly this slowness.) In the present case of *discrete* states ($S_i = \pm 1$), a much more acute difficulty is that this simple algorithm will in all probability get trapped in one or other "*local minimum*", that is, a configuration for which no single spin flip can reduce the energy, but which is not a ground state.

For example, consider the uniform ($J_{ij} \equiv 1$) and homogeneous ($h_i \equiv 0$) Ising model, with periodic boundary conditions. The only ground states are clearly the uniform spin up configuration $\{S_i \equiv 1\}$ and the uniform spin down configuration $\{S_i \equiv -1\}$. A clear *local* minimum is any configuration having some of its rows with spins up and the rest with spins down. Clearly in such a configuration, flipping any individual spin, or even any small group of spins, would only raise the energy.

3.2 Stochastic minimization. Annealing

One way to escape local minima is to introduce a certain "*artificial temperature*" $1/\beta$ into the system. Using the Metropolis algorithm with that β will still flip in its turn any spin S_i for which that flip reduces the energy (i.e., for which $\delta E_i < 0$; cf. (2.4)), but will assign the positive probability (2.3) for flipping, even when the energy is thereby increased ($\delta E_i > 0$). In the context of the minimization (zero temperature) problem such a process is called *point-by-point stochastic minimization* or *point-by-point stochastic relaxation*.

To approach the real minimum one cannot of course keep the fluctuations. Thus, the artificial temperature should gradually be decreased toward 0 (β gradually raised to infinity). This process is called *(simulated) annealing*, and has been introduced into the context of minimization problems by S. Kirkpatrick [16], [17]. The rate at which the temperature is lowered, or the *annealing schedule*, is crucial and sensitive. Although the algorithm will most likely escape bad local minima, it is still very likely to be trapped in larger-scale attraction basins.

The situation can more specifically be understood in terms of the lattice geometry. Suppose that a certain, rather large block of spins is *reversed*, i.e., its flipping would decrease the energy, while flipping any of its individual spins or any smaller sub-block, would increase the energy. In case the block is "narrow" (e.g., one or two spin wide), stochastic relaxation has a very good chance of breaking it into smaller blocks. On the other hand, if the block is many spin wide, it can be broken only by the simultaneous flipping of a large number of spins. The probability for this to happen in any stretch of point-by-point stochastic relaxation sweeps (where higher probability is assigned to the original spin values) decreases exponentially with the width of the block, and thus becomes extremely unlikely – unless the artificial temperature is so large that the entire block structure is broken into pieces. Since eventually the temperature must get down, and since wide reversed blocks are very likely to form exactly at that stage (on large grids, especially on such sub-regions where the average local field runs counter to the spins of the ground state), the process gets easily trapped (cf. Sec. 6.2).

3.3 Block relaxation

Having observed that the main difficulty is associated with the need to flip wide blocks, multi-level approaches readily suggest themselves. The general idea of such approaches is to supplement the point-by-point minimization process with *block-by-block minimization* sweeps, on various scales. In such sweeps, *blocks* of spins are scanned, flipping each block in its turn (i.e., flipping simultaneously all its spins) if this decreases the energy. The energy difference in flipping any block Q is easily calculated by the formula

$$\delta E(Q) = 2 \sum_{<i,j>\in\partial Q} J_{ij} S_i S_j + 2 \sum_{i\in Q} h_i S_i \tag{3.1}$$

where ∂Q is the set of *bonds* (pairs of nearest neighbors) $< i, j >$ *around the boundary of* Q (i.e., i being inside Q and j outside Q), and S_i and S_j are the spin values *before* flipping. The main question, of course, is which blocks to scan. It is impractical to try all possible blocks, since their number grows exponentially with the number of spins.

The first approach, a natural extension to familiar multigrid PDE solvers, could be to simply divide the grid into 2×2 (then 4×4, etc.) disjoints blocks, and scan them in some prescribed order (lexicographic, or red-black, etc.). Such a naive approach is doomed to failure. Experiments will immediately show that no significant improvement, in fact very little change at all, is contributed by the block sweeps once the processed configurations are highly evolved (i.e., have undergone many point-by-point minimization sweeps, reaching, for example, a local minimum). Indeed, in such highly evolved configurations most bonds $< i, j >$ are *unviolated* (i.e., $J_{ij}S_iS_j > 0$). Hence, the flipping of an *arbitrary* block of spins is likely to violate most of the bonds around its boundary, and thus almost certainly *increase* the energy. There are only very *special* blocks whose flipping would reduce the energy, and these blocks will seldom coincide with any of those arbitrary square blocks chosen for processing.

Nor would it change much if "annealing" were used with these arbitrary blocks, that is, if *block stochastic relaxation* were tried on them. Such stochastic relaxation is easy to produce: The flipping probability of any block Q at a given artificial temperature $1/\beta$ is

$$P_f(Q) = 1/(1 + e^{\beta \delta E(Q)}) \tag{3.2}$$

with $\delta E(Q)$ given by (3.1). But again, since the blocks are arbitrary, flipping them would almost always increase the energy so much as to make them highly improbable ($P_f \ll 1$; unless the artificial temperature is raised so much that *any* block will be flipped with sizeable probability, rendering the procedure useless).

3.4 Revised block relaxation

These observations directly lead to the central algorithmic idea of this research. It is clear that before deciding whether or not to flip any given block, the block itself should be modified so as to become a better candidate for flipping. That is, the flipped region should undergo a certain optimization. This can be done by point-by-point (possibly stochastic) minimization in the vicinity of the flipped region. This optimization will reduce locally, as far as possible, the number of violated bonds around the boundary of the block, thus making its boundary less arbitrary. The new boundary is in fact likely to be the best possible one in the neighborhood of the original, highly artificial boundary. We call this optimization process the *block revision process*; its result is called the *revised block*.

Only *after* this revision, should δE of the (revised) block be examined, and the decision whether actually to adopt this flipping be accordingly made. The decision may be either stochastic (based on (3.2) with some finite β) or deterministic ($\beta = \infty$). In case it is decided to adopt the flip, it is the *revised* block, not the original square, which is flipped. If it is decided to reject the flip, the system returns to the configuration that existed before the whole process (the process of flipping the original square and then revising it) has started. For this purpose, the block revision process is done on a separate "*work area*". Only if and when it is decided to flip, the resulting configuration in the work area actually replaces the original configuration.

A full relaxation sweep on blocks of a linear scale of b is performed this way: The entire lattice is divided into squares of $b \times b$ sites. The squares are usually, but not necessarily, chosen so as to have no or minimal overlap. These squares are scanned in some prescribed order. Each one in its turn is transferred to the center of the work area, where it is flipped, then revised, and then either adopted or rejected, based on (3.2). (The choice of the artificial β, which may depend on the revised block, is discussed in Sec. 5.2.) Such a relaxation sweep is called *revised* $b \times b$ *block* (*stochastic* or *deterministic*) *relaxation sweep*, where the adjectives "stochastic" or "deterministic" stand for finite or infinite artificial β, respectively.

3.5 Multi-level recursion

The revised block relaxation, as described so far, works very well as long as the blocks one needs to flip are not too large. Revised 2×2, and then 4×4, block relaxation sweeps are likely to locate and flip all reversed blocks of comparable sizes. This is no longer true when the flipping of larger blocks is required. At such sizes the differences between the standard squares and the reversed blocks are no longer so local, and thus cannot efficiently be realized by point-by-point optimization. They can, however, be realized by supplementing the point-by-point optimization with a block-by-block optimization. Thus, the whole process should be *recursive*: The large-scale block relaxation employs, in its revision processes, relaxation sweeps over smaller-scale blocks, which in turn may employ still smaller scales, etc.

In the next section this idea is restated from more general points of view. Further important rules and algorithmic ideas are then described in Sec. 5.

4. Principle of Multi-Level Discrete-State Optimization

The optimization problems to which multi-level processes have traditionally been applied (e.g., discretized PDE problems in variational formulation) are *continuous-state* problems; that is, each of their unknowns (states) can assume any real value, or at least (in constrained optimization problems) any real value within certain bounds. In such problems the solution can be regarded as a linear combination of components, each having a specific scale; e.g., Fourier components, each having a specific wavelength. The point-by-point minimization relaxation is efficiently used to converge the small-scale components (those with wavelengths comparable to the meshsize), while block relaxation is used to effect larger-scale convergence: each relaxation sweep over blocks of a given size is efficient in converging components of comparable scale (e.g., Fourier components with wavelength comparable to the block width). Thus, upon completing the block relaxation, the remaining errors are small-scale errors, representing fine-grid details which are invisible to block processing. Owing to their local nature, these small-scale errors can effectively be reduced by subsequent point-by-point relaxation.

By contrast, in case of discrete-state problems, where each unknown can assume only some *discrete* values (such as $+1$ and -1 in Ising spin problems), the solution cannot be described as a sum of components of different scales. Hence, the fine-scale details cannot be *added* to coarse-level changes; at each point they either completely *cancel* those changes, or leave them exactly as they were. More importantly, in discrete-state relaxation the question at each point or block is not *how much* change to introduce, but *whether or not* to change. In this situation the finer-scale details accompanying a coarse-level change cannot be added *later*, *after* that change has been introduced, because they may well affect the very decision whether to introduce the change at all. Hence, in discrete-state minimization (or optimization) problems, multi-level processing should be governed by the following principle.

Decide on a large-scale change only after calculating its effects at all finer scales.

Employ this principle recursively.

This principle, we believe, can successfully be applied to many discrete-state optimization problems, including combinatorial problems such as the travelling salesman, problems in discrete-state physics, in electronic chip optimization, in image restoration and pattern recognition, and others. Moreover, this principle is likely to be very useful also to various *continuous*-state minimization problems, especially those which share with the discrete-state ones the property of having many local minima.

Indeed, the above multi-level approach can readily be motivated in terms of the typical topology of local minima. The role of large-scale changes is to take us out and away from a local minimum, sufficiently away that we do not return there; take us, in other words, out of the current attraction basin. This would bring us into a new basin. Now to decide whether to prefer the new basin to the old one, it is first necessary to reach its bottom, so that it can be compared to the bottom of the old basin. The large-scale change itself is not likely to land exactly at the bottom; in fact, it is most likely to land far from it, so arbitrarily high on the new basin's walls that comparison to the old basin, where the bottom had already been reached, is completely meaningless. By calculating the finer-scale effects of the large-scale change we basically attain the new bottom, where comparison to the old bottom is meaningful. Hence the principle as stated above. Furthermore, this principle should be employed recursively, because usually the energy function is made basins within basins within basins. The different scales of changes should, very roughly, correspond to moving between the different scales of basins.

One additional philosophical comment. Since the multi-level optimization described here proves to be a very powerful algorithm, it is likely to have been used in biological evolution. This suggests that evolution itself has not been accomplished by merely accumulating many small-scale improvements. Once in a while, a larger-scale (i.e., a more profound) change must have occured which, in all probability, did not immediately produce more viable individuals; but, if these individuals were just fit enough and lucky enough to survive several generations so as to evolve the smaller-scale (less profound) features needed to support the large-scale change, then the advantages (if any) of the overall change could come to play toward its preferred selection. Once in a much longer while a still more profound change could have emerged this way, lucky to survive enough generations to see the several recursive levels of improvements needed to show its profound advantages.

5. Further Multi-Level Rules

The central algorithmic concept introduced above can be implemented in various different ways and degrees of sophistication. Our experience with diversified types of examples has led us to several further observations and rules, which have subsequently been incorporated into the algorithm, correcting its performance in important ways. The main rules of multi-levelling are described below, while various other techniques are reported in the Appendix.

5.1 Level scheduling: the basic algorithm

5.1.1 General rule. An important rule, applicable recursively at all levels of the algorithm, is the following.

LEVEL SCHEDULING RULE: Do not relax on a coarser level (i.e., with larger blocks) before your configuration has been optimized (i.e., relaxed) on finer levels.

The reason for this rule is simple. In the block relaxation as described above, any flipped block is optimized on finer levels before its energy is being compared to the energy of the old (pre-flipping) configuration. Now clearly, this comparison can be meaningful only if the old configuration has itself previously been optimized on the finer levels.

Moreover, for the same reason, this rule should also be applied in scheduling levels *within each revision process.* This leads to a multi-level algorithm which should basically have the following structure.

5.1.2 Recursive algorithm. Let $1 = b_0 < b_1 < \cdots < b_M$ be the block sizes; that is, at level k, the unrevised blocks are $b_k \times b_k$ (mostly disjoint) squares of spins. We have usually chosen $b_k = 2^k$ for $k = 0, 1, \ldots, M-1$, and $b_M = L \leq 2^M$, where $L \times L$ is the entire lattice. (In many problems, some of the coarser levels can be omitted. Also, other choices of block sizes can as easily be implemented in the present program.) Denote by RELAX(k, D) a relaxation pass on level k over a domain D, where $D = \bigcup_{j=1}^{J} D_j$ is a union of (usually disjoint) $b_k \times b_k$ squares D_j. For $k = 0$ these blocks are single spins, and the relaxation pass is the usual point-by-point relaxation described in Secs. 3.1 and 3.2 (the scheduling of the artificial β is discussed in Sec. 5.2). For $k > 0$ we recursively define RELAX(k, D) as performing, for each $j = 1, 2, \ldots, J$, the following steps A to C.

A. Transfer D_j and a sufficiently large neighborhood of it to the work area of level k, and there flip D_j.

B. Revise D_j by performing RELAX(ℓ, D_j^ℓ) $n_{\ell,k}$ times, for $\ell = 0, 1, \ldots, k-1$, *in that order.* (Sometimes a reversed sequence is added, so that the entire order is $\ell = 0, 1, \ldots, k-2, k-1, k-2, \ldots, 1, 0$.) Each of the domains D_j^ℓ is usually a $b_{\ell,k} \times b_{\ell,k}$ square (sometimes a $\bar{b}_{\ell,k} \times \bar{b}_{\ell,k}$ square at its center is subtracted from it), whose center coincides with the center of D_j. Parameters like $n_{\ell,k}$, $b_{\ell,k}$ and $\bar{b}_{\ell,k}$ are chosen apriori.

C. Based on δE (cf. (3.1)) of the revised block, decide whether or not to accept its flipping (using the prescription detailed in Secs. 5.2.2 and A.3). In case of acceptance, replace the original D_j and its neighborhood by the content of the work area of level k.

Note that the algorithm uses only one work area per level, the size of which is comparable to the corresponding block size. The total storage requirement is therefore at most several times the size of the problem's lattice. In many problems the coarsest levels are not needed, in which case the required storage is just a fraction more than that of the given lattice.

A complete multigrid cycle consists of performing RELAX(k, the entire lattice) n_k times, for $k = 0, 1, \ldots, M$, *in that order.*

5.1.3 Complexity considerations. The size of the relaxation areas should carefully be checked, otherwise the total computational work may grow too fast as a function of N, the total number of sites.

To see the essential relations, denote by w_k the work *per spin* involved in RELAX(k,D). Denote also by $a_{\ell,k}$ the number of spins in D_j^ℓ (in Step B of RELAX(k,D)) divided by b_k^2; that is, $a_{\ell,k}$ is the size of the level-ℓ relaxation domain relative to the size of the square it serves to revise. Usually $a_{\ell,k} = b_{\ell,k}^2/b_k^2$. Neglecting the small work of Steps A and C, one gets

$$w_k = \sum_{\ell=1}^{k-1} w_\ell \, n_{\ell,k} \, a_{\ell,k} \tag{5.1}$$

This implies that w_k grows geometrically with k, and is therefore best restrained by choosing $n_{k-1,k} = 0$, i.e., by skipping level $k-1$ altogether. Indeed, *revising level-k blocks by relaxation on level $k-1$ is not really needed*, especially when stochasticity (which takes care of blocks far from square-like shapes; cf. Sec. 5.2) rectangular enlargements (cf. Sec. A.6) and adaptivity (on level $k-2$; see Sec. A.4) are used. It may in fact even be *undesired* to relax on level $k-1$, because its scale is too close to the scale of the revised block, hence the relaxation may completely destroy the block (flip it back) instead of revising it.

The growth of w_k is further restrained by choosing $n_{k-2,k}$ and $a_{k-2,k}$ as small as possible. Our usual choice has been $n_{k-2,k} = 1$ and $a_{k-2,k} = 25/16$; namely we chose $b_{k-2,k} = 5b_{k-2}$, just enough to cover a little more than $b_k = 4b_{k-2}$. Necessary exceptions to these prescriptions are $n_{0,1} = 1$, $b_{0,1} = 4$, $b_{0,2} = 6$ and $b_{\ell,M} = L$, yielding $a_{0,1} = 4$, $a_{0,2} = 9/4$ and $a_{\ell,M} = 1$.

Note, on the other hand, that the growth of w_k is not much affected if somewhat larger values of $n_{\ell,k}$ are chosen for $k - \ell > 2$. Similarly, if the entire lattice is large, the work of a complete cycle does not significantly increase if many more sweeps are made on the finest levels: e.g., if $n_0 = 20$ and $n_1 = 5$ in problems with $L \geq 50$. *It is important to use such extra fine-level sweeps when L is large*, because it is exactly for large L that very special local situations (e.g., special snake-like reversed blocks of width 1) become more probable, which requires (e.g., for flipping those blocks) a longer sequence of point-by-point stochastic relaxation sweeps.

In our current algorithm, the work is somewhat larger than indicated by (5.1), due to adaptation processes (Sec. A.4). A typical relation in practice may for example turn out to be $w_k \approx 2\sum_{\ell=1}^{k-2} w_\ell$. This implies $w_k \sim 2^k$, hence the overall work of a cycle is $W \sim 2^M N = N^{3/2}$. With other choices of the parameters a larger W may emerge, but it need never exceed $O(N^2)$.

5.2 Artificial stochasticity: how much when

An important question of course is how much stochasticity (i.e., how large artificial annealing temperature $1/\beta$) should be used at various stages of the algorithm. Since the multi-level process is itself a mechanism for escaping from local minima, it is not clear that artificial temperature is at all required (especially when the approach of Sec. A.2 below is adopted). Indeed, in some cases we have found that strict minimization ($\beta = \infty$ at all stages) multi-level algorithms (especially when equipped with the devices explained in Secs. A.3 and A.4) converge to a ground state in less than one cycle, thus achieving best performance without *any* stochasticity.

Still, at least at the present stage of algorithmic development, stochasticity was found to be very helpful in most cases. More specifically, its usefulness is typically pronounced whenever a deterministic algorithm gets stuck with a reversed block which is *long and narrow.* Such a block is not approximable, even roughly, by any square at any level. Hence a deterministic algorithm may miss it, while a stochastic relaxation, on a scale comparable to the *width* of that block, can quite easily break it into shorter blocks.

5.2.1 Annealing rule. Since the goal is to reach strict ground states, not to just wander stochastically around them, the process cannot be left at finite artificial temperatures. It should always be terminated with a strict minimization ($\beta = \infty$) pass.

This, moreover, is again recursively true at all levels. That is, each block revision process which employs a stochastic relaxation pass on any finer level, should follow it up with strict minimization on that same finer level, before switching to a coarser level. This is because, as explained in Sec. 5.1.1, there is no point in going to a coarser level with a configuration which has not first been optimized on the finer level. Hence the following rule:

ANNEALING RULE: At every stage and every level of the algorithm any stochastic relaxation should be followed up by strict minimization on the same level.

Observe that strict minimization at a certain level may of course itself employ *stochastic* passes (followed by strict minimization passes) at *finer* levels.

Also note that the annealing could be *gradual*: at any level one could start with some lower β and gradually increase it. In practice, however, the best approach is usually to employ just one finite β (see Sec. 5.2.2) before increasing it to ∞. This is in sharp contrast to one-level annealing processes, which should be done in a careful, gradual manner.

5.2.2 Size of β. In principle there is a specific value of β for each block Q. That is that value which would make it probable to flip Q whenever Q is likely to be part of a longer (but not wider) reversed block. Such a value of β should be different at different levels. Moreover, it can actually be tuned to the very block Q whose flipping is currently to be (stochastically) decided.

Thus, more precisely, the decision whether or not to flip a revised block Q is done as follows. If $\delta E(Q) \leq 0$, the flipping is always accepted. If $\delta E(Q) > 0$ and relaxation is deterministic, it is rejected. If $\delta E(Q) > 0$ but relaxation is stochastic, then the algorithm measures the likelihood of Q to be part of a longer block whose flipping would lower the energy, and randomly decides accordingly. In our present algorithm, the flipping probability $P_f(Q)$ is based on the following two easily-measured quantities:

1) The energy difference $\delta E(Q)$ (see (3.1)).

2) $B_\lambda(Q)$, the bond violation, at size λ, around the boundary of Q, where λ is comparable to the linear size of Q. $B_\lambda(Q)$ is defined as follows. Let $\langle i_\nu, j_\nu \rangle$, for $\nu = 1, \ldots, \mu$, be the μ bonds around the boundary ∂Q, ordered in their (cyclic) geometrical order. Set $b_\nu = J_{i_\nu j_\nu} S_{i_\nu} S_{j_\nu}$, where S_i is the spin value at site i *before* flipping Q. Also, because of the cyclic ordering, for $\nu > \mu$ define $b_\nu = b_{\nu-\mu}$. Then

$$B_\lambda(Q) = \max \left[0, \max_{\substack{1 \leq \nu_1 \leq \mu \\ 0 \leq \nu_2 - \nu_1 < \lambda}} \sum_{\nu=\nu_1}^{\nu_2} b_\nu \right] . \tag{5.2}$$

Thus, $B_\lambda(Q)$ measures how much bond violation can be cancelled by flipping, together with Q, a neighboring block coupled to Q by at most λ consecutive bonds. We have usually chosen $\lambda = \mid Q \mid^{1/d}$, where $\mid Q \mid$ is the number of spins in Q, and d is the problem's dimension ($d = 2$ in our present description). The amount of work in calculating $B_\lambda(Q)$ is negligible compared with the revision work that produced Q.

With those two quantities, the block flipping probability $P_f(Q)$ is defined by (3.2), with

$$\beta = \gamma / B_\lambda(Q), \tag{5.3}$$

where γ is a global control constant; $\gamma = 1$ may be a general good value (when used together with the device explained in Appendix A.1).

Note that at the finest (single spin) level, $\lambda = \mid Q \mid = 1$, and almost always $B_1(Q) = 1$, hence normally $\beta = \gamma$ and the block stochastic relaxation described here reduces to the usual point-by-point stochastic relaxation described in Sec. 3.2. The only exception is when $\delta E(Q) \geq 0$ and $B_1(Q) = 0$, that is, when all the bonds of the unflipped spin are violated but in spite of that its flipping would not lower the energy. This is possible when a strong counter field exists at that site. In many problems the field cannot be that strong, so that stochastic relaxation on the finest grid can be programmed in the traditional way ($\beta \equiv \gamma$), saving the work of calculating B_1. Nevertheless, note that in case $\delta E(Q) \geq 0$ and $B_1(Q) = 0$ does occur, rejecting the flipping (as dictated by the present scheme, but not by the usual stochastic relaxation) is indeed the correct step, because there can be no ground state where this spin is flipped.

6. Numerical Minimization Tests

6.1 Test classes

The minimization algorithm defined above, with its supplementary techniques described in the Appendix, was trained and tested on the following five classes of two dimensional Ising spin problems.

(1) The uniform ($J_{ij} \equiv 1$) and homogeneous ($h_i \equiv 0$) model, with periodic boundary conditions on various sizes of square lattices. The two ground states are known ($S_i \equiv 1$ and $S_i \equiv -1$), and the purpose of testing is to see how fast these states are obtained from various initial configurations.

(2) The same, except that the field h_i is non-constant, but still constructed so that the ground states are known. For example, $h_i = H_1 > 0$ at all sites i inside some convex domain and $h_i = H_2 < 0$ outside that domain, yielding three possible ground states ($S_i \equiv 1$, $S_i \equiv -1$ and $S_i \equiv$ sign h_i), depending on the values of H_1 and H_2. Of particular interest are of course those special values that give two, or even all three, ground states; an efficient solver should then easily move back and forth between those states. More complicated geometries were also tried.

(3) The same, except that each h_i is a real random number, uniformly distributed in the interval $(-H, H)$. In this case the ground states are not apriori known, but we can compare the minima reached from different initial configurations, by different algorithms. In most experiments we have taken $H = 2$, which is small enough to produce long-range interactions, but not as small as to make them trivial: typically, the ground state has a sea of spins of one sign, with large randomly shaped islands of the opposite sign.

(4) $J_{ij} \equiv 1$, h_i is randomly either $-H$ or $+H$, with equal probabilities, and the boundary conditions are free (no periodicity). For one particular distribution of signs on a 50×50 lattice, and for four different values of H (73/26, 73/27, 73/30 and 73/32), exact ground states, calculated by a special graph-theoretic method (cf. Sec. 3.1), were supplied to us by the Grenoble group [3].

(5) Spin-glass models: $h_i \equiv 0$, J_{ij} is randomly $+1$ or -1, with probabilities p and $1 - p$,

respectively, and the boundary conditions are periodic. For three cases ($p = .12$, $p = .146$ and $p = .5$) on a 20×20 lattice, exact ground states are described in [6]. The difficult case here is $p = .146$, nearly the critical value, at which large blocks of aligned spins tend to form. For $p = .12$ and $p = .5$ the correlations are short range. Moreover, for $p = .5$ there are many ground states.

Note that in all cases $|J_{ij}| \equiv 1$. The present program is not developed for cases of *strong* local variations in $|J_{ij}|$ (see Sec. A.3), but in principle could handle general couplings.

6.2 Comparison to simple annealing

For each of these classes we have compared multi-level solutions with a single-level solution by simulated annealing. The latter turned out as effective as the former for Class 4 problems, at the above-cited values of H. These problems are indeed dominated by difficulties at the finest level. Namely, the reversed blocks that tend to form are thin, mostly one spin wide. Only for smaller H (e.g., $H \approx 1$) wider blocks would become likely, making multi-levelling necessary. But, even in multi-level processing, thin reversed blocks in isotropic problems should be flipped by relaxation at the single-spin level. In fact, these Class 4 problems – especially those with the lower H, which produce long snake-like reversed blocks (length 4 for $H = 73/30$ and length 7 for $H = 73/32$) – served as important test beds for our single level processing, leading to several of the techniques described in the Appendix.

For Class 1 problems, simple annealing still performed reasonably well. The reason is that reversed blocks are in a sense still local: even if a wide reversed block is formed (e.g., an extensive island of -1 spins in a sea of $+1$ spins), it is not necessary to flip this whole block to see a decrease in energy. It decreases each time one of the end rows or columns is flipped. Moreover, if an annealing process starts with a sufficiently small β – or equivalently: if one starts with a random first approximation – the chance is that the reversed blocks are not very wide, hence the rows to be flipped, one at a time, are not very long. Unlike the previous class, however, multi-levelling did accelerate convergence of Class 1 problems, typically reducing simple annealing solution times by one order of magnitude for moderate-size grids (e.g., 32×32).

The real strength of multi-levelling is shown in various problems of Classes 2, 3 and 5. Here, in many cases, simple annealing fails even to *approach* the ground energy, no matter how slowly β grows, how many iterations are made, or what supplementary techniques are tried. It is doomed to fail whenever local convergence contradicts global convergence; e.g., whenever there exist some wide subdomains where the magnetic field is mostly in a direction opposing the ground-state spins.

For example, taking a Class 2 problem with a sufficiently wide convex domain and values of H_1 and H_2 not far from the values that give two ground states (e.g., a 5×5 square with $H_1 \approx .8$ and $H_2 = -.1$), annealing could reach only the configuration $S_i \equiv$ sign h_i, which (e.g., for $H_1 < .8$) is not necessarily the true ground state. For several Class 3 problems (with $H = 2$) annealing never even approached the lowest energy produced by the multi-level algorithm, no matter how gradually the artificial temperature was decreased. In Class 5, annealing did reach ground states in the easier cases, but failed for $p = .146$.

6.3 Multi-level performance: current status

The current set of multi-level minimization programs is not fully streamlined. It is still a patchwork. Some of the supplementary techniques (see Appendix) are not yet fully implemented: some of them were introduced after most of the experiments were done, some are programmed only for the finest level, others only for coarser ones. (For necessary technical reasons, the finest level is treated by different routines, unlike the traditional multigrid practice.) Also, the programs are still far from being optimized with respect to CPU time.

Instead of timing, we have measured performance by counting *point decisions (PDs)*. One PD is consumed whenever a decision (to flip or not) is made at the single-spin level. This involves calculating (2.4) and, for a probabilistic decision, also (2.3). The main work at coarser levels is the block revision, which ultimately always leads to finest-level sweeps, hence can also be measured in PDs.

For any one of the test problems the algorithm produced the minimal energy in at most few cycles, always costing less than $3N^2$ PDs. In some cases the algorithm jumps several times within one cycle back and forth between several approximate ground states. For example, in Class 2 problems with a 5×5 convex domain with $H_1 = .8$ and $H_2 = -.1$, the ground state $S_i \equiv$ sign h_i was produced in the first relaxation sweep over 4×4 blocks, the other ground state ($S_i \equiv -1$) was then produced in relaxing over 8×8 blocks, and two additional jumps between the two states occurred still within that 8×8 block relaxation (due to the adaptivity feature). A proper use of LCC (see § A.3) would in this case determine the existence of two equivalent ground states, or would choose the lower of them in case H_1 is slightly different from the transition value .8. Similarly, in one of Class 3 examples, there were two widely different configurations with almost the same energy, one of them minimal. Approximate transition between them always occurs at the coarsest scale (revising the entire grid flip), with exact transition seen upon using the suitable LCC.

Many more tests are of course needed to establish the exact efficiency of the multi-level algorithms, especially since the present tests have not been entirely "fair": they have been performed with the same classes of problems used in the training of the algorithm.

6.4 Summary. One-cycle algorithms

The present minimization algorithm is not "perfect". To any one of its versions it may be possible to construct "counter examples" which would require exponential solution time, and this may remain true for future versions as well. But the more developed is the algorithm, the less likely are the counter examples. At the present state, the probability of such examples seem already to be small enough so that average solution times is $O(N^2)$ or better. To maintain this efficiency in the future, with new *types* of problems, we may of course find, as we have found in the past whenever new types appeared, that some additional rules should be understood and implemented.

What is important to realize is that, for most purposes, it is not the ground states that are required. All that is needed are *approximate* ground states, which approximate ground state statistical properties. We have observed that the present algorithm easily yields such approximations in just one cycle.

Indeed, whenever the configuration obtained by one multi-level cycle is not itself a ground state, the difference turns out to be insignificant: the energy is very close to the true minimum, and, more importantly, slight changes of data (e.g., very small changes in the random magnetic field) can turn the obtained configuration into a ground state; hence its *apriori* chance of being itself a ground state is likely to have been about the same as that of the current ground states themselves.

We thus conjecture that *for many statistical purposes, one cycle of the multi-level minimization algorithm is enough for each set of data.* We further conjecture that even a relatively "light" cycle will often do, lighter even than the $O(N^{3/2})$ cycle mentioned in Sec. 5.1.3. Moreover, in each additional cycle many more approximate ground states may be encountered, which may similarly serve in calculating the desired statistics. We plan to test these conjectures on some classes of problems.

One should of course be careful in using this approach. It cannot be used when the desired statistics are strongly affected exactly by those special rare reversed blocks which the algorithm takes longer to flip.

7. Finite Temperature: Preliminary Observations

7.1 Continuous-state problems

For zero temperature (minimization problems), it has been shown above that discrete-state multi-level processing is considerably more involved than continuous-state multi-level processing. The same is expected at finite temperatures, as long as they are not as high as to have correlation lengths comparable to the meshsize. The first step in developing multi-level processes at positive temperatures is therefore made in the context of continuous-state problems, such as the XY model or the Heisenberg model. Multi-level Monte-Carlo processes can then be developed along lines similar to familiar multigrid techniques. In particular, since the problems are nonlinear, the inter-grid transfers are made in the *Full Approximation Scheme* (*FAS*; cf. [10, §8]), whose conventions will be used below.

For simplicity, assume first that only two levels are involved. The coarse level is typically made of every other column and every other row of the fine level. Two basic transfer operators should be defined between these two levels: the coarse-to-fine interpolation I_c^f, and the fine-to-coarse local averaging I_f^c. The interpolation can be simple bilinear (in terms of angles, in case of Heisenberg or XY models, so as to preserve the unit size of the spins), and the local averaging can be its transpose.

We assume that for every fine-grid configuration u^f, an energy function $E^f(u^f)$ is defined. This induces, for every coarse-grid configuration u^c, the energy function $E^c(u^c) = E^f(I_c^f u^c)$. Since this coarse-grid energy function will be defect-corrected below, we can actually replace it by any convenient function $\bar{E}^c(u^c)$, as long as $\bar{E}^c(u^c)$ approximates $E^c(u^c)$ for smooth u^c; i.e.,

$$\| \nabla^c \bar{E}^c(e^{i\omega x}) - \nabla^c E^c(e^{i\omega x}) \| \,/\, \| \nabla^c E^c(e^{i\omega x}) \| \to 0 \quad \text{as} \quad |\omega| \to 0 \tag{7.1}$$

where $x = (x_1, \ldots, x_d)$ are the space coordinates, $\omega = (\omega_1, \ldots, \omega_d)$, $\omega x = \omega_1 x_1 + \cdots + \omega_d x_d$, $\nabla^c = (\partial_1^c, \partial_2^c, \ldots)$, $\partial_i^c E^c(u^c) = (\partial/\partial u_i^c) E^c(u^c)$, and $\| \cdot \|$, $| \cdot |$ are any finite norms. This normally allows using $\bar{E}^c$ which has the same functional form as E^f; e.g., nearest neighbor couplings only.

For any given fine-grid configuration $\tilde{u}^f$ we then define the corrected interpolation $\tilde{I}_c^f$ (the FAS interpolation), given by

$$\tilde{I}_c^f u^c = \tilde{u}^f + I_c^f (u^c - I_f^c \tilde{u}^f) \tag{7.2}$$

and the corrected coarse-grid energy

$$\tilde{E}^c(u^c) = \bar{E}^c(u^c) + (u^c, \tilde{\tau}_f^c) \tag{7.3}$$

where $(,)$ is the inner product and

$$\tilde{\tau}_f^c = (I_c^f)^T \nabla^f E^f(\tilde{u}^f) - \nabla^c \bar{E}^c(I_f^c \tilde{u}^f), \tag{7.4}$$

$(I_c^f)^T$ denoting the transpose of I_c^f. These corrected forms interpret any coarse-grid configuration u^c as describing only the change from $I_f^c\tilde{u}^f$, with $\tilde{u}^f$ (the current fine-grid configuration) still describing its fine-grid details.

A typical cycle of the multi-level algorithm starts with several sweeps of a usual point-by-point Monte-Carlo process (see Sec. 2.2) on the fine grid, bringing the system to a *local* statistical equilibrium. The resulting configuration $\tilde{u}^f$ is then used in a coarse-grid Monte-Carlo, based on $\tilde{E}^c$ and on the starting configuration $I_f^c\tilde{u}^f$. This coarse-grid Monte-Carlo can (and for full efficiency usually should) itself use still coarser grids, in a similar manner. The cycle terminates with the final coarse-grid configuration u^c being used to update the fine grid, replacing $\tilde{u}^f$ by $\tilde{I}_c^f u^c$.

Note that in each cycle, throughout the coarse-grid processing, $\tilde{\tau}_f^c$ is fixed, representing a fixed field-like fine-to-coarse defect correction. This allows the coarse-grid Monte-Carlo to be done without constantly using the fine grid, hence to consume relatively short CPU times. The main work per cycle is the few fine-grid sweeps. Their number is small since they need to equilibrate only on the smallest scale (the scale invisible to the coarse grid).

When enough grids, to the coarsest possible scale, are recursively used in such a manner, this algorithm has fast transition times. In one cycle it almost equilibrates at all scales, hence also almost decorrelates at all scales. Hence few cycles could be enough for calculating statistical averages, provided the slow balancing of deviations (cf. Sec. 2.3) is also treated at all levels. This is indeed possible, using suitable inter-grid transfers.

As a simple example, consider the calculation of $\langle u^f \rangle$, the average of u^f over the domain and over all configurations, weighted by their physical probabilities. If the fine-to-coarse transfer I_f^c is sum-preserving (or "full weighting", in the usual multigrid terminology – cf. [10, §4.4]), then $\langle I_f^c u^f \rangle = \langle u^f \rangle$. To average out local deviations, $I_f^c u^f$ should be a local sum-preserving averaging not of u^f itself, but of $\bar{u}^f$, where $\bar{u}_i^f$ is defined at each site i as the expected value of u_i^f given its neighborhood. (Using suitable pre-calculated tables, somewhat larger neighborhoods can quickly be taken into account – making the averaging out of local deviations even better.) Employing similar fine-to-coarse transfers at all levels, the configuration u^M on the coarsest level will represent averaging of u^f over all scales except for the coarsest. Hence $\langle u^M \rangle$, calculated for example by averaging $\bar{u}^M$ over a sequence of coarsest-grid Monte-Carlo steps over several multigrid cycles, will give us a good approximation to $\langle u^f \rangle$.

An important feature of this algorithm is that its statistics are as accurate at large scales as at small ones. Hence, even rather crude approximations have reliable large-scale figures. Moreover, and for a similar reason, one need not calculate $\langle u^f \rangle$ to a high precision in order, for example, to be able to calculate its derivative ∂ with respect to the temperature; $\langle \partial u^f \rangle$ can itself be calculated, along with $\langle u^f \rangle$, and to a comparable accuracy, by transfering it from finer levels to coarser ones in an analogous manner. The local computation of $\overline{\partial u}^f$ is as easy as that of $\bar{u}_i^f$, but, unlike the latter, it gives meaningful approximations to $\langle \partial u^f \rangle$ only when repeated at all scales.

At scales where large local deviations are statistically important, if there are such scales, the algorithm should be modified (or not allowed to reach such scales), since the coarse-level moves are no longer independent of their statistical effects at the fine level. To see how such effects can be taken into account, consider the more extreme case of nonlinearity discussed in the next section.

7.2 Discrete-state problems

Returning to Ising spins as a model for discrete-state problems, it is clear from our zero temperature investigations (Sec. 3), that coarse-grid moves cannot correctly be decided without calculating the details of their fine-grid effects, in a certain recursive manner (see Sec. 5.1). The general outline for such an algorithm (not yet implemented) is as follows.

A multi-level cycle should start on the finest (single spin) level, and move to increasingly coarser levels. At each level Monte-Carlo sweeps over the entire domain should be carried out. Each step of a sweep at scale b level should consist of flipping a $b \times b$ block, then revising it by finer level Monte-Carlo sweeps in its vicinity, then deciding whether to accept or reject the revised block, employing for example the Metropolis rule. This rule should use the *physical* β, of course. When this β is large (substantially larger than critical), though, it should be useful to prefix such "physical sweeps" with some "artificial sweeps", using lower artificial β. Enough physical sweeps should in principle be made at each level (also within the revision processes) to reach local equilibrium, but a reasonable approximation to such an equilibrium should in practice be obtained in just a couple of sweeps.

Two important requirements that should (approximately) be satisfied by the revision processes seem uncertain: (a) Maintaining detailed statistical balance, i.e., ensuring the stationarity of the physical distribution (2.2). (b) Balancing statistical deviations: the revision may, especially at critical and smaller β, create deviations unrelated to (e.g., not opposing) those in the original configuration, hence weighting together the revised and the original configurations may not be fully efficient at averaging deviations out.

The two requirements can concurrently be met by the following three devices:

(A) The same revising process which is applied to the configuration with the flipped $b \times b$ square is also applied to the original (pre-flip) configuration. The two processes are correspondingly called the *flipped revision* and the *unflipped revision.*

(B) All revising processes are designed so as to preserve a certain property, such as sum total (average magnetization) or certain majorities (e.g., for I_j^c defined by a majority rule). Only moves which preserve that property should be considered by the revising Monte-Carlo. In addition, the revising processes are of course *local*, hence they preserve the configuration outside (a certain neighborhood of) the flipped square. Thus each revision is done within a certain set of permissible configurations. For the unflipped revision, that set is denoted by R; for the flipped revision – R'. Starting with any member of such a set, all other members of the same set should be accessible by the revising Monte-Carlo.

(C) The end products of the two revising processes should be weighted against each other (either for the purpose of choosing one of them to continue the process, or for calculating suitably weighted averages) *not* according to their relative probabilities (2.3), but according to the relative probabilities of the corresponding *sets*, R and R'. The relative probability of R (R') can be computed during the unflipped (flipped) revising process as the inverse of the average of $e^{\beta E(c)}$, calculated along with other averages of interest. The averages are calculated using single-spin averaging of deviations (see Sec. 2.3), and also (for $b > 4$) similar *block* averaging – the whole process being recursive, like that described in Sec. 5.1.

These three devices need not always fully be implemented. Device (A) may not be necessary when the level scheduling rule (Sec. 5.1.1) is adopted, ensuring that the original configuration is statistically the same as the one produced by the unflipped revision. Device (B) may perhaps be skipped when a crude approximation is desired and only very few sweeps are therefore performed at each level, approximately preserving the desired property anyway. Device (C) may be avoided by (crudely) taking the relative probabilities of the end configurations as representing those of the corresponding sets. As noted before, owing to the multi-levelling, crude approximations may already yield reliable long-range statistics.

Appendix: Supplementary Minimization Techniques

The development of the multi-level energy minimization algorithm for Ising spin systems has led to the introduction of several new techniques, reported below. All these techniques could in principle be developed also for single-level minimization algorithms, but the multi-level framework has been important in their origination, for two related reasons: First, once multi-level processing has overcome the most fundamental sources of algorithmic slowness, less profound sources come into clear view and must be dealt with. (The history of this work is thus itself an illustration of multi-level evolution, as pictured in Sec. 4.) Secondly, in the multi-level context the task of relaxation at each level is more sharply focused (to the localization and flipping of only those reversed blocks whose width is comparable to that level's scale), hence easier to study and devise.

A.1 Lower starting β

The above value of β (5.3) has been chosen so as to make it probable to flip a block Q (correspondingly: a single spin) whenever it sits *at one end* of a reversed longer block. Sometimes, especially at the finest levels, reversed blocks can (usually in low probability) appear which are considerably long (compared with their width, which itself is comparable to the meshsize of the level). On these finest levels, several (even many) sweeps are usually performed (cf. Secs. 5.1.3 and A.4), and it is desired to make it highly probable that the first sweep will break any long reversed block, whose width is comparable to the meshsize, into shorter ones. It is desired, in other words, to make it probable to flip a block Q even when it sits *in the middle* of a reversed longer block. For this purpose, this first sweep should employ, instead of (5.3), a lower value of β, given by

$$\tilde{\beta} = \gamma / [B_\lambda(Q) + \tilde{B}_\lambda(Q)] \tag{A.1}$$

where, similarly to (5.2)

$$\tilde{B}_\lambda(Q) = \max\left[0, \max_{\substack{1 \le \nu_1 \le \mu \\ 0 \le \nu_2 - \nu_1 < \lambda}} \sum_{\nu=\nu_1}^{\nu_2} \tilde{b}_\nu\right]. \tag{A.2}$$

Here $\tilde{b}_\nu = b_\nu$ except that $\tilde{b}_\nu = 0$ for $\bar{\nu}_1 \le \nu \le \bar{\nu}_2$, where $\bar{\nu}_1$ and $\bar{\nu}_2$ are the values of ν_1 and ν_2 for which the max in (5.2) has been attained.

Observe that on the finest (the single spin) level, where $\lambda = 1$, the above prescriptions reduce to choosing at site i

$$\beta = \gamma / \min(1, B^i), \qquad \tilde{\beta} = \gamma / \min(2, B^i) \tag{A.3}$$

where B^i is the number of bonds $< i, j >$ violated upon flipping S_i.

A.2 Deterministic trials

Another possible strategy is to replace the above stochastic relaxation with deterministic trials. That is, whenever $P_f(Q)$ is not small, *tentatively* flip Q and then flip and revise its neighboring block (neighboring on that part of the boundary at which the bond violation $B_\lambda(Q)$ has appeared). The two flipped blocks together are now forming one longer block Q'. In case $\delta E(Q') \leq 0$, then accept this combined flipping. If $\delta E(Q') > 0$ and $\delta E(Q')/B_\lambda(Q')$ is not small (still using the original $\lambda = |Q|^{1/d}$), then reject the whole flipping, including the original tentative flipping of Q. If $\delta E(Q')/B_\lambda(Q')$ is positive but small, then continue the trial, now adding a third block. And so on. The criterion for smallness of $\delta E/B_\lambda$ can be made stricter (smaller) at each additional step. This strategy has not been tested yet.

A.3 Lowest Common Configuration (LCC)

The revised block Q is not always connected. Whenever it is disconnected, an important modification to the above algorithm is to make a separate flipping decision for each of its disconnected components. In other words, instead of choosing between the two configurations (the flipped Q vs. the original configuration), their "lowest common configuration" should be constructed.

Generally, given two spin configurations $\{S_i^1\}$ and $\{S_i^2\}$, their *lowest common configuration (LCC)* $\{S_i\}$ is defined as follows. Let Q be the set of sites i for which $S_i^1 \neq S_i^2$. And let $Q = \bigcup_{j=1}^{J} Q_j$ be the decomposition of Q into its disconnected components; i.e., the sites of each Q_j are connected to each other by a chain of sites belonging to Q, but are not so connected to the sites of any other $Q_{j'}$. A *chain of sites* is defined as a sequence of sites where each pair of subsequent sites α and β are *strongly coupled*, i.e., $J_{\alpha\beta} \neq 0$ and $|J_{\alpha\beta}|$ is not much smaller than $\min[\max_\gamma |J_{\alpha\gamma}|, \max_\gamma |J_{\gamma\beta}|]$. (In current programs, α and β are considered strongly coupled if and only if they are nearest neighbors. The programs are thus inapplicable to cases of strong anisotropy or other large local variations in $|J_{\alpha\beta}|$. The sign of $J_{\alpha\beta}$ may however change arbitrarily.) Now, with this decomposition, for $i \notin Q$ define $S_i = S_i^1 = S_i^2$, while for each $1 \leq j \leq J$ define $S_i = S_i^{\nu(j)}$ for all $i \in Q_j$, where $\nu(j)$ is either 1 or 2, whichever yields the lower energy.

Given Q, it is easy to calculate the LCC in $O(|Q|)$ operations, where $|Q|$ is the number of sites in Q. In our algorithms this work will be negligible.

A *stochastic lowest common configuration (SLCC)* of two configurations can similarly be calculated, the only difference being that the $\nu(j)$ are chosen stochastically, as prescribed in Sec. 5.2.2, using $\delta E(Q_j)$ and $B_\lambda(Q_j)$, with $\lambda = | Q_j |^{1/d}$.

The use of LCCs in other parts of the algorithm is sometimes also very important. In particular it is important when, typically on a large grid, the minimization processes at some particular regions of the grid are weakly coupled to each other. (For example, a block whose flipping changes the energy very little is weakly coupled to similar blocks far from it). In such a situation, each cycle is likely to attain the correct minimum at many of these regions, but the chances of getting all these minima simultaneously (hence the global minimum) may be slim, hence requiring many cycles. If, on the other hand, at the end of each cycle the attained configuration is replaced by its LCC with the configuration at the end of the previous cycle, each of those regions, once minimized, will remain so, and the global minimum will soon be reached.

The LCC can similarly further be used *within the revision processes*, especially for large blocks. Generally, *any* subprocess can be strengthened by accompanying it with a procedure of keeping in a special memory the configuration with lowest energy attained by it so far, replacing that configuration by its LCC with a newly encountered configuration whenever the latter shows promisingly low energy. At the end of the subprocess, its end product should be replaced by the LCC of this end product and the configuration in that special memory, before being used by an outer process.

Generalization. When a general number of configurations $\{S_i^1\}, \ldots, \{S_i^m\}$ is given, their simultaneous LCC is defined as follows. Let Q be the set of all sites i such that $S_i^k \neq S_i^\ell$ for at least one pair (k, ℓ). Let as before $Q = \cup_{j=1}^J Q_j$ be the decomposition of Q into its disconnected components. Then

$$\{S_i\} = \text{LCC}(\{S_i^1\}, \ldots, \{S_i^\ell\})$$

is defined by $S_i = S_i^1 = \cdots = S_i^\ell$ for $i \notin Q$, while for each $1 \leq j \leq J$ define $S_i = \mu(j) S_i^{\nu(j)}$ for all $i \in Q_j$, where the values of $\mu(j) = \pm 1$ and $1 \leq \nu(j) \leq \ell$ are chosen so as to yield the lowest possible energy. This generalization can be very useful (see end of Sec. A.4).

A.3.1 Minimized Common Configuration (MCC). A further generalization,

$$\{S_i\} = \text{MCC}(\{S_i^1\}, \ldots, \{S_i^\ell\})$$

is similarly defined by $S_i = S_i^1 = \cdots = S_i^\ell$ for $i \notin Q$, whereas, unlike in LCC, the values of S_i in each Q_j are determined by a separate minimization process, trying to reach the minimal Q_j configuration with all spins outside Q_j held fixed. This separate minimization can use very many sweeps, since Q_j would normally be small; it can use decimation procedures, in case Q_j is long but mostly one spin wide; Or it can itself use multi-level procedures, when Q_j is wider.

The generalized techniques can again be used within the revision processes. They should normally use only configurations obtained at a large β stage, lest they contain too much thermal noise.

A.4 Adaptive relaxation

The basic algorithm described above (Sec. 5.1.2) assumes pre-assigned relaxation domains D_j^ℓ and fixed numbers $n_{\ell,k}$ of sweeps at level ℓ. Efficiency can be enhanced by adapting these parameter to the local situation. In other words, following the $n_{\ell,k}$ *fixed sweeps* (over the fixed domain D_j^ℓ), further *adaptive sweeps* can be made, each of which scans only the nearest neighbors of the sites (or blocks, if $\ell > 0$) which were flipped in the former sweep. In fact, for each block Q that was formerly flipped, the adaptive sweeps scan only its *violated neighbors*, i.e., those nearest neighbors which show large bonds violation (comparable to $B_\lambda(Q)$) with the flipped Q.

When the adaptive sweeps are stochastic, they should be followed up by deterministic adaptive sweeps (see Sec. 5.2.1). Each of the latter should then scan, in addition to the former-sweep's violated neighbors, also all those blocks whose last flipping was probabilistic (energy increasing).

Such adaptive schemes serve several purposes. First, they save scanning sub-domains of D_j^ℓ where no further improvement occurs. More importantly, they allow the relaxation region to expand, when necessary, beyond the originally allotted D_j^ℓ. They thus enable using D_j^ℓ of minimal size (see Sec. 5.1.3). Similarly, they permit the use of small $n_{\ell,k}$, with more passes performed only when and where needed.

A general rood procedure, in relaxing either the entire domain or within any revision process, is to first make one fixed sweep (employing usually the lower starting β; cf. Sec. A.1), then stochastic adaptive sweeps (with the usual β), and then deterministic adaptive sweeps, continuing each of these two adaptive successions until it becomes empty. On those levels where more sweeps are allowed (and desired – cf. Sec. 5.1.3), this whole sequence can be repeated several times, the end product of each such sequence being replaced by its LCC with the previous end product. Or, better still, several such end products at a time can be accumulated and then replaced by their *simultaneous* LCC or MCC.

A.5 Position shifts

Certain troublesome reversed blocks stand a better chance of being flipped when the unrevised blocks of a certain level are suitably positioned. Hence, whenever returning to relaxing the entire lattice on a given level, it is advisable to shift the positioning of that level's squares. In our power-two scheme ($b_k = 2^k$), most important are half-size shifts (shifting level-k blocks by $b_k/2$ in either or both directions). Similarly, D_j^ℓ can be shifted (e.g., replacing $b_{\ell,n}$ by $b_{\ell,n} - b_\ell$) in between relaxation passes.

In the present code only simpler shifts are implemented, produced by shifting the entire problem at the beginning of each new cycle, alternating between one-position-upward and one-position-leftward shifts.

A.6 Rectangular enlargements of flipped squares

Another device that was found useful is to start the revision process of a flipped square with rectangular enlargements. This means to add, one by one, rows and/or columns to the flipped region as long as this reduces the energy of the entire configuration. Having reached in this way the optimal rectangular flipping, the regular revision process then starts to work on it.

No detailed experiments were conducted to determine the effect of this device. It seems useful especially when skipping level $k-1$ in revising level k (see Sec. 5.1.3). Similar enlargements make also sense on the finest (single spin) level, and might in fact be very useful there (similar to the tentative trials; cf. Sec. A.2), but are not yet implemented.

Acknowledgement

We are deeply indebted to the Grenoble group of R. Rammal and J.C. Angles d'Auriac for providing us with invaluable detailed information on exact ground states of two dimensional spin-glasses and Ising models with random fields, before during and after publication. One of us DJA has also benefitted from discussions with G. Parisi, who has brought to his attention previous work on improving high temperature statistics.

References

[1] D. Andelman, H. Orland and L.C.R. Wijewardhana: Lower critical dimension of the random-field Ising model: a montecarlo study. *Phys. Rev. Lett.* **52** (1984), 145.

[2] J.C. Angles D'Auriac and R. Maynard: On the random antiphase state in the $\pm J$ spin glass model in two dimensions. *Solid State Commun.* **49** (1984), 785.

[3] J.C. Angles D'Auriac, M. Preissmann and R. Rammal: The random field Ising model: algorithmic complexity and phase transition. *J. Physique Lett.* **46** (1985), 173.

[4] J.C. Angles D'Auriac and R. Rammal: Phase Diagram of the Random Field Ising Model. CNR Grenoble, preprint, 1986.

[5] F. Barahona: On the computational complexity of Ising spin glass models. *J. Phys.* **A15** (1982), 3241.

[6] F. Barahona, R. Maynard, R. Rammal and J.P. Uhry: Morphology of ground states of the two-dimensional frustration model. *J. Phys.* **A15** (1982), 673.

[7] G. Bhanot, M. Creutz and H. Neuberger: Microcanonical simulation of Ising systems. *Nucl. Phys. FS* **B235** (1984), 417.

[8] I. Bieche, R. Maynard, R. Rammal and J.P. Uhry: On the ground states of the frustration model of a spin glass by a matching method in graph theory. *J. Phys.* **A13** (1980), 2553.

[9] E. Bonomi and J-L Lutton: The N-city travelling salesman problem: statistical mechanics and the metropolis algorithm. *SIAM Review* **26** (1984), 551. See also [16], [17] and [18].

[10] A. Brandt: Multigrid Techniques: 1984 Guide with Applications to Fluid Dynamics. Monograph, 191 pages. Available as GMD Studien 85 from GMD-AIW, Postfach 1240, D-5205, St. Augustin 1, W. Germany.

[11] M. Falcioni, E. Marinari, M.L. Paciello, G. Parisi, B. Taglienti and Y.S. Zhang: Large distance correlation functions for an $SU(2)$ lattice gauge theory. *Nucl. Phys. FS* **B215** (1983), 265.

[12] K.H. Fisher: Spin glasses I. *Phys. Stat. Solidi* **B116** (1983), 357.

[13] K.H. Fisher: Spin glasses II. *Phys. Stat. Solidi* **B130** (1985), 13.

[14] R. Graham and F. Haake: Quantum Statistics in Optics and Solid-State Physics. Springer Tracts in Modern Physics, Vol. 66, Springer-Verlag, Berlin, 1973.

[15] A. Hasenfratz and P. Hasenfratz: Lattice Gauge Theories. Florida State University, preprint

FSU-SCRI 85-2, 1985.

[16] S. Kirkpatrick: Models of disordered systems. Lecture Notes in Physics 149 (C. Castellani et al., eds.), Springer-Verlag, Berlin, 280.

[17] S. Kirkpatrick, C.D. Gelatt Jr. and M.P. Vecchi: Optimization by simulated annealing. *Science* **220** (1983), 671.

[18] S. Kirkpatrick and G. Toulouse: Configuration space analysis of travelling salesman problems. *J. Physique* **46** (1985), 1277.

[19] S.K. Ma: Renormalization group by Monte-Carlo methods. *Phys. Rev. Lett.* **37** (1976), 461.

[20] N.D. Mackenzie and A.P. Young: Lack of ergodicity in the infinite range Ising spin-glass. *Phys. Rev. Lett.* **49** (1982), 301.

[21] N.D. Mackenzie and A.P. Young: Statics and dynamics of the infinite range Ising spin glass model. *J. Phys.* **C16** (1983), 5321.

[22] E. Marinari, G. Parisi and C. Rebbi: Computer estimates of meson masses in $SU(2)$ lattice gauge theory. *Phys. Rev. Lett.* **47** (1981), 1795. For a recent review see [15].

[23] N. Metropolis, A.W. Rosenbluth, M.N. Rosenbluth, A.H. Teller and E. Teller: Equation of state calculations by fast computing machines. *J. Chem. Phys.* **21** (1953), 1087.

[24] A.T. Ogielski and I. Morgenstern: Critical behavior of three-dimensional Ising model of spin glass. *J. Appl. Phys.* **57** (1985), 3382.

[25] G. Parisi, R. Petronzio and F. Rapuano: A measurement of the string tension near the continuum limit. *Phys. Lett.* **128B** (1983), 418.

[26] G. Parisi and Y.S. Wu: Perturbation theory without gauge fixing. *Scientia Sinica* **24** (1981), 483.

[27] H. Sompolinsky: Time dependent order parameters of spin-glasses. *Phys. Rev. Lett.* **47** (1981), 935.

[28] H. Sompolinsky and A. Zippelius: Relaxational dynamics of the Edwards-Anderson model and the mean-field theory of spin-glasses. *Phys. Rev.* **B25** (1982), 6860.

[29] R.L. Stratonovich: Topics in the Theory of Random Noise, Vol. 1, Gordon and Breach, New York, 1963.

[30] R.H. Swendsen: Monte-Carlo renormalization group. *Phys. Rev. Lett.* **42** (1979), 859.

MULTIGRID METHOD FOR NEARLY SINGULAR AND SLIGHTLY INDEFINITE PROBLEMS

A. Brandt*
Weizmann Institute of Science, Rehovot, Israel

S. Ta'asan**
Institute for Computer Applications in Science and Engineering

Abstract

This paper deals with nearly singular, possibly indefinite problems for which the usual multigrid solvers converge very slowly or even diverge. The main difficulty is related to some badly approximated smooth functions which correspond to eigenfunctions with nearly zero eigenvalues. A modification to the usual coarse-grid equations is derived, both in Correction Scheme and in Full Approximation Scheme. With this modification, the algorithm exhibits the usual multigrid efficiency.

INTRODUCTION

Usual multigrid for indefinite problems is sometimes found to be very inefficient. A strong limitation exists on the coarsest grid to be used in the process. This limitation is not so much a result of the indefiniteness (existence of eigenvalues with different signs) itself, but of the nearness to singularity, that is, the existence of nearly zero eigenvalues. These eigenvalues are badly approximated (e.g., they may even have a different sign) on coarse grids, hence the

---

*Sponsored by the Air Force Wright Aeronautical Laboratories, Air Force Systems Command, Unites State Air Force under Grant AFOSR 84-0070.

**Research was supported by the National Aeronautics and Space Administration under NASA Contract Nos. NAS1-17070 and NAS1-18107 while the author was in residence at ICASE, NASA Langley Research Center, Hampton, VA 23665-5225.

corresponding eigenfunctions, which are usually smooth ones, cannot efficiently converge. As a remedy, one could avoid using grids which are too coarse, but in many cases this would degrade efficiency.

This trouble of the coarse-grid approximation has been resolved by introducing a modification to the usual coarse-grid equations, based on the observation that there are just few smooth eigenfunctions which are not well represented on the coarse-grid, and these can be controlled by specially added relations. This modifiction removes the restriction on the coarseness of the grids that can be used.

Another issue when dealing with indefinite problems is the choice of relaxation. Mode analysis shows that the Gauss-Seidel relaxation is suitable for such problems if fine enough grids are considered. Indeed, even though some smooth components diverge with this relaxation, on fine enough grids this divergence is slow and can, therefore, easily be corrected by the coarse-grid corrections. On coarser grids, however, the divergence of smooth components in Gauss-Seidel relaxation is faster, hence, another relaxation scheme is needed. We have used for that purpose the Kaczmarz relaxation, which always converges.

The multigrid algorithm obtained here has good asymptotic convergence rates for problems in which the indefiniteness is not too high, i.e., the number of eigenvalues with the "wrong" sign (positive in our text) is small. For higher indefiniteness another method has been developed and will be reported elsewhere.

2. RELAXATION

Generally, in order to achieve good multigrid performances, the relaxation involved need to have good smoothing properties on one hand, and at most slow divergence on the other hand. We discuss below Gauss-Seidel and Kaczmarz relaxations and their proper use in our context.

2.1. Gauss-Seidel

Fourier analysis of Gauss-Seidel relaxation even for slightly indefinite problems shows that on fine grids high frequencies converge very fast. The reason for this is that the principal part for indefinite problems is the same as that of definite ones. Smooth components may diverge on such grids, but slowly enough to be handled by the coarse-grid correction. For example, in case of the operator $\Delta + k^2$ in two dimensions the worst divergence factor per sweep of smooth components is $1/(1 - \frac{1}{2} k^2 h^2)$, h being the mesh size. On coarse grids, typically when this factor becomes larger than 1.2 or so, Gauss-Seidel relaxation can no longer be used.

2.2. Kaczmarz Relaxation

Given an equation

$$Ax = b \tag{2.1}$$

where A is an $n \times n$ nonsingular matrix, define a new unknown y such that

$$A^* y = x \tag{2.2}$$

where A^* is the adjoint of A. The equation obtained for y is

$$AA^* y = b. \tag{2.3}$$

The matrix AA^* is symmetric positive definite for any A which is nonsingular. Hence, Gauss-Seidel relaxation for equation (2.3) will converge. It induces a relaxation on equation (2.1) via the relation (2.2). This relaxation of (2.1) is called Kaczmarz relaxation. Its i-th step is

$$x_j \leftarrow x_j + \overline{a}_{ij}\, \delta_i \qquad (j = 1,\ldots,n)$$

$$\delta_i = (b_i - \textstyle\sum_j a_{ij} x_j)/\sum_j |a_{ij}|^2$$

where $\overline{a}_{ij}$ is the complex conjugate of a_{ij}. For general smoothing properties of this relaxation, see [1] Section 1.1 and [2].

Kaczmarz relaxation converges whenever solution exists, and can therefore be used on coarse grids. Moreover, when more than one solution exists the convergence is to the one closest to the initial approximation (Tanabe [3]). Hence, this relaxation would not allow the growth of error eigenfunctions corresponding to $\lambda = 0$, and would similarly allow only very slow change in eigenfunctions corresponding to λ close to 0.

3. TROUBLES WITH THE COARSE-GRID APPROXIMATION

Having settled the question of relaxation, another difficulty is encountered: some coarse grids do not well approximate some smooth components. To understand this situation, suppose the error on the fine-grid (grid h) contains a smooth eigenfunction ϕ^h, so that the corresponding residual is $L^h \phi^h = \lambda^h \phi^h$. The corresponding equations on the coarse-grid (grid H) are

$$L^H V^H = \lambda^h I_h^H \phi^h$$

where I_h^H is the fine-to-coarse transfer, i.e., some local averaging. Since ϕ^h is a smooth eigenfunction of L^h, $I_h^H \phi^h$ is approximately an eigenfunction of L^H, but with slightly different eigenvalue λ^H. The solution of the coarse-grid equations is approximately

$$\tilde{V}^H = \frac{\lambda^h}{\lambda^H} I_h^H \phi^h.$$

After interpolating the $\tilde{V}^H$ as a correction to the fine-grid solution, the new error is approximately

$$\phi^h - I_H^h \tilde{V}^H \approx \left(1 - \frac{\lambda^h}{\lambda^H}\right)\phi^h$$

where I_H^h is the interpolation operator and since ϕ^h is a smooth function, we assume that $I_H^h I_h^H \phi^h \approx \phi^h$. Thus, due to the coarse-grid correction the error is reduced by the factor $|1 - \lambda^h/\lambda^H|$; hence a condition for good convergence is

$$|1 - \frac{\lambda^h}{\lambda^H}| \ll 1 \tag{3.1}$$

for any eigenfunction ϕ^h which has poor convergence by relaxation.

When λ^h, λ^H are close to zero, relation (3.1), even if it holds on fine enough grids, it may strongly be violated on coarse grids. Such coarse grids cannot then be used in the multigrid process. Without them, however, efficiency may very much degenerate. We will therefore present a new method in which restriction (3.1) is removed.

4. MODIFIED COARSE-GRID EQUATIONS: Two-Grid Case

The modification described here is based on the assumption that there are only few smooth eigenfunctions for which relation (3.1) is violated. Denote by H_0 the subspace spanned by these badly approximated eigenfunctions. We assume for the description below that H_0 is known. In Section 6 we present a method for approximating H_0.

4.1. Correction Scheme (CS) Version

Assume first that H_0 is spanned by one function ϕ^h, and let u^h be the exact solution of the fine-grid (grid h) equation

$$L^h u^h = F^h. \tag{4.1}$$

Suppose the current approximation $\tilde{u}^h$ to u^h satisfies

$$\langle u^h, \phi^h\rangle = \langle \tilde{u}^h + \eta\phi^h, \phi^h\rangle. \tag{4.2}$$

If η were known we would have the approximation $\tilde{u}^h + \eta\phi^h$ on the fine-grid instead of $\tilde{u}^h$. This would yield the coarse-grid (grid H) equation

$$L^H v^H = I_h^H \left[F^h - L^h \tilde{u}^h - \eta L^h \phi^h\right] \tag{4.3}$$

where v^H approximates the error $v^h = u^h - \tilde{u}^h - \eta\phi^h$. Since by (4.2) the latter does not have components in H_0, equation (4.3) could be used to accelerate fine-grid convergence. However, since η is not known, we need to add another equation on the coarse grid which will enable us to solve also for η. A reasonable choice for such an equation is an approximation of equation (4.2), namely

$$\langle v^H, \bar{I}_h^H \phi^h\rangle = 0 \tag{4.4}$$

where $\bar{I}_h^H$ is some fine-to-coarse transfer, not necessarily identical with I_h^H. Equations (4.3), (4.4) form the modified CS equations.

Suppose now that H_0 is spanned by $\{\phi_1^h,\cdots,\phi_N^h\}$ and $\langle\phi_j^h, \phi_k^h\rangle = \delta_{jk}$. Because of linearity, the corrected CS equations for this case will be

$$L^H v^H = I_h^H R^h - \sum_{j=1}^{N} \eta_j I_h^H L^h \phi_j^h \tag{4.5a}$$

$$\langle v^H, \phi_j^H \rangle = 0 \qquad j = 1,\cdots,N \tag{4.5b}$$

where $R^h = F^h - L^h \tilde{u}^h$, $\tilde{u}^h$ being the current fine-grid approximation, and $\phi_j^H = \bar{I}_h^H \phi_j^h$. The coarse-grid correction will finally be done either by

$$\tilde{u}^h \leftarrow \tilde{u}^h + I_H^h \left[\tilde{v}^H + \sum_{j=1}^{N} \tilde{\eta}_j \phi_j^H \right] \tag{4.6a}$$

or by

$$\tilde{u}^h \leftarrow \tilde{u}^h + I_H^h \tilde{v}^H + \sum_{j=1}^{N} \tilde{\eta}_j \phi_j^h \tag{4.6b}$$

depending on whether or not ϕ_j^h are stored on the fine grid, $\tilde{u}$ and $\tilde{\eta}_j$ being the computed (approximate) solutions to equations (4.5). The difference between (4.6a) and (4.6b) is usually unimportant, so ϕ_j^h need not be stored. The only case where (4.6b) must be used is when the fine-grid problem is much closer to singularity than the coarse-grid one. In that case, $\tilde{\eta}_j$ may be large; therefore, $I_H^h \tilde{\eta}_j \phi_j^H$ may have large high frequency components, which will magnify the residuals on the fine grid. By doing (4.6b) one avoids introducing high frequency components that arise from interpolating $\tilde{\eta}_j \phi_j^H$, and therefore the mentioned difficulty is removed. See Section 7, Tables 5 and 6.

4.2. Full Approximation Scheme (FAS) Version

The Full Approximation Scheme is essential for nonlinear problems or when local refinement is used. It is important, therefore, to derive the modified equations in that formulation too. This derivation can be done directly from (4.5), but to gain an additional insight we do it independently. The usual FAS equation on the coarse grid is

$$L^H u^H = F^H + \tau_h^H(\tilde{u}^h) \tag{4.7}$$

where $\tau_h^H(\tilde{u}^h) = L^H \bar{I}_h^H \tilde{u}^h - I_h^H L^h \tilde{u}^h$ is the "fine-to-coarse (defect) correction," $F^H = I_h^H F^h$, $\tilde{u}^h$ is the current approximation on the fine grid, and $\bar{I}_h^H$ is the fine-to-coarse solution transfer (see [1], Sections 8.1-8.2).

In the present case, however, we wish to approximate on the coarse-grid only the part of the error which is free of H_0 components; hence, the H_0 components of the correction, $\sum \eta_j \phi_j^h$, should be considered part of the fine grid approximation, replacing (4.7) by

$$L^H u^H = F^H + \tau_h^H\left(\tilde{u}^h + \sum_{j=1}^{N} \eta_j \phi_j^h\right). \tag{4.8}$$

The additional conditions, ensuring that the coarse-grid correction is indeed approximtely free of H_0 components, can be written as

$$\langle u^H, \phi_j^H\rangle = \langle \bar{I}_h^H \tilde{u}^h, \phi_j^H\rangle + \eta_j\langle \phi_j^H, \phi_j^H\rangle, \quad (j = 1,\cdots,N). \tag{4.9}$$

Equations (4.8) and (4.9) together should determine u^H and $\eta_1,\cdots,\eta_N$. Once an approximate solution $(\tilde{u}^H, \tilde{\eta}_1,\cdots,\tilde{\eta}_N)$ has been calculated, the correction to the fine grid solution can be done analogously to either (4.6a) or (4.6b), but the former option yields here a particularly simple formula, namely

$$\tilde{u}^h \leftarrow \tilde{u}^h + I_H^h\left(\tilde{u}^H - \bar{I}_h^H \tilde{u}^h\right) \tag{4.10a}$$

which is just the usual FAS correction formula. The latter option, which must be used in some extreme cases, reads

$$\tilde{u}^h \leftarrow \tilde{u}^h + I_H^h \left(\tilde{u}^H - I_h^H \tilde{u}^h - \sum_{j=1}^{N} \eta_j \phi_j^H\right) + \sum_{j=1}^{N} \eta_j \phi_j^h . \tag{4.10b}$$

Equation (4.8) is not in a form convenient for calculations on grid H. In case the problem is linear, a more convenient form is

$$L^H u^H = \bar{F}^H + \sum_{j=1}^{N} \eta_j \psi_j^H \tag{4.11}$$

where

$$\bar{F}^H = F^H + \tau_h^H\left(\tilde{u}^h\right) = I_h^H R^h + L^H\left(\bar{I}_h^H \tilde{u}^h\right) \tag{4.12}$$

and $\psi_j^H = \tau_h^H(\phi_j^h)$. The solution of (4.9) and (4.11) thus involves the 2N+1 input functions $\bar{F}^H$, $\psi_1^H,\cdots,\psi_N^H$, $\phi_1^H,\cdots,\phi_N^H$, of which $\bar{F}^H$ should be calculated and stored whenever the algorithm switches from level h to level H, while the other 2N functions can be calculated and stored once for all. The same equations can be used also for nonlinear problems, but with ψ_j^H generally calculated by

$$\psi_j^H = \varepsilon^{-1}\left\{\tau_h^H(\tilde{u}^h + \varepsilon\phi_j^h) - \tau_h^H(\tilde{u}^h)\right\}, \tag{4.13}$$

with sufficiently small positive ε. The dependence of ψ_j^H on $\tilde{u}^h$ is very crude (e.g., no dependence at all when L^h is linear); hence it will usually be unnecessary to update them on a new switch to level H.

5. GENERAL MULTIPLE GRID EQUATIONS

Suppose a sequence of discretization with mesh sizes $h_1 > h_2 > \cdots > h_M$ is given, where $h_j = 2h_{j+1}$. Let the h_k-grid equations be

$$L^k u^k = F^k$$

where L^k approximates L^{k+1} for $k < M$, and L^M approximate some differential operator L.

Usually, if level M well approximates the differential equation (even in terms of H_0 components) then level M - 1 will approximate level M well enough for acceleration purposes. Hence, modified coarse-grid equations may not be needed on level M - 1. Denote by ℓ the finest level on which modified equations are needed. We describe now the modified equations on levels $k \leq \ell$, assuming the subspace of bad components, H_0, is spanned on level ℓ by the orthogonal set $\{\phi_1^\ell, \cdots, \phi_N^\ell\}$.

5.1. CS Version

For $k \leq \ell$ the equations to be solved for v^k, η_j^k on level k are

$$L^k v^k = f^k - \sum_{j=1}^{N} \eta_j^k I_{\ell+1}^k L^{\ell+1} \phi_j^{\ell+1} \tag{5.1a}$$

$$\langle v^k, \phi_j^k \rangle = \rho_j^k \qquad (j = 1, \cdots, N) \tag{5.1b}$$

where

$$f^k = I_{k+1}^k (\tilde{f}^{k+1} - L^{k+1} \tilde{v}^{k+1}) \qquad (k < \ell) \tag{5.2a}$$

$$\tilde{f}^k = f^k - \sum_{j=1}^{N} \tilde{\eta}_j^k I_{\ell+1}^k L^{\ell+1} \phi_j^{\ell+1} \qquad (k \leq \ell) \tag{5.2b}$$

$$\tilde{f}^{\ell+1} = f^{\ell+1} \tag{5.2c}$$

$$\rho_j^k = \rho_j^{k+1} - \langle \tilde{v}^{k+1}, \phi_j^{k+1} \rangle \qquad (k < \ell,\ j = 1, \cdots, N) \tag{5.2d}$$

$$\rho_j^{\ell} = 0 \qquad (j = 1,\cdots,N) \qquad (5.2e)$$

$$\phi_j^k = \bar{I}_{k+1}^k \, \phi_j^{k+1} \qquad (k < \ell;\ j = 1,\cdots,N) \qquad (5.2f)$$

$$\bar{I}_{\ell+1}^k = \bar{I}_{k+1}^k \, \bar{I}_{\ell+1}^{k+1}, \qquad I_{\ell+1}^k = I_{k+1}^k \, I_{\ell+1}^{k+1} \qquad (5.2g)$$

$\bar{I}_{k+1}^k$, I_{k+1}^k are fine-to-coarse grid transfers, not necessarily the same. $\tilde{v}^k$, $\tilde{\eta}_j^k$ are the current approximation to v^k, η_j^k respectively. Initial approximations are $\tilde{v}^k = 0$, $\tilde{\eta}_j^k = 0$. The input functions for level k are thus f^k, ϕ_j^k and $I_{\ell+1}^k \, L^{\ell+1} \, \phi_j^{\ell+1}$, $(j = 1,\cdots,N)$, of which only f^k should be updated on every new switch from level k + 1.

For efficient relaxation, instead of storing f^k one should store $\tilde{f}^k$ and update it whenever the $\tilde{\eta}_j^k$ are changed.

Note that η_j^{k-1} is designed to be a correction to η_j^k. Thus, the coarse-grid corrections for $2 \le k \le \ell$ will be done by the replacements

$$\tilde{\eta}_j^k \leftarrow \tilde{\eta}_j^k + \tilde{\eta}_j^{k-1} \qquad (j = 1,\cdots,N) \qquad (5.3a)$$

$$\tilde{f}^k = \tilde{f}^k - \sum_{j=1}^{N} \tilde{\eta}_j^{k-1} \, I_{\ell+1}^k \, L^{\ell+1} \, \phi_j^{\ell+1} \qquad (5.3b)$$

$$\tilde{v}^k \leftarrow \tilde{v}^k + I_{k-1}^k \, \tilde{v}^{k-1} \qquad (5.3c)$$

while for $k = \ell + 1$ use

$$\tilde{v}^{\ell+1} \leftarrow \tilde{v}^{\ell+1} + I_{\ell}^{\ell+1}\left(\tilde{v}^{\ell} + \sum_{j=1}^{N} \tilde{\eta}_j^{\ell} \, \phi_j^{\ell}\right) \qquad (5.3d)$$

or

$$\tilde{v}^{\ell+1} \leftarrow \tilde{v}^{\ell+1} + I_{\ell}^{\ell+1} \, \tilde{v}^{\ell} + \sum_{j=1}^{N} \tilde{\eta}_j^{\ell} \, \phi_j^{\ell+1}, \qquad (5.3e)$$

(see discussion in Section 4.1 for the use of (5.3d) versus (5.3c)).

5.2. FAS Version

For $k \le \ell$ the equations to be solved for u^k, η_j^k on level k are given by

$$L^k \, u^k = \bar{F}^k + \sum_{j=1}^{N} \eta_j^k \, \psi_j^k \qquad (5.4a)$$

$$\langle u^k, \phi_j^k \rangle = \sigma_j^k + \eta_j^k \alpha_j^k \qquad (j = 1,\cdots,N) \tag{5.4b}$$

where

$$\psi_j^k = \varepsilon^{-1}\left\{\tau_{k+1}^k(\tilde{u}^{k+1} + \varepsilon\phi_j^{k+1}) - \tau_{k+1}^k(\tilde{u}^{k+1})\right\} + I_{k+1}^k \psi_j^{k+1} \quad (k \leq \ell) \tag{5.5a}$$

(hence $\psi_j^k = \tau_{k+1}^k(\phi_j^{k+1}) + I_{k+1}^k \psi_j^{k+1}$ in the linear case)

$$\psi_j^{\ell+1} = 0 \qquad (j = 1,\cdots,N) \tag{5.5b}$$

$$\tau_{k+1}^k(\tilde{u}^{k+1}) = L^k \bar{I}_{k+1}^k \tilde{u}^{k+1} - I_{k+1}^k L^{k+1} \tilde{u}^{k+1} \tag{5.5c}$$

$$\phi_j^k = \bar{I}_{k+1}^k \phi_j^{k+1} \qquad (j = 1,\cdots,N) \tag{5.5d}$$

$$\bar{F}^k = L^k \bar{I}_{k+1}^k \tilde{u}^{k+1} + I_{k+1}^k(\tilde{F}^{k+1} - L^{k+1}\tilde{u}^{k+1}) \tag{5.5e}$$

$$\tilde{F}^k = \bar{F}^k + \sum_{j=1}^{N} \tilde{\eta}_j^k \psi_j^k \qquad (k \leq \ell) \tag{5.5f}$$

$$\tilde{F}^{\ell+1} = \bar{F}^{\ell+1} \tag{5.5g}$$

$$\sigma_j^k = \langle \bar{I}_{k+1}^k \tilde{u}^{k+1}, \phi_j^k \rangle + \begin{cases} 0 & (k = \ell) \\ \tilde{\sigma}_j^{k+1} - \langle \tilde{u}^{k+1}, \phi_j^{k+1} \rangle & (k < \ell) \end{cases} \tag{5.5h}$$

$$\tilde{\sigma}_j^k = \sigma_j^k + \tilde{\eta}_j^k \alpha_j^k \qquad (j = 1,\cdots,N) \tag{5.5i}$$

$$\alpha_j^k = \langle \phi_j^k, \phi_j^k \rangle \qquad (j = 1,\cdots,N) \tag{5.5j}$$

and $\tilde{\eta}_j^k$, $\tilde{u}^k$ are the current approximations to η_j^k, u^k respectively. Intital approximations are $\tilde{\eta}_j^k = 0$, $\tilde{u}^k = \bar{I}_{k+1}^k \tilde{u}^{k+1}$. The input functions are $\bar{F}^k$, $\phi_1^k,\cdots,\phi_N^k$, $\psi_1^k,\cdots,\psi_N^k$, of which only $\bar{F}^k$ must be recalculated each time the level k problem is formulated.

For efficient relaxation, instead of storing $\bar{F}^k$, and σ_j^k $(j = 1,\cdots,N)$ one should store $\tilde{F}^k$ and $\tilde{\sigma}_j^k$ and update them whenever $\tilde{\eta}_j^k$ is updated.

Initially (when the level k problem is set up) $\tilde{F}^k = \bar{F}^k$ and $\tilde{\sigma}_j^k = \sigma_j^k$.

The coarse-grid corrections will be done by the replacements

$$\tilde{\eta}_j^k \leftarrow \tilde{\eta}_j^k + \tilde{\eta}_j^{k-1} \qquad (2 \le k \le \ell;\ j = 1,\cdots,N) \qquad (5.6a)$$

$$\tilde{F}^k \leftarrow \tilde{F}^k + \sum \tilde{\eta}_j^{k-1}\ \psi_j^k \qquad (2 \le k \le \ell) \qquad (5.6b)$$

$$\tilde{\sigma}_j^k \leftarrow \tilde{\sigma}_j^k + \tilde{\eta}_j^{k-1}\ \alpha_j^{k-1} \qquad (2 \le k \le \ell;\ j = 1,\cdots,N) \qquad (5.6c)$$

$$\tilde{u}^k \leftarrow \tilde{u}^k + I_{k-1}^k(\tilde{u}^{k-1} - \bar{I}_k^{k-1}\ \tilde{u}^k) \qquad (2 \le k \le M). \qquad (5.6d)$$

In case, for some $2 \leqslant k \leqslant \ell+1$, the grid k problem is much closer to singularity than grid k+1 problem, (5.6d) should be replaced by

$$\tilde{u}^k \leftarrow \tilde{u}^k + I_{k-1}^k\left(\tilde{u}^{k-1} - \bar{I}_k^{k-1}\ \tilde{u}^k - \sum_{j=1}^N \tilde{\eta}_j^{k-1}\ \phi_j^{k-1}\right) + \sum_{j=1}^N \tilde{\eta}_j^{k-1}\ \phi_j^k \qquad (5.6e)$$

which is the analogue of both (5.3c) and (5.3e). Of course, (5.6e) can always be used, but (5.6d) is somewhat simpler (cf. end of Section 4.1).

Observe, indeed, that in the linear case

$$\psi_j^k = \tau_{k+1}^k\left(\phi_j^{k+1}\right) + I_{k+1}^k\ \psi_j^{k+1} = L^k\ \phi_j^k - I_{\ell+1}^k\ L^{\ell+1}\ \phi_j^{\ell+1}$$

and by identifying u^k with $\bar{I}_{k+1}^k\ \tilde{u}^{k+1} + v^k + \sum \eta_j^k\ \phi_j^k$ the equivalence of the FAS and the CS is easily seen.

5.3. Solution Process for Modified Equations

We refer in this section to the FAS version, namely the equation

$$L^k\ u^k = \bar{F}^k + \sum_{j=1}^N \eta_j^k\ \psi_j^k \qquad (5.7a)$$

$$\langle u^k,\ \phi_j^k\rangle = \sigma_j^k + \eta_j^k\ \alpha_j^k \qquad (j = 1,\cdots,N), \qquad (5.7b)$$

where the unknowns are the function u^k and the constants $\eta_1^k,\dots,\eta_N^k$. The CS version is treated similarly. As before, $\tilde{u}^k$, $\tilde{\eta}_j^k$, $\tilde{F}^k$ and $\tilde{\sigma}_j^k$ will denote the current (stored) approximations to u^k, η_j^k, the right-hand side of (5.7a) and the right-hand side of (5.7b), respectively.

In relaxing equations (5.7) we distinguish between the following:

(i) <u>a local relaxation sweep</u>

Relax $L^k u^k = \tilde{F}^k$ for $\tilde{u}^k$ by either Gauss-Seidel or Kaczmarz, keeping $\tilde{\eta}_j^k$, and therefore also $\tilde{F}^k$, fixed.

(ii) <u>a global step</u>

This will be the step for updating $\eta_1^k,\dots,\eta_N^k$, and the H_0 components in u^k by using (5.7b) together with (approximatley) the H_0 components of (5.7a). Most generally this is done by solving simultaneously for β_j, $\hat{\eta}_j$ $(j = 1,\dots,N)$ the system of 2N equations:

$$\langle L^k\left(u^k + \sum_{i=1}^{N} \beta_i \phi_i^k\right), \phi_j^k\rangle = \langle \tilde{F}^k + \sum_{i=1}^{N} \hat{\eta}_i \psi_i^k, \phi_j^k\rangle \quad (j = 1,\dots,N) \qquad (5.8a)$$

$$\langle u^k + \beta_j \phi_j^k, \phi_j^k\rangle = \tilde{\sigma}_j^k + \hat{\eta}_j \alpha_j^k \quad (j = 1,\dots,N) \qquad (5.8b)$$

and then introducing the following changes

$$\tilde{u}^k \leftarrow \tilde{u}^k + \sum_{j=1}^{N} \beta_j \phi_j^k \qquad (5.9a)$$

$$\tilde{\eta}_j^k \leftarrow \tilde{\eta}_j^k + \hat{\eta}_j \quad (j = 1,\dots,N) \qquad (5.9b)$$

$$\tilde{F}^k \leftarrow \tilde{F}^k + \sum_{j=1}^{N} \hat{\eta}_j \psi_j^k \qquad (5.9c)$$

$$\tilde{\sigma}_j^k = \tilde{\sigma}_j^k + \hat{\eta}_j \alpha_j^k \quad (j = 1,\dots,N). \qquad (5.9d)$$

The local relaxation is used to smooth the error in u^k and therefore should be done at all levels. On the other hand, it may be enough to do step (ii) on the coarsest grids <u>only</u>, since it deals with global variables $\left(\eta_j^k\right)$ and with global changes to u^k. Thus, step (ii) will be done on grids $k \le m$, where usually $m < \ell$. This will usually reduce the storage requirement of the algorithm, since there is no need to store ϕ_j^k on levels $m < k \leq \ell$. In fact, it is often unnecessary to store even ψ_j^k for $m < k \leq \ell$. Indeed for $m < k \leqslant \ell$ these functions are only used in the interpolation step (5.6b), which can be skipped in case of a V cycle, because as a smooth change to $\tilde{F}^k$, its effect on the subsequent relaxation on level k is negligible. On the other hand, step (5.6b) cannot be skipped in case it is followed by a switch back to the coarser $(k - 1)$ grid, since in this case the smooth update to $\tilde{F}^k$ is essential. Thus, in case of W cycles, ψ_j^k must be stored for all levels $k \leq \ell$. Generally, $m < \ell$ can be used only if no intermediate level k $(m < k \leqslant \ell)$ is much closer to singularity than level k+1.

5.4. Summary. Work and Storage

A cycle for improving $\tilde{u}^k$ and $\tilde{\eta}^k = (\eta_1^k,\cdots,\eta_N^k)$ $(k \leq \ell)$ is denoted by

$$(\tilde{u}^k, \tilde{\eta}^k) \leftarrow CMG(\tilde{u}^k, \tilde{\eta}^k, L^k, \tilde{F}^k, \tilde{\sigma}^k)$$

and is defined recursively by the following steps (A) through (D).

(A) Make the following $\nu_1(k)$ times

(a) a local step for $L^k u^k = \tilde{F}^k$

(b) for $k \leq m$ make a global step defined in (5.8), (5.9). For $k = 1$, choose $\nu_1(k)$ to guarantee convergence to small residuals, or solve the equations directly, and then terminate the cycle. If $k > 1$, continue.

(B) Starting with $\tilde{u}^{k-1} = \bar{I}_k^{k-1} \tilde{u}^k$, $\tilde{\eta}_j^{k-1} = 0$ $(j = 1,\cdots,N)$ make the cycle

$$(\tilde{u}^{k-1},\tilde{\eta}^{k-1}) \leftarrow CMG(\tilde{u}^{k-1}, \tilde{\eta}^{k-1}, L^{k-1}, \tilde{F}^{k-1}, \tilde{\sigma}^{k-1})$$

$\gamma(k)$ times, where $\tilde{F}^{k-1}$, $\tilde{\sigma}_j^{k-1}$ are defined by (5.5) with k replaced by $k - 1$.

(C)
$$\tilde{\eta}_j^k \leftarrow \tilde{\eta}_j^k + \tilde{\eta}_j^{k-1} \qquad (k \leq m;\ j = 1,\cdots,N)$$

$$\tilde{F}^k \leftarrow \tilde{F}^k + \sum_{j=1}^{N} \tilde{\eta}_j^{k-1} \psi_j^k \qquad (k \leq m)$$

$$\tilde{\sigma}_j^k \leftarrow \tilde{\sigma}_j^k + \tilde{\eta}_j^{k-1} \alpha_j^{k-1} \qquad (k \leq m)$$

and interpolation is done either by

$$\tilde{u}^k \leftarrow \tilde{u}^k + I_{k-1}^k\left(\tilde{u}^{k-1} - \bar{I}_k^{k-1}\,\tilde{u}^k\right)$$

or

$$\tilde{u}^k \leftarrow \tilde{u}^k + I_{k-1}^k\left(\tilde{u}^{k-1} - \bar{I}_k^{k-1}\,\tilde{u}^k - \sum_{j=1}^{N} \tilde{\eta}_j^{k-1}\,\phi_j^{k-1}\right) + \sum_{j=1}^{N} \tilde{\eta}_j^{k-1}\,\phi_j^k.$$

The second option is necessary in case the grid k problem may be almost singular.

(D) Make steps (a), (b) of (A) $\nu_2(k)$ times.

$\gamma(k) = 2$ corresponds to a W cycle; hence if $\gamma(k) = 2$, the global step has to be done on level k, which implies $m \geq k$. If $\gamma(k) = 1$ we can choose m to be smaller than k. Since equations (5.7) are for $k \leq \ell < M$, this cycle is part of a bigger cycle for $k = M$.

The storage and work required by this algorithm are essentially the same as in the usual multigrid algorithm, since all extra work and storage involved are made on very coarse grids, often only on level 1, sometimes also on level 2. In fact, $\ell = M-1$ should be used only when the finest (grid M) problem is itself a rather poor approximation to the differential problem, so usually $\ell < M-1$, in which case the extra work is negligible compared to the work of relaxing grid M.

6. APPROXIMATION OF SUBSPACE H_0

In the preceding discussion it is assumed that H_0 is accurately known. This section deals with how accurately H_0 needs to be known and how to approximate it.

6.1. Accuracy Needed for H_0

Let ϕ_i^h $(i \geq 1)$ be the smooth eigenfunctions on the finest grid

$$L^h\,\phi_i^h = \lambda_i^h\,\phi_i, \qquad \langle\phi_i, \phi_j\rangle = \delta_{ij}, \tag{6.1}$$

and let H_0 be for simplicity spanned by ϕ_1^h alone. Suppose that ϕ_1^h is not known to the algorithm, and instead ϕ^h is used, where

$$\phi^h = \sum_{i\geq 1} a_i\,\phi_i^h. \tag{6.2}$$

Suppose an error $v^h = e_1\,\phi_1^h$ has emerged on the fine grid so that

$$L^h V^h = e_1 \lambda_1 \phi_1^h, \tag{6.3}$$

and the corresponding modified CS coarse-grid equations are

$$L^H V^H + \eta I_h^H L^h \phi^h = e_1 \lambda_1 \phi_1^H \tag{6.4a}$$

$$\langle V^H, \phi^H \rangle = 0 \tag{6.4b}$$

where $\phi^H = I_h^H \phi^h$, $\phi_i^H = I_h^H \phi_i^h$. For smooth eigenfunctions ϕ_i^h, we can assume that ϕ_i^H are again eigenfunctions

$$L^H \phi_i^H = \lambda_i^H \phi_i^H, \qquad \langle \phi_i^H, \phi_j^H \rangle = \delta_{ij}, \tag{6.5}$$

neglecting changes in eigenfunctions since important to our discussion are only changes in the eigen<u>values</u>. If we write the solution to (6.4) as

$$V^H = \sum_{i \geq 1} E_i \phi_i^H \tag{6.6}$$

then (6.4) gives

$$\lambda_i^H E_i + \eta \lambda_i^h a_i = 0 \qquad i \geq 2 \tag{6.7a}$$

$$\lambda_1^H E_1 + \eta \lambda_1^h a_1 = \lambda_1^h e_1 \tag{6.7b}$$

$$\sum_{i \geq 1} a_i E_i = 0. \tag{6.7c}$$

Equations (6.7) imply

$$\eta = \frac{q_1 a_1 e_1}{\beta + q_1 a_1^2} \tag{6.8a}$$

$$E_1 = \eta\beta/a_1, \qquad E_i = -\eta q_i a_i \quad (i \geq 2) \tag{6.8b}$$

where

$$\beta = \sum_{i \geq 2} q_i a_i^2, \qquad q_i = \frac{\lambda_i^h}{\lambda_i^H}. \tag{6.9}$$

Hence, the coarse-grid correction is

$$V^H + \eta\phi^H = \sum_{i\geq 1} (E_i + \eta a_i)\phi_i^H$$

$$= \frac{\beta + a_1^2}{\beta + q_1 a_1^2} q_1 e_1 \phi_1^H + \sum_{i\geq 2} \frac{q_1 a_1 a_i}{\beta + q_1 a_1^2} (1 - q_i)e_1 \phi_i^H. \qquad (6.10)$$

Extra errors have thus been introduced in the directions of $\{\phi_2^H, \phi_3^H,\cdots,\}$; but these should be small (relative to e_1), since q_i should be close to 1 for ϕ_i^H not in H_0 (and also a_i will be small compared to a_1 by the condition below) and, more importantly, these errors can efficiently be reduced by the next coarse-grid correction. Our focus here should thus be the behavior of the ϕ_1^h component. Assuming $I_H^h \phi_1^H \approx \phi_1^h$ by smoothness, the coarse-grid correction, when interpolated to the fine grid and subtracted from the old fine grid error, gives in this component the new error $\bar{e}_1 \phi_1^h$, where

$$\bar{e}_1 = (1 - q_1)\beta(\beta + q_1 a_1^2)^{-1} e_1. \qquad (6.11)$$

The main condition for convergence is therefore

$$\left|\frac{(\lambda_1^h - \lambda_1^H)\beta}{\lambda_1^h a_1^2 + \lambda_1^H \beta}\right| < 1, \qquad (6.12)$$

and the convergence factor per cycle is bounded (below) by the left-hand side of (6.12). This bound is indeed small when λ_1^H is a good approximation to λ_1^h, i.e., when $|\lambda_1^h - \lambda_1^H| << \lambda_1^h$. But if that is not the situation (which is why one should want to include ϕ_1^h in H_0 in the first place) then the necessary condition for fast convergence is that both $|\lambda_1^h \beta|$ and $|\lambda_1^H \beta|$ are small compared to $\lambda_1^h a_1^2$. Since $q_i \approx 1$ for $i \geq 2$, the condition for fast convergence can be summarized as

$$\sum_{i\geq 2} a_i^2 << \min\left(1, \frac{|\lambda_1^h|}{|\lambda_1^H|}\right) a_1^2. \qquad (6.13)$$

This condition implies in particular that, if $\lambda_1^h = 0$, that is, if the given problem is singular, then $a_2 = a_3 = \cdots = 0$, i.e., the eigen-function ϕ_1^h must be known exactly. This seems to be too stringent, but in fact, the increase in accuracy for ϕ_1^h can be obtained as the algorithm proceeds, by doing for each cycle of the original problem, a cycle for improving ϕ_1^h. (See Section 7, Tables 8 and 9.)

On the other hand,(6.13) implies that there is no complication when the coarse-grid problem is singular (see Section 7, Tables 1-4).

Generally, condition (6.13) gives a precise idea as to how closely ϕ_1^h should be approximated.

6.2 Algorithm for Approximating H_0

We motivate the algorithm by considering the case where H_0 is spanned by one function. It is assumed in this discussion that the finest grid problem is well-posed. This implies that errors in components which belong to H_0 show sizeable average residuals on the finest grid.

The method for approximating H_0 is based on the following observation: components which belong to H_0 are spanned by eigenfunctions whose eigenvalues are much closer to zero than others, and exactly such eigenfunctions will converge in Kaczmarz relaxation much slower than other eigenfunctions. Hence, if the coarse-grid equation

$$L^k W^k = 0 \qquad \text{with homogeneous boundary conditions} \qquad (6.14)$$

is relaxed, starting with a random approximation, then when convergence has slowed down, the dominant part in the resulting $\tilde{W}^k$ must be a component in H_0; therefore, $\tilde{W}^k$ at this stage can serve as an approximation to a function in H_0 on the coarse grid. H_0 is needed on finer grids. A first candidate will be just an interpolation of $\tilde{W}^k$ to these grids. However, since interpolation introduces high frequency errors which will leave large residuals in equation (6.14) on finer grids, and therefore give the wrong ψ_j^k, one needs to smooth somehow the interpolated $\tilde{W}^k$ from coarser grids. A reasonable way to do it is to relax (6.14) on fine grids after obtaining a first approximation from coarser grids. This is summarized in the following algorithm.

Algorithm

Repeat the following for $i = 1,\cdots,N$ ($N = \dim H_0$)

(A) Set $k = 1$

(B) $\tilde{W}^k$ = random function if $k = 1$

$\tilde{W}^k = I^k_{k-1}\, \tilde{W}^{k-1}$ if $k > 1$.

(C) Relax (6.14), starting with initial approximation $\tilde{W}^k$, keeping $\tilde{W}^k$ orthogonal to $\{\phi^k_1,\cdots,\phi^k_{i-1}\}$, until convergence becomes slow

(D) $k \leftarrow k + 1$

(E) if $k \leq \ell+1$ go to (A), else $\phi^{\ell+1}_i = \tilde{W}^{\ell+1}$.

(F) Define $\phi^k_i = I^k_{k+1}\, \phi^{k+1}_i$ ($k = \ell,\ell-1,\cdots,1$).

In case N is not known in advance, stop the above procedure when step (C) no longer reaches slow convergence in just a few sweeps. If, after few cycles of solving the original problem, convergence rate still deteriorates, repeat (A) through (F) once more, replacing the random function in (B) by the residuals left by the original problem on the coarsest grid. If the addition of the new function $\phi^{\ell+1}_i$ to the set $\{\phi^{\ell+1}_1,\cdots,\phi^{\ell+1}_{i-1}\}$ does not improve convergence rate significantly, it means that the accuracy of $\{\phi^{\ell+1}_1,\cdots,\phi^{\ell+1}_i\}$is not enough and this can be improved by inverse iteration on the grid $k = \ell+1$ (using standard multigrid for doing the inverse iteration). The improvement of the functions $\phi^{\ell+1}_j$ by inverse iteration is done by one multigrid cycle before each multigrid cycle of the original problem. Such an improvement is needed when the original finest grid probelm is much closer to singularity than the next level; see Section 7, Tables 8 and 9.

Finally, if cycles on level $\ell+1$ converge satisfactorily, but not on finer levels, then ℓ should be increased (by 1). The above algorithm can then, of course, be shortened, starting with the known H_0 on level ℓ.

7. NUMERICAL RESULTS

Experiments were performed with the new algorithm using the model problem

$$\begin{cases} (\Delta + k^2)U = F & \text{in } \Omega = (0,1) \times (0,1) \\ U = g & \text{on } \partial\Omega. \end{cases}$$

The tables below show the residual history on the finest level. We denote by M, m, ℓ, h_1 the following:

M -- the finest level,

ℓ -- the finest level on which corrected equations are needed,

m -- the finest level on which the global step is performed,

h_1 -- the mesh size of the coarsest grid (grid 1).

The subspace H_0 was calculated by the algorithm of Section 6.2, where in step (B) 40 relaxations were done on the coarsest grid (k = 1) and two on every finer grid ($1 < k \leq \ell$). The algorithm CMG of Section 5.3 was used with

$\nu_1(k) = 2$, $\nu_2(k) = 1$ when Gauss-Seidel relaxation is used,

$\nu_1(k) = \nu_2(k) = 3$ when Kaczmarz relaxation is used,

$\nu_1(k) = 13$ for $k = 1$,

$\gamma(k) = 1$ for $k > 2$, $\gamma(k) = 2$ for $k = 2$.

We give below the discrete first two eigenvalues of the Laplacian for the grids used in the examples of this section. This will enable us to see how far from singularity is each of the different levels used in the process.

h	λ_1^h	λ_2^h
.25	-18.74516600406	-41.37258300203
.125	-19.48683967711	-47.23375184668
.0625	-19.67587286709	-48.81161578777
.03125	-19.72335955067	-49.21342550952

In Tables 1-4 interpolation of corrections was made according to (5.6b); in other tables the interpolation is specified. Residuals were transferred by 9-point full weighting and the local relaxation was Gauss-Seidel for $kh \leq .5$ and Kaczmarz for $kh > .5$. In all examples $M = 4$, $h_1 = .25$, $\ell = 3$, $m = 2$.

Tables 3, 4 show a case in which the second eigenvalue is very close to zero, and its corresponding eigenspace is two-dimensional. Therefore, only two functions were used in spanning H_0. The algorithm for finding functions in H_0 finds first these eigenfunctions whose eigenvalues are closest to zero. Therefore, the eigenfunction belonging to the first eigenvalue was not used in these computations, and it was not needed as can be seen from the fast convergence shown by these tables. This clearly shows that H_0 is related to almost-singularity, not to indefiniteness.

In Tables 5, 6 we show that in case the finest grid problem is too close to a singularity one must use interpolation of correction according to (5.6e) (or (5.3e) in the CS version) and not the usual FAS interpolation. In these two tables the exact H_0 was used.

Table 7 shows that if ϕ is not known accurately enough, poor results are obtained. ϕ in this example is found by the same procedure described in the beginning of this section.

In Table 8 inverse iteration (done by usual multigrid) was used to improve the accuracy of ϕ. Starting with ϕ obtained as in Table 7, one multigrid cycle of inverse iteration was done to improve ϕ before each multigrid cycle for the original problem. Results are identical to the ones obtained with the exact ϕ (Table 5).

Table 9 shows a case in which the distance of the closest eigenvalue to zero is about 1.10^{-8}. As seen from this table, improving ϕ by only one cycle of inverse iteration per cycle of the original problem is not quite enough to maintain the full speed of the algorithm. Once in few cycles the residuals are magnified, and this happens whenever the L_2-norm of the error in the approximation is reduced significantly. This reduction of the error is due to a correction of the approximate solution by $\eta\phi$. If ϕ is not accurate enough, components other than the desired ones enter to u^h and since their residuals are much higher than those of $\eta\phi$, a magnification of the residuals occurs. (A similar phenomenon can be seen also in Table 7, where the distance from singularity is larger, but ϕ is not improved at all by inverse iterations.) This would not have happened if we allowed the speed of convergence of the inverse-iteration cycles to be slightly faster than that of the main cycles, e.g., by adding an extra inverse-iteration cycle once per several cycles. But this is not really needed, because all that may happen is a minor slowdown at high-accuracy solutions (much below truncation errors) for cases of extreme closeness to singularity.

In fact, we believe that, to obtain solutions with errors smaller than trunction error, all one has generally to do is a one-cycle FMG algorithm for calculating H_0 (meaning one inverse-iteration cycle for level k after step (C) in the algorithm of Section 6.2), followed by a one-cycle FMG algorithm for solving the original problem.

REFERENCES

[1] A. Brandt, Multigrid Techniques: 1984 Guide With Applications to Fluid Dynamics. Monograph available as GMD-Studie No. 85, from GMD-FlT, Postfach 1240, D-5205, St. Augustin 1, W. Germany.

[2] A. Brandt, Algebraic multigrid theory: the symmetric case. Preliminary Proceedings of International Multigrid Conference, Copper Mountain, Colorado, April 1983. Applied Math. Comp., to appear.

[3] S. Ta'asan, Multigrid Methods for Highly Oscillatory Problems. Ph.D. Thesis, The Weizmann Institute of Science, Rehovot, Israel 1984.

[4] K. Tanabe, Projection methods for solving a singular system of linear equations and its applications, Numer. Math., 17 (1971), pp. 203-214.

Table 1: $k^2 = 18.745166$, $\dim H_0 = 1$

cycle #	$\|residuals\|_2$
1	.363(+3)
2	.172(+2)
3	.114(+1)
4	.891(-1)
5	.762(-2)
6	.685(-3)
7	.652(-4)
8	.658(-5)
9	.684(-6)
10	.744(-7)

Table 2: $k^2 = 19.486839$, $\dim H_0 = 1$

cycle #	$\|residuals\|_2$
1	.363(+3)
2	.174(+2)
3	.114(+1)
4	.892(-1)
5	.763(-2)
6	.687(-3)
7	.654(-4)
8	.661(-5)
9	.688(-6)
10	.749(-7)

Table 3: $k^2 = 41.372583$, $\dim H_0 = 2$

cycle #	$\|residuals\|_2$
1	.363(+3)
2	.172(+2)
3	.112(+1)
4	.938(-1)
5	.864(-2)
6	.832(-3)
7	.820(-4)
8	.815(-5)
9	.796(-6)
10	.811(-7)

Table 4: $k^2 = 47.233752$, $\dim H_0 = 2$

cycle #	$\|residuals\|_2$
1	.363(+3)
2	.171(+2)
3	.110(+1)
4	.910(-1)
5	.824(-2)
6	.778(-3)
7	.755(-4)
8	.740(-5)
9	.682(-6)
10	.673(-7)

Table 5: $k^2 = 19.723368$, $\dim H_0 = 1$, interpolation according to (5.6e)

cycle #	$\|residuals\|_2$
1	.363(+3)
2	.172(+2)
3	.114(+1)
4	.893(-1)
5	.765(-2)
6	.687(-3)
7	.691(-4)
8	.685(-5)
9	.759(-6)
10	.768(-7)

Table 6: $k^2 = 19.723368$, $\dim H_0 = 1$, interpolation according to (5.6d)

cycle #	$\|residuals\|_2$
1	.363(+3)
2	.172(+2)
3	.114(+1)
4	.893(-1)
5	.125
6	.102(-1)
7	.162(-1)
8	.132(-2)
9	.909(-3)
10	.742(-4)

Table 7: $k^2 = 19.72336843$, dim $H_0 = 1$, interpolation according to (5.6e), ϕ_1 crudely computed.

cycle #	$\|residuals\|_2$
1	.363(+3)
2	.174(+2)
3	.114(+1)
4	.879(-1)
5	.116
6	.550(-2)
7	.134
8	.427(-1)
9	.465(-2)
10	.534(-3)

Table 8: $k^2 = 19.72336843$, dim $H_0 = 1$, interpolation according to (5.6e), ϕ_1 successively improved.

cycle #	$\|residuals\|_2$
1	.363(+3)
2	.174(+2)
3	.114(+1)
4	.893(-1)
5	.765(-2)
6	.687(-3)
7	.691(-4)
8	.683(-5)
9	.759(-6)
10	.768(-7)

Table 9: $k^2 = 19.72335955955$, dim $H_0 = 1$, interpolation according to (5.6e).

cycle #	$\|Residuals\|_2$	$\|\tilde{u}^h - u^h\|_2$
1	.363(+3)	.555
2	.174(+2)	.392
3	.114(+1)	.392
4	.893(-1)	.392
5	.764(-2)	.392
6	.687(-3)	.392
7	.655(-4)	.392
8	.359(-3)	.268(-1)
9	.284(-4)	.268(-1)
10	.219(-5)	.268(-1)

AN ADAPTIVE MULTI-GRID SCHEME FOR SIMULATION OF FLOWS

Laszlo Fuchs
Department of Gasdynamics,
The Royal Institute of Technology,
S-100 44 Stockholm, SWEDEN.

SUMMARY

An adaptive MG scheme has been applied to the computation of certain incompressible flows. The scheme uses a basic (low) order solver of the Navier-Stokes equations, on a system of zonal subgrids. These subgrids may be defined independently, and may contain locally refined regions. The MG scheme is used to solve efficiently the discrete equations, even on such systems of grids. Local mesh refinements are done dynamically, in regions where the estimated truncation errors are larger than the average. When a final grid system is found (with almost uniformly distributed truncation errors) the order of approximation is improved by a few additional MG cycles using a defect correction type scheme. The adaptive scheme is also used to find regions where certain simplified approximations to the governing equations (e.g. PNS and potential equations) are valid. Such approximations are than applied to produce, rapidly, solutions that are valid in these regions. In this way, boundary conditions may be applied with controlled accuracy. In regions where such approximations are not valid, the approach may produce (natural) block relaxation schemes.

The scheme has been applied to the solution of the flow in a channel with symmetric sudden enlargement. For Reynolds numbers larger than some certain value, the solution is not unique. The symmetry breaking bifurcation that occurs can be traced easily by the method.

1. INTRODUCTION

Adaptive numerical techniques address classically the problem of defining , dynamically, different subregions where the different numerical scales (grid size) are such that the length scales of the solution, are resolved. Such an adaptive procedure may be approached in several ways. In simple cases, when the physical scales are known approximately, one may use a proper (non-uniform and non-rectangular) mesh (e.g. [1]). This mesh may be then changed slightly during the solution procedure. The modification may be done directly on the existing mesh, or by introducing a modified (locally defined) mesh that is patched to the other existing mesh. This type of mesh embedding/patching technique have been applied to the computation of some transonic flows (see e.g. [2-3]). A more general way of resolving local scales is using a zonal technique. The zonal mesh is composed of several overlaping (possibly independently defined) grids. The different subgrids are 'connected' through the subregions in common to at least two of the zones. This method requires information exchange among the zones. For hyperbolic systems the method have been used successfully by Benek et al [4], Berger and Oliger [5] and Berger [6]. The zonal technique in the Multi-Grid (MG) context have been applied for transonic flows by Gu and Fuchs [7-9]. The information exchange among the zones must satisfy in some cases certain global constrains. The treatment of these condition for viscous incompressible flows has been investigated by Fuchs [9-11].

Dynamical (i.e. during the solution process) grid construction is only one aspect of more general adaptive numerical schemes. A less frequently considered element is the use of adaptively defined discrete approximation. For one-dimensional problems one could produce rather efficient adaptive algorithms where the local order of approximation and the mesh-size are defined simultaneously (see e.g. [12-14]). The application to multi-dimensional cases is much more complicated and only very limited work is reported in the literature [16]. Our approach is not to try to optimize the order of approximation together with the grid generation, but rather to improve numerical accuracy once the scales of the problem are resolved, (using lower order approximations). That accuracy improving step shall be described in the following.

An important aspect of numerical simulation of physical phenomena, and which is very often neglected during the numerical treatment, is the question of modelling. It is a practice that one assumes certain properties of the physics of the problem and than solves, numerically, the resulting simplified set of equations. In our adaptive scheme we do not try to implement the assumptions on the governing equations a priori, but rather to fined different subregions where different approximations may be adopted without affecting the global accuracy of the numerical solutions. Different approximations to the governing equations may also be used to device proper relaxation schemes and also to define some of the boundary conditions.

The application of the adaptive scheme for a simple (but nevertheless interesting) viscous flow problem, is given in the following. The techniques that are used for this problem, may not be applicable for all cases of viscous flows, but the basic principles may be generalized to include more complex cases as well. These aspects and the basic elements of the adaptive scheme are described in the following.

2. THE ADAPTIVE SCHEME

The basic elements of our adaptive scheme include adaptive modeling, dynamical definition of grids and finally, an accuracy improving step. At the current stage, and for reasons given below, we do not use adaptively defined discrete approximations to the model equations.

The MG scheme can be regarded as a successive approximation process. As such, it provides information about the convergence rate of the numerical solution, the discrete approximation to the model equations and also on the accuracy of model equations (and in some cases the accuracy of the boundary approximations). General successive approximation processes can be written as sequences in some parameters that approach an asymptotical value. That is, $L\ q = R$ is approximated by $L_\varepsilon\ q = R$. ε may represent different mesh spacings, h (as in the classical MG scheme) or different orders of finite-differences (as in the case of successive defect correction scheme). In general, L_ε may also represent a differential approximation to the governing equations. Examples for such approximations are the boundary layer equations [or in more general form the so called PNS ('Parabolized Navier-Stokes') equations]. An other example is the potential approximation to inviscid flows. In the former case the 'small' parameter, ε, is the length scales ratio in two perpendicular directions whereas in the later case the 'small' parameter is the vorticity of the flow field. Unfortunately, these kinds of approximations are not valid generally. However, when the approximations are not too bad they can still be used in several ways to improve the efficiency of the numerical schemes without loosing accuracy.

2.1 SUCCESSIVE APPROXIMATION

Consider L_ε to be an approximation (or a sequence of approximations for a sequence of ε values) to L. The approximation (modelling or 'truncation') error is given by $(L - L_\varepsilon)\, q = O(\varepsilon^t)$. If $L_\varepsilon q_\varepsilon = R$ and $||q - q_\varepsilon|| = O(\varepsilon^p)$, one may improve the accuracy of the solution by the following (defect correction) steps:

$$L_\varepsilon q_\varepsilon^{(1)} = R$$

$$L_\varepsilon q_\varepsilon^{(n+1)} = R + (L_\varepsilon - L)\, q_\varepsilon^{(n)} \tag{1}$$

where $n > 1$.

The error in the solution ($E^{(n)} = ||q - q_\varepsilon^{n)}||$) at the n-th step is $O(\varepsilon^{n \cdot t})$ (when $p=t$). Thus, by repeated steps, the modelling error can be reduced to any desired level. (In practice, the L's in (1) are replaced by an s-th order finite-difference approximation to L. The error in the n-th step is then: $\min\{O(\varepsilon^{n \cdot t}), O(h^s)\}$.) For numerical calculations the desired level should be such that the modelling and the truncation errors are of the same order. By using such a criterion one may find that when the approximative model is a good one, it is enough with a single step in (1). When the approximation is not that good, few iterative step may be required. In the general case, when the error in the first step is O(1), many steps must be done. In the last case the scheme is equivalent to a (block) iterative method.

Based on the modeling error, the approximative equations may be used to determine the regions where they should be used iteratively, and those regions where a desired level of approximation is achieved by using the 'few step' approach. In this way, fully iterative techniques has to be used only in parts of the computational domain. Furthermore, if the boundaries of the computational domain are placed inside the region where a single step is adequate, the approximative model may be used to define some of the boundary conditions (see below). It should be emphasized that one may use several approximative models in the computation of a single problem. This approach leads to a rational way of detecting different regions, where the simplified equations are valid. In some cases, this leads also to the reduction in the number of dependent variables, and a further improvement in numerical efficiency.

For the computation of viscous incompressible flows, we use one or two approximative models. One approximation states that the flow is irrotational and therefore there exists a velocity potential. The flow can satisfy only slip boundary conditions, and in most cases this approximation leads to an iterative scheme. The iterative scheme results in computing the irrotational and the rotational parts of the velocity field. This type of scheme have been applied for solving the incompressible Navier-Stokes equations in primitive variable form [14,16-18]. The use of the non-linear full potential equation in approximating the Euler equations (both as relaxation operator and as 'far-field' solver) for transonic flow, is being currently studied.

2.2 ADAPTIVE CONSTRUCTION OF GRID-SYSTEMS

We use the zonal technique to discretize the space. This technique implies that one or more basic grid systems are used, such that the union of all the grids covers the whole domain, and the section of the grids is non-empty. The mesh in different zones, may be constructed independently of the mesh in the other zones. From a given system of zones, one may construct a MG structured grid by doubling the mesh spacing in each step of coarsening. Beside this zonal grid construction, one may easily refine the mesh locally, by adding finer subdomains, derived from the given grid, by halving the mesh spacing. The management of such local mesh refinement technique (in the MG procedure) is simple and it requires the storage of a few additional scalars [20].

The flexibility of the algorithm in refining the grid wherever it is necessary, can be utilized only if the scheme is used in an adaptive manner. That is, the grid is refined when and where it should be refined. Some simple criteria for such adaptive processing are described in [10] both for regular flows, and for flows with singularities. The adaptive criteria are based on the estimates of the truncation errors (the right hand side in (1) when ε corresponds to the different grid-spacings of the MG procedure).

2.3 THE ACCURACY INCREASING STEP

The current adaptive scheme does not try to adapt the order of finite-difference approximation to the behaviour of the solution and the grid that is being used. The reasons for our approach are the following:

* The scheme is much simpler and the operations involved in computing the controlling criteria are negligible.

* Higher order approximations are used only when the scales of the problem are resolved (by using lower order approximation) and when the lower order truncation errors are (almost) uniformly distributed. Thus, increasing the order of approximation guarantees also improved accuracy of the numerical solution.

* The accuracy correcting step involves only few computations of the higher order operators and therefore the increase in the total amount of operations increases only slightly. There is no need to introduce changes in the basic MG solver to maintain efficiency. Furthermore, the ('few' steps defect corrections scheme as in eq. (1)) have better stability properties compared to the use of higher order approximations in the relaxation steps.

The basic accuracy increasing step is described in [21]. Basically, one uses the same scheme as in (1) with the exception that the L_ε operator represents the low order relaxation operator of the basic MG scheme, and the L operator is replaced by L_n. The order of approximation of the operators L_n in the sequence is increasing by at least the order of the approximation of the low order relaxation operator. This way of computing higher order approximations has improved stability properties and requires only one additional MG cycle, for each defect correction step, once the basic solution has been computed. The basic scheme has been developed in [21] and it has been extended, in a straightforward manner, to the zonal grid-system with local mesh refinements.

3. COMPUTED EXAMPLE

Consider the flow in a channel with a sudden, symmetrical, enlargement. This problem is interesting, since it exhibits, even for relatively low Reynolds number (Re ≈ 150), a symmetry breaking bifurcation. The geometry is very simple and can be treated by using some type of transformation of coordinates. Here, we use two cartesian grid system, defined independently, to describe the inflow and the outflow sections. The fact that the coordinates in the two zones are aligned is not used by the scheme. The 2-D Navier-Stokes equations are written (for the computed cases below) in terms of the streamfunction and the vorticity. The primitive variable form of the zonal-MG code has also been developed leading to similar results.

The inflow boundary condition is given by specifying the velocity profile (free-stream parabolic profile). The outflow boundary conditions are given by one of the two ways: fully developed velocity profile or by using the parabolized equations.

The last type of boundary condition can be placed at a distance where the errors due to parabolization are as small as the discretization errors. The same parabolized operator is used (in some of the computations) in the iterative mode (in regions where the approximation is not so good). To exemplify the adaptive criteria in using the different approximations (modeling and discretization) we study these errors for the computed results.

Figures 1.a, 2.a and 3.a show the streamline pattern in the channel (with an inflow section width of one unit and symmetrical steps of half units) for Reynolds number, Re= 50, 100 and 200, respectively. As seen, despite the symmetry in geometry, the flow becomes asymmetrical for Re=100 and larger. The behaviour of the

Figure 1.a: The streamline pattern in the channel with a sudden symmetrical expansion. Re=50.

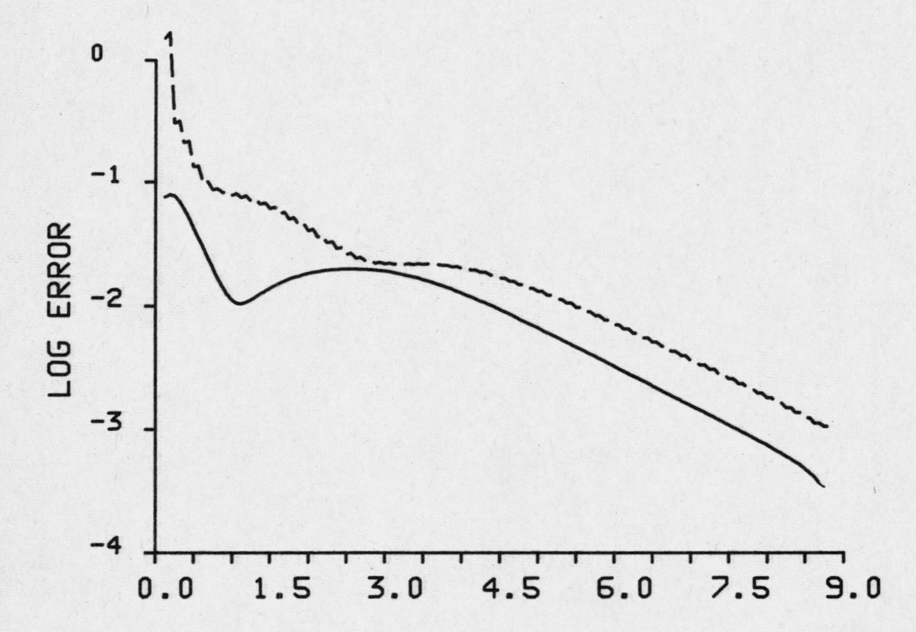

Figure 1.b: The relative mean (PNS) modeling error (solid line) and the relative mean truncation error (dashed line) corresponding to Fig 1.a.

relative mean parabolization error ($||u_{xx}||/||u_{yy}||$) and the relative mean discretization error, is shown in figures 1.b, 2.b and 3.b, respectively. For Re=50, both errors decrease at the same rate, once the separated region is passed. (Note that the truncation error also goes to zero asymptotically since the finite-difference approximation and the model equations are both exact for free-stream conditions). For higher Re, where the separated region is larger, and where the parabolization approximation is not good, the errors decrease at much slower rate and at greater distance from the entrance. By estimating the relative parabolization error, one can determine also where to place the outflow (parabolized) boundary conditions without affecting the global accuracy.

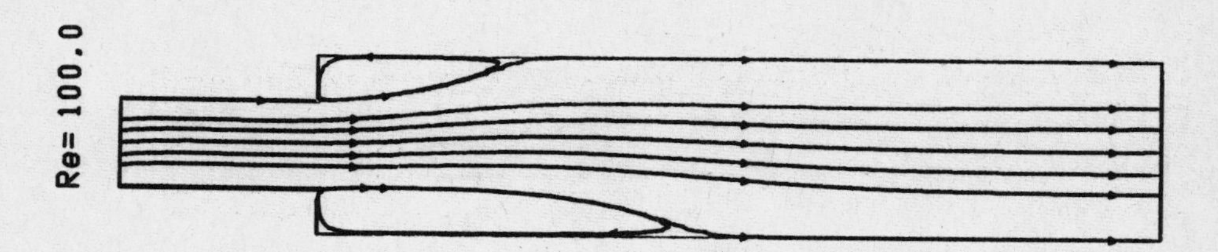

Figure 2.a: The streamline pattern in the channel with a sudden symmetrical expansion. Re=100.

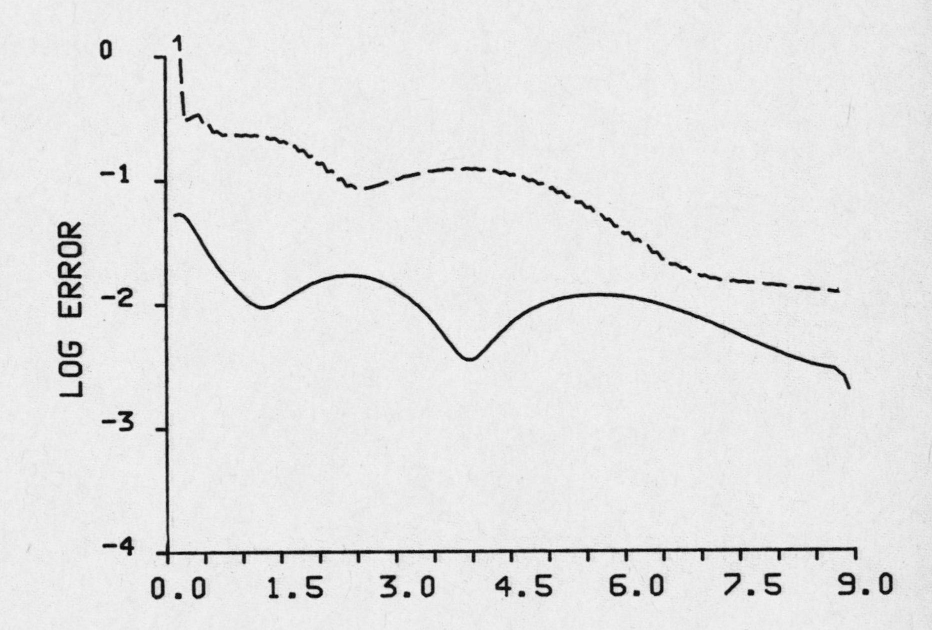

Figure 2.b: The relative mean (PNS) modeling error (solid line) and the relative mean truncation error (dashed line) corresponding to Fig 2.a.

The asymmetry in the numerical solution at the higher Re may be surprising at first glance. However, this type of behaviour is not unexpected in light of the experimental results of Durst et al. [22]. To study more closely the effects of the inflow boundary conditions on the flow, we considered two (mass conserving) perturbed inflow conditions: by 0.15% and 1%. Both these perturbed inflow conditions result in asymmetrical states. Figures 4 and 5 show the computed solutions with 0% and a 1% perturbation, respectively, in the inflow velocity profile, on a system of (locally) refined grids for Re=100. The solution seems to be, qualitatively, very sensitive to perturbations in the velocity at the expansion plane. This sensitivity increases with Re. The numerical solution, with a 4-th

Figure 3.a: The streamline pattern in the channel with a sudden symmetrical expansion. Re=200.

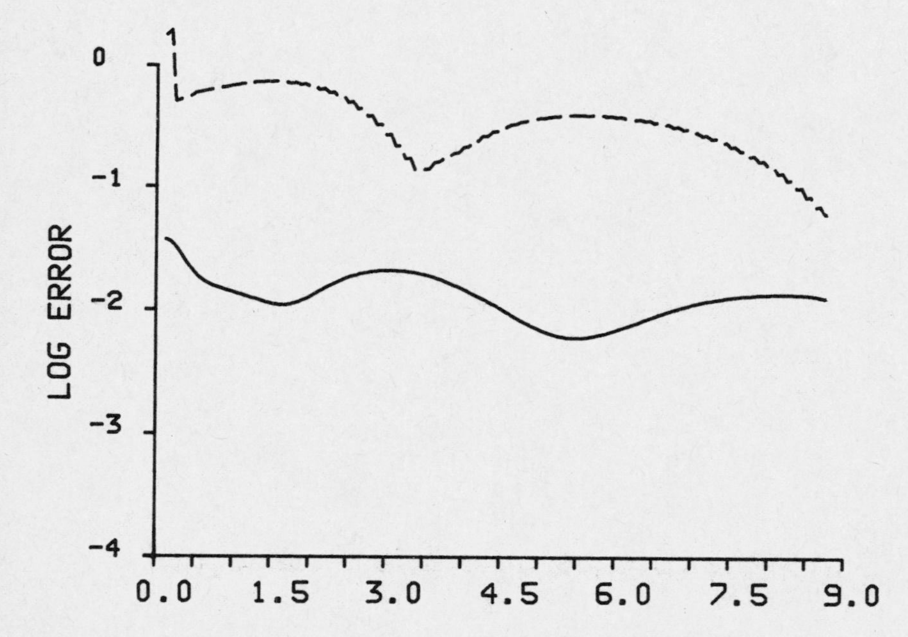

Figure 3.b: The relative mean (PNS) modeling error (solid line) and the relative mean truncation error (dashed line) corresponding to Fig 3.a.

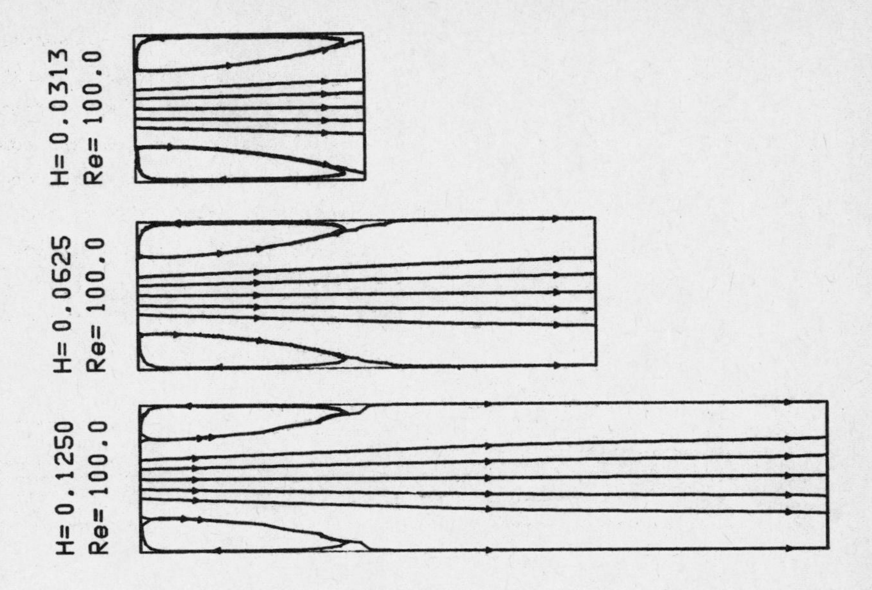

Figure 4: The streamline pattern in the outflow section, using locally refined grids. Re=100. 0% inflow velocity perturbation.

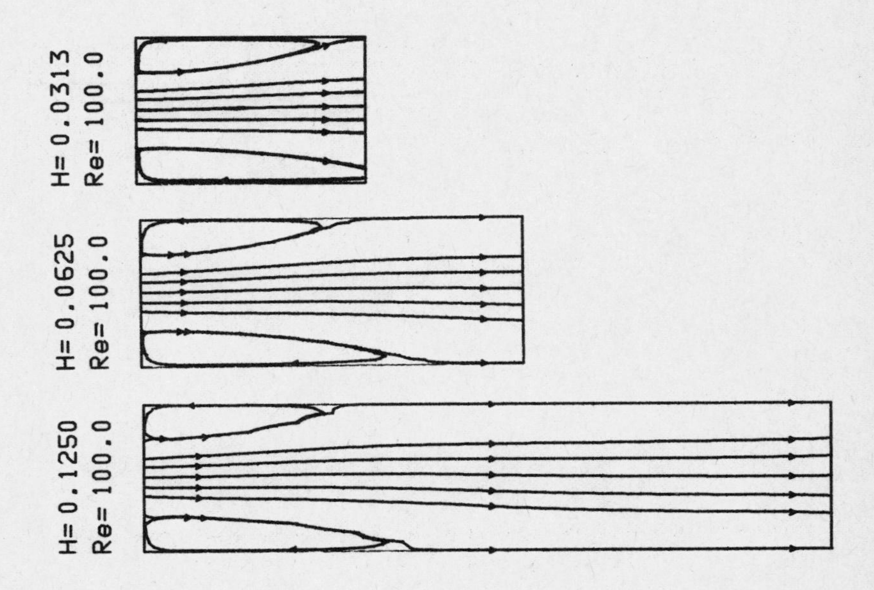

Figure 5: The streamline pattern in the outflow section, using locally refined grids. Re=100. 1% inflow velocity perturbation.

order scheme on a similar mesh (Re=100) results in a symmetrical solution for the unperturbed case, and is asymmetrical for the perturbed case (Fig. 6). The difference between the low order (first order upwind) and the 4-th order schemes is some few percents. The local grids in Fig. 4-6 are constructed using truncation error estimates. The truncation error fields, on the sequence of the local grids, are shown in Fig. 7. As seen, the extend of the regions with the large truncation errors decrease very fast as the mesh is refined (even locally).

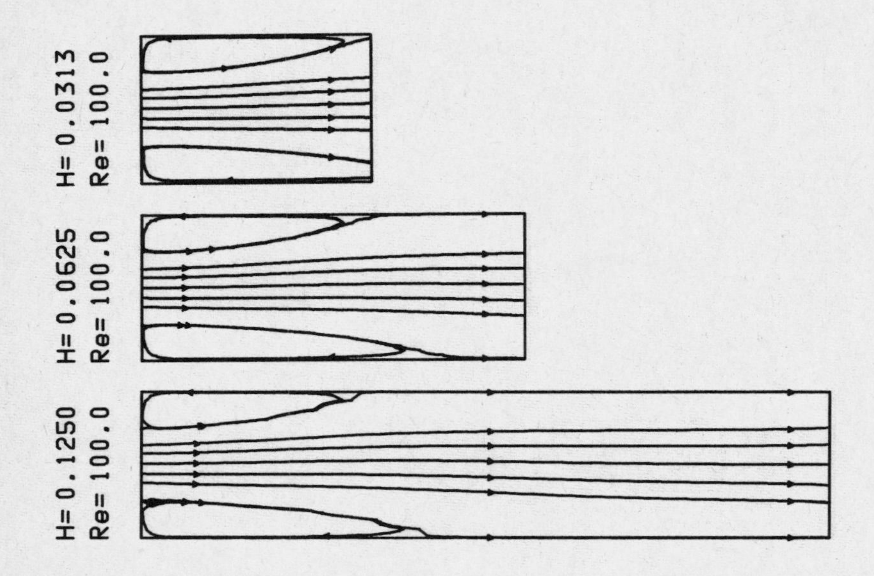

Figure 6: The same case as Fig. 5, computed by a 4-th order finite-difference scheme.

4. CONCLUDING REMARKS

An adaptive scheme that contains three main features, have been developed. These feature include the use of different differential approximation to the governing equations. These approximations are used to reduce the computational domain (by placing boundary condition based on the approximation at a place that does not reduce global accuracy), or as a (block) relaxation operators. This approach, when done in a controlled manner results in increased computational efficiency. Other elements of the current scheme include the use of a zonal-local MG solver, that can be applied in a second stage to increase accuracy in a simple and stable manner. Here, we have demonstrated some of the basic elements of the scheme on the flow in a channel with a symmetric sudden expansion. The basic elmements of the scheme are being implemented, currently, for more complex problem than the one given here.

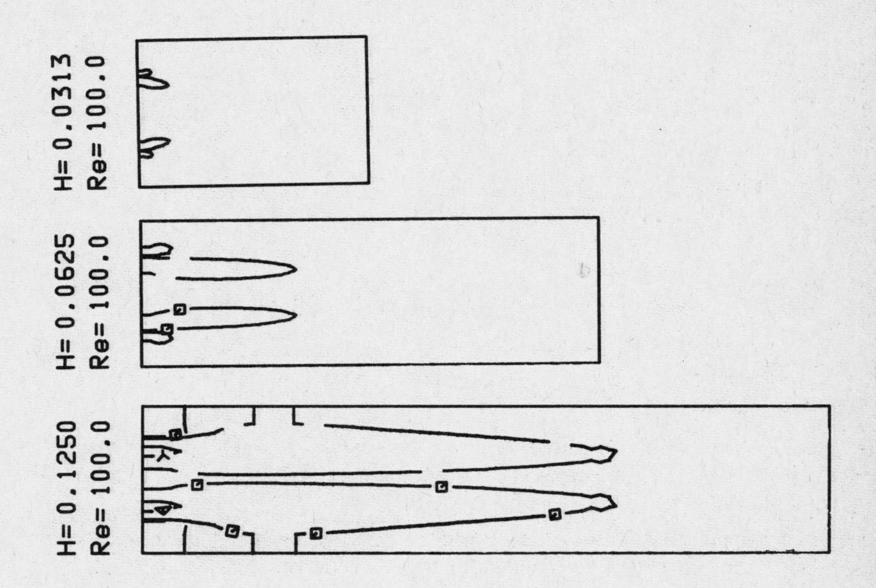

Figure 7: The truncation error field corresponding to the case in Fig. 4.

REFERENCES

1. C.W. Mastin and J.F. Thompson - Adaptive Grids Generated by Elliptic Systems. AIAA 83-451 (1983).

2. M.M. Rai - A Conservation Treatment of Zonal Boundaries for Euler Equation Calculation. AIAA 84-164 (1984).

3. W.J. Usab and E.M. Murman - Embedded Mesh solution of the Euler Equations Using Multiple-Grid Method. AIAA 83-1946 (1983).

4. J.A. Benek, J.L. Steger and F.C. Dougherty - A Flexible Grid Embedding Technique with Application to the Euler Equations. AIAA 83-1944 (1983).

5. M.J. Berger and J. Oliger - Adaptive Mesh Refinement for Hyperbolic Partial Differential Equations. J. Comp. Phys. vol 53, p. 484 (1984).

6. M.J. Berger - On Conservation at Grid Interface. ICASE No. 84-43 (1984).

7. C-Y. Gu and L. Fuchs - Numerical Computation of Transonic Airfoil Flows. Proc. 4-th Int. Conf. on Numerical Methods in Laminar and Turbulent Flow. Eds. C. Taylor, M.D. Olson, P.M. Gresho and W.G. Habashi, Pineridge Press, pp. 1501-1512, (1985).

8. C-Y. Gu and L. Fuchs - Transonic Potential Flow: Improved Accuracy by Using Local Grids. IMACS Conference, Oslo (1985).

9. L. Fuchs - Multi-Grid Solutions on Grids with non-aligning Coordinates. 2-nd Copper-Mountain Conference on Multigrid Methods, (1985).

10. L. Fuchs - Adaptive Construction of Grid-Systems for Flow Simulations. Proc. ICFD Conference. Oxford University Press. (1985).

11. L. Fuchs - Numerical Flow Simulation using Zonal Grids. AIAA 85-1518. (1985).

12. T.H. Chong - A Variable Mesh Finite-Difference Method for Solving a Class of Parabolic Differential Equations in one Space Variable. SIAM J. of Numer. Anal. vol 15. p. 835, (1978).

13. S.F. Davis and J.E. Flaherty - An Adaptive Finite-Difference Method for Initial-BOundary Value Problems for PDE. SIAM. J. Sci. Stat. vol 3, pp. 6-27, (1982).

14. A. Brandt - Multigrid Techniques: 1984 Guide. CFD Lecture Series at von-Karman Institute, March (1984).

15. H.A. Dwyer, R.J. Kee and B.R. Sanders - Adaptive Grid Methods for Problems in Fluid Mechanics and Heat Transfer. AIAA J. vol 18. pp. 1205-1212 (1980).

16. T. Thunell and L. Fuchs - Numerical Solution of the Navier-Stokes Equations by Multi-Grid Techniques. Numerical Methods in Laminar and Turbulent Flow. Eds: C. Taylor and B.A. Schrefler. Pineridge Press. pp. 141-152, (1981).

17. L. Fuchs - Multi-Grid Schemes for Incompressible Flows. Proc. GAMM-Workshop on Efficient Solvers for Elliptic Systems. Ed: W. Hackbusch. Notes on NUmerical Fluid MEchanics. Vieweg. Vol. 10 pp. 38-51, (1984).

18. L. Fuchs and H-S. Zhao - Solution of Three-Dimensional Incompressible Flows by a Multi-Grid Method. Int. J. Numerical Methods in Fluids. vol 4. pp. 539-555.

19. L. Fuchs - A Local Mesh Refinement Technique for Incompressible Flows. Computers & Fluids. In press (1985).

20. L. Fuchs - Defect Corrections and Higher Numerical Accuracy. Proc. GAMM-Workshop on Efficient Solvers for Elliptic Systems. Ed: W. Hackbusch. Notes on NUmerical Fluid MEchanics. Vieweg. Vol. 10 pp. 52-63, (1984).

21. F. Durst, A. Melling and J.H. Whitelaw - Low Reynolds Number Flow Over a Plane Symmetric Sudden Expansion. J. Fluid Mech. Vol 64. pp. 111-128, (1974).

MULTIGRID METHODS FOR CALCULATING THE LIFTING POTENTIAL INCOMPRESSIBLE FLOWS AROUND THREE-DIMENSIONAL BODIES

W. Hackbusch
Institut für Informatik
und Praktische Mathematik
Universität Kiel
2300 Kiel, Olshausenstraße 40

Z.P. Nowak
Institute of Applied Mechanics
and Aircraft Technology
Warsaw Technical University
00-665 Warsaw, Nowowiejska 24

1. Formulation of the problem

The problem of determining the non-lifting incompressible irrotational flow around an impermeable body can be mathematically expressed as the Neumann boundary value problem for the perturbation velocity potential:

(1a) $\Delta\varphi = 0$

in the region R_e outside the body, with the boundary condition

(1b) $\frac{\partial\varphi}{\partial n}(p) = - \bar{V}_\infty \cdot \bar{n}(p)$

at the points p on the body surface S, where $\bar{V}_\infty$ is the velocity of the undisturbed flow far from the body and $\bar{n}(p)$ is the outer normal at p. Additionally,

(1c) $\varphi(p) \to 0 \quad \text{when} \quad |p| \to \infty\ ,\ p\in R_e.$

The zero value is obtained when the solution φ of this Neumann problem is used for calculating the total force on S. To obtain lift, one must introduce additional impermeable surfaces, immersed in the flow region. These surfaces, called *vortex sheets* or *wakes*, originate at the sharp edges on S, such as the leading or the trailing edge of a wing.

The wake sheet is an idealization of a thin layer present in a real viscous flow downstream of a wing. The inclusion of these sheets prevents the derivatives of the velocity potential, i.e., the velocity components, from becoming infinite at the edge points.

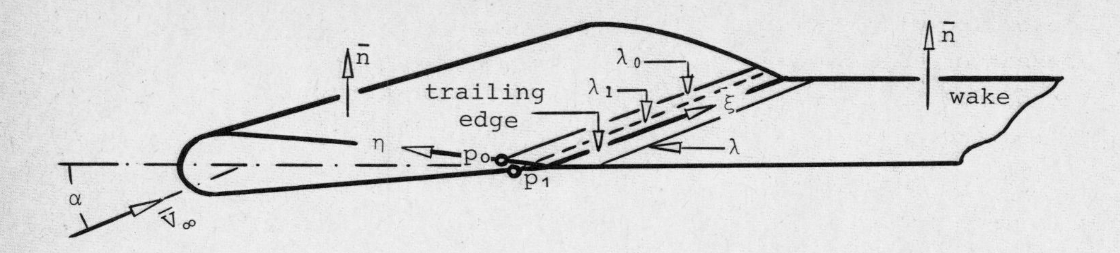

Figure 1. The wake originating at the trailing edge of a wing

Let S' denote the sum of all the vortex surfaces present in the flowfield and let $S'' = S + S'$. The Neumann problem (1a-c) could be reformulated by replacing S with S'' and R_e with $R_e'' = R_e - S'$. The shape of S' could be determined from the additional nonlinear condition of the pressure equality on both sides of S'. Usually, however, the shapes of the sheets are simply assumed and only the *Kutta condition* of the finite velocity at the edges is enforced.

For the boundary integral method the solution of the lifting flow problem is sought in the form of the combination of the surface potentials:

$$(2) \qquad \varphi(p) = -\frac{1}{4\pi}\int_S \frac{\sigma(q)}{r(p,q)}\, dS_q + \frac{1}{4\pi}\int_{S+S'} \frac{\partial}{\partial n_q}\left(\frac{1}{r(p,q)}\right)\mu(q)\, dS_q, \quad p\in R_e'',$$

where $r(p,q) = \| p - q \|$ is the distance between the points $p\in R_e''$ and $q\in S''$. Substituting (2) into the Neumann condition (1b) on S, we arrive at an integral equation

$$(3) \qquad \sigma(p) = f(p)+(A_{11}\sigma)(p)+(A_{12}'\mu)(p), \quad p\in S,$$

where $f(p) = -2\circ\bar{V}_\infty\cdot\bar{n}(p)$, and A_{11}, A_{12}' are the integral operators:

$$(4) \qquad (A_{11}\sigma)(p) = \frac{1}{2\pi}\int_S \frac{\partial}{\partial n_p}\left(\frac{1}{r(p,q)}\right)\sigma(q)\, dS_q,$$

$$(A_{12}'\mu)(p) = -\frac{1}{2\pi}⨍_{S+S'} \frac{\partial}{\partial n_p}\left(\frac{\partial}{\partial n_q}\left(\frac{1}{r(p,q)}\right)\right)\mu(q)\, dS_q.$$

In the last formula $⨍$ denotes the principal value at the surface integral.

We see that one of the functions σ or μ can be chosen arbitrarily on S. The various boundary integral methods of aerodynamics differ in the choice of these functions (for a review see e.g. [4]). Here, we shall present one of the successful choices, which serves as a basis for a method due to Hess [3]. We shall confine our attention to the case of a flow around a wing.

Let ξ, η denote the arc-length coordinates: ξ measured along the trailing edge and η measured along the contours of the wing cross-sections, starting from the trailing edge (Fig. 1). The coordinate η will be scaled by the total length of the current cross-section, so that its maximum value is always 1.

In the method of Hess, the distribution of μ on S is assumed in the form

(5) $$\mu(\xi,\eta) = (1-2\eta)\ \hat{\mu}(\xi),$$

where $\hat{\mu}$ is the function to be determined. Let $\nu(\xi)$ be the value of μ on the wake surface, close to the trailing edge. We must set

(6) $$\nu(\xi) = 2\hat{\mu}(\xi).$$

Otherwise, the contact of the three doublet sheets distributed on the upper surface of the wing, the lower surface and the wake, with the respective local densities: $\hat{\mu}(\xi)$, $-\hat{\mu}(\xi)$ and $\nu(\xi)$ would produce a line vortex along the trailing edge, resulting in the arbitrarily large velocities forbidden by the Kutta condition.

It will be assumed that the wake surface is generated by the trailing edge bisectors or the straight lines parallel to $\overline{V}_\infty$, issuing from the trailing edge, and that μ is constant along these generators. Hence, μ on the wake surface is equal to the appropriate values $\nu(\xi)$ near the trailing edge. By (5) and (6) also the distribution of μ on S can be expressed with the aid of $\nu(\xi)$. Consequently , (3) can be rewritten in the form

(7) $$\sigma(p) = f(p)+(A_{11}\sigma)(p)+(A_{12}\nu)(p), \quad p\in S ,$$

where A_{12} is an integral operator transforming the function $\nu(\xi)$ into a function defined on S.

Since $\nu(\xi)$ can still be chosen arbitrarily, one can additonally require that the impermeability condition (1b) should be satisfied along a certain line λ on S', placed near the trailing edge (Fig. 1). Let $p(\xi) \in \lambda$ lie on the wake generator, originating at the trailing edge point with the coordinate ξ. Using (7) for $p(\xi)$, with $\sigma(p(\xi)) = 0$, we obtain the additional linear relation

(8) $$(A_{21}\sigma)(\xi) + (A_{22}\nu)(\xi) = g(\xi), \quad 0\le\xi\le\xi_{max} ,$$

where

$$(A_{21}\sigma)(\xi) = (A_{11}\sigma)(p(\xi)),$$

$$(A_{22}\nu)(\xi) = (A_{12}\nu)(p(\xi)),$$

$$g(\xi) = - f(p(\xi)),$$

and ξ_{max} is the total length of the trailing edge. Often, instead of the linear impermeability condition on λ, the pressure equality condition is imposed at all the pairs of points $p_o(\xi)$ and $p_1(\xi)$ along the two lines λ_o and λ_1, lying on the upper and lower surface near the trailing edge (Fig. 1). Since by the Bernoulli equation the pressure value depends only on the modulus of the local velocity vector, this requirement can be put into the form

(9) $$\bar{\tau}_o \cdot \bar{V}(p_o(\xi)) - \bar{\tau}_1 \cdot \bar{V}(p_1(\xi)) = 0,$$

where $\bar{\tau}_o$ and $\bar{\tau}_1$ are the unit vectors pointing in the directions of the velocities $\bar{V}(p_o(\xi))$ and $\bar{V}(p_1(\xi))$, respectively. Let us denote

$$\bar{t}_o = \bar{V}_\infty - (\bar{V}_\infty \cdot \bar{n}(p_o(\xi))) \cdot \bar{n}(p_o(\xi)).$$

The vector $\bar{t}_o$ is the projection of $\bar{V}_\infty$ on the tangent plane at $p_o(\xi)$. We shall asssume that $\bar{\tau}_o$ and $\bar{t}_o$ have approximately the same direction. Hence,

$$\bar{\tau}_o \approx \bar{\tau}_o' = \bar{t}_o/|\bar{t}_o|.$$

Similarly we can define $\bar{t}_1$ at $p_1(\xi)$, and write

$$\bar{\tau}_1 \approx \bar{\tau}_1' = \bar{t}_1/|\bar{t}_1|.$$

Now, (9) can be approximated by the linear condition

$$\bar{\tau}_o' \cdot \bar{V}(p_o(\xi)) - \bar{\tau}_1' \cdot \bar{V}(p_1(\xi)) = 0,$$

which, in view of the equality $\bar{V} = \bar{V}_\infty + \nabla\varphi$, is the same as

(10) $$\bar{\tau}_o' \cdot \nabla\varphi(p_o(\xi)) - \bar{\tau}_1' \cdot \nabla\varphi(p_1(\xi)) = (\bar{\tau}_1' - \bar{\tau}_o') \cdot \bar{V}_\infty$$

Substituting (2) into (10) we obtain another additional condition of the form (8).

2. Splitting of the integral operator

Solving (8) with respect to ν and substituting the result into (7) we obtain

(11) $$\sigma(p) = u(p) + (A\sigma)(p), \quad p \in S,$$

where

(12) $$u = f + A_{12}A_{22}^{-1}(g)$$

is a given function, and

$$A = A_{11} - A_{12}A_{22}^{-1}A_{21}.$$

In the general case one should not expect the operator A to be compact, e.g., on the space C^o of the piecewise continuous functions on S with the maximum norm. Therefore, the Fredholm theory cannot be applied to (11). Moreover, the multigrid method of the second kind (cf. [2]) for a discretized form of (11) need not converge.

As proposed in [5] (cf. [2, p. 325]), in such cases the operator A could be split into the two parts:

(13) $$A = C + D$$

where C is a compact part. For example, the compact part could be extracted from the integral operator A_{11}, defined by (4). For this purpose the kernel of A_{11} could be split as follows

$$\frac{\partial}{\partial n_p}\left(\frac{1}{r(p,q)}\right) = c(p,q) + d(p,q),$$

where $c(p,q)=0$ whenever the modulus of the kernel exceeds a chosen bound M. The operators C and D can be defined as

(14) $$(C\sigma)(p) = \frac{1}{2\pi}\int_S c(p,q)\sigma(q)dS_q \quad , \quad p\in S \quad ,$$

(15) $$(D\sigma)(p) = \frac{1}{2\pi}\int_S d(p,q)\sigma(q)dS_q - (A_{12}A_{22}^{-1}A_{21}\sigma)(p), \quad p\in S \quad .$$

3. The discrete problem

The following general theorems can be proved on the basis of the theory of the collectively compact operators (cf. [1]).

Theorem I (On the convergence of the discrete solutions)

Let us suppose that

(i) $(I-A)^{-1}$ exists and D is a contraction on C^o.

(ii) R_n is a restriction operator acting from C^o into the space $C^{o,n}$ of the discrete functions (n-element sequences with the maximum norm), such that

$$\| R_n \|_{C^{o,n} \leftarrow C^o} \le 1.$$

(iii) P_n is a prolongation operator, acting in the reverse direction, such that

$$R_n P_n = I_n \quad ,$$

$$M \| x_n \|_{C^{o,n}} \leq \| P_n x_n \|_{C^o} \leq \| x_n \|_{C^{o,n}} ,$$

$$\lim_{n \to \infty} \| (I - P_n R_n) x \|_{C^o} = 0 ,$$

where I_n is the identity on $C^{o,n}$, M is a positive constant, and x_n, x are any elements of $C^{o,n}$ and C^o, respectively.

(iv) The discrete forms of C and D from (13) are given by the formulae

(16a) $C_n = R_n C P_n$,

(16b) $D_n = R_n D P_n$,

and the discrete form of u is

$$u_n = R_n u.$$

Then for sufficiently large n the discrete scheme

(17) $\sigma_n = u_n + (C_n + D_n) \sigma_n$

is stable, i.e.,

$$\| (I_n - C_n - D_n)^{-1} \|_{C^{o,n} \leftarrow C^{o,n}} \leq M < \infty$$

and the discrete solutions σ_n satisfy the convergence property

$$\lim_{n \to \infty} \| \sigma - P_n \sigma_n \|_{C^o} = 0 ,$$

where σ is the solution of the problem $\sigma = u + A\sigma$.

Theorem II (On the convergence of a two-grid iteration)

Let the two-grid iteration consist of the smoothing step

(18a) $(I_n - D_n) \hat{\sigma}_n = u_n + C_n \sigma_n^j$,

and the correction step (18b, c, d):

(18b) $\rho_n := u_n - (I_n - C_n - D_n) \hat{\sigma}_n$,

(18c) solve $(I_m - \hat{C}_m - \hat{D}_m) \delta_m = R_{m \leftarrow n} \rho_n$,

(18d) $\sigma_n^{j+1} := \hat{\sigma}_n + P_{n \leftarrow m} \delta_m$,

where $R_{m \leftarrow n}$ is a discrete restriction operator acting from $C^{o,n}$ into $C^{o,m}$ (m is a given function $m(n) < n$, $m(n) \to \infty$ when $n \to \infty$), $P_{n \leftarrow m}$ is a discrete prolongation operator acting in the reverse direction, and

$$\hat{C}_m = R_{m \leftarrow n} C_n P_{n \leftarrow m} ,$$

$$\hat{D}_m = R_{m \leftarrow n} D_n P_{n \leftarrow m} .$$

If, in addition to the assumptions of the Theorem I, the restriction and prolongation operators satisfy the conditions

$$\| R_{m\leftarrow n} \|_{C^{o,m} \leftarrow C^{o,n}} \leq 1 \quad ,$$

$$\| P_{n\leftarrow m} \|_{C^{o,n} \leftarrow C^{o,m}} \leq 1 \quad ,$$

$$R_{m\leftarrow n} P_{n\leftarrow m} = I_m \quad ,$$

$$\lim_{n\to\infty} \| (R_m - R_{m\leftarrow n} R_n) x \|_{C^{o,m}} = 0 \quad ,$$

$$\lim_{n\to\infty} \| (I_n - P_{n\leftarrow m} R_{m\leftarrow n}) R_n x \|_{C^{o,n}} = 0 \quad ,$$

for $m = m(n)$ and any fixed $x \in C^o$, then the multigrid iteration (18a-d) is stable, i.e.,

$$\| (I_n - D_n)^{-1} \|_{C^{o,n} \leftarrow C^{o,n}} \leq M_1 < \infty \quad ,$$

$$\| (I_m - \hat{C}_m - \hat{D}_m)^{-1} \|_{C^{o,m} \leftarrow C^{o,m}} \leq M_2 < \infty \quad , \qquad m = m(n)$$

for sufficiently large n, and

$$\lim_{j\to\infty} \sigma_n^{\,j} = \sigma_n \quad \text{in} \quad C^{o,n} \quad ,$$

where σ_n is the solution of (17). More particularly,

$$\sigma_n^{\,j+1} - \sigma_n = Q_n\,(\sigma_n^{\,j} - \sigma_n), \qquad j=1,2,3,\ldots,$$

where

$$\lim_{n\to\infty} \| Q_n \|_{C^{o,n} \leftarrow C^{o,n}} = 0 \quad .$$

The proofs of these theorems will not be given here due to lack of space.

The numerical calculations indicate that if the pressure equality condition (10) is used for the derivation of (8), then the operator D, defined by (15), may not be a contraction on C^o, thus violating the assumption (i) of Theorem I.

4. The discretization method

Let us note that the discretizations (16a, b) which require, among other things, the integrations over the pieces of the curved surface S, can only be regarded as idealizations of the procedures used in practice.

The simplest practically useful discretization is obtained when S is approximated by a polyhedron, composed of the flat "panels" (Fig. 2). The wake is replaced by a system of flat strips of finite length.

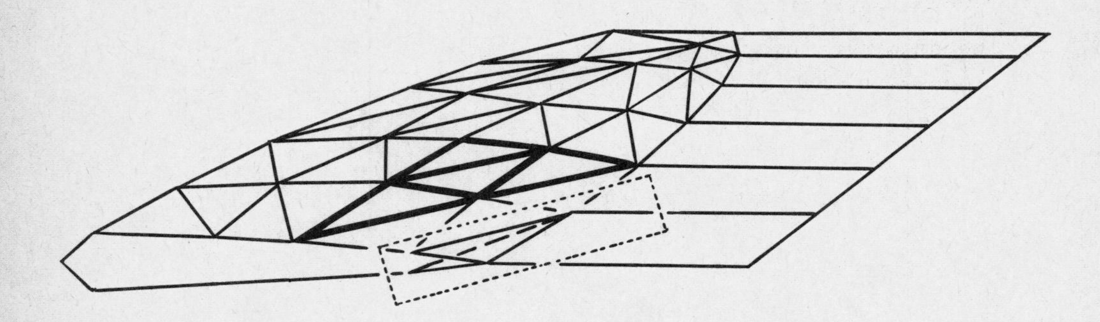

Figure 2. The approximating surface

The numerical solution σ_n is assumed to be constant over the n panel areas (i.e., the piecewise constant prolongation is chosen for P_n). Also $\nu(\xi)$ is approximated by a stepwise function ν_l, constant on each of the l strips. These approximations are substituted into condition (7), which is enforced at the n panel centres (i.e., the pointwise restriction is chosen for R_n). Similarly, condition (8) is enforced at the l locations near the centres of the trailing edge segments. As a result, a linear system of n+l algebraic equations with n+l unknown components of σ_n and ν_l is obtained.

Let $A_{n\cdot n}$, $A_{n\cdot l}$, $A_{l\cdot n}$, $A_{l\cdot l}$ denote the discrete counterparts of the respective operators A_{11}, A_{12}, A_{21} and A_{22}, and let u_n be calculated from a discrete form of (12). The expensive calculation of the discrete counterpart of the operator A will not be necessary. The matrix $A_{n\cdot n}$ is split into the two parts $C_{n\cdot n}$ and $\tilde{D}_{n\cdot n}$, corresponding to (14) and the first term on the right-hand side of (15)(here, we prefer to write $C_{n\cdot n}$ instead of C_n). Hence, $C_{n\cdot n}$ is a dense part of $A_{n\cdot n}$, composed of the small coefficients due to the influence of the distant panels, while $\tilde{D}_{n\cdot n}$ is a sparse part.

5. The multigrid solution procedures

The panel system on the body surface has a nested structure: it can be divided into the groups of the four panels, one of which is shown in Fig. 2, bordered by bold lines.

5.1. Method I. The discrete restriction operator $R_{m\leftarrow n}$, where $m = n/4$, is defined as the average value of the constant distributions over the four panels in each group. The discrete prolongation operator $P_{n\leftarrow m}$ assigns the equal values to the four panels in a group.

The smoothing step (18a) requires the expensive multiplication $C_{n\cdot n}\sigma_n^j$, and then solving the system

$$(19) \qquad (I_{n\cdot n} - \tilde{D}_{n\cdot n} + A_{n\cdot l}\, A_{l\cdot l}^{-1}\, A_{l\cdot n})\hat{\sigma}_n = v_n \quad ,$$

where v_n is a given discrete function.

When the impermeability condition along a line λ in the near wake was used for the derivation of (8), it was found that the sufficiently exact solution of (19) can be obtained in a few iterations of the form

$$(20) \qquad (I_{n\cdot n} - \tilde{D}_{n\cdot n})\hat{\sigma}_n^{\,k} = v_n - (A_{n\cdot l}\, A_{l\cdot l}^{-1}\, A_{l\cdot n})\hat{\sigma}_n^{\,k-1}, \qquad k=1,2,\ldots,$$

starting from $\hat{\sigma}_n^{\,o} = \sigma_n^{\,j}$. The matrix-vector multiplications, involved in the right-hand side of (20), are not expensive, since the appropriate matrices contain a small number of rows or columns. The QR-decomposition of the matrix $A_{l\cdot l}$ was used to obtain the product involving $A_{l\cdot l}^{-1}$.

It was also found that the subsequent inversion of $I_{n\cdot n} - \tilde{D}_{n\cdot n}$ can be accomplished with the aid of the Jacobi iteration

$$(21) \qquad \hat{\sigma}_n^{\,i} = t_n + \tilde{D}_{n\cdot n}\, \hat{\sigma}_n^{\,i-1}, \qquad i=1,2,\ldots,$$

$$\hat{\sigma}_n^{\,o} = \bar{\sigma}_n^{\,k-1},$$

where t_n represents the current right-hand side of (20). Since $\tilde{D}_{n\cdot n}$ is a sparse matrix, iteration (21) is relatively inexpensive. However, for thin bodies with edges of a small angle, such as the wing of an aircraft, more than twenty iterations (21) may be necessary. Therefore, $\tilde{D}_{n\cdot n}$ was split into the two parts

$$(22) \qquad \tilde{D}_{n\cdot n} = \tilde{D}'_{n\cdot n} + \tilde{D}''_{n\cdot n} \quad ,$$

where $\tilde{D}'_{n\cdot n}$ contains only the influence coefficients for the pairs of the mutually most sensitive panels (such panels usually lie on the opposite sides of the wing surface; (cf. [2, p. 326]). Now, instead of (21) we can use the so-called paired Jacobi iteration (cf. [5], [2, p. 326]):

$$(23) \qquad (I_{n\cdot n} - \tilde{D}'_{n\cdot n})\, \hat{\sigma}_n^{\,i} = t_n + \tilde{D}''_{n\cdot n}\, \hat{\sigma}_n^{\,i-1}, \qquad i=1,2,\ldots$$

Each iteration (23) requires the solution of at most n/2 independent pairs of the linear algebraic equations. If the opposite panels are approximately symmetric near the trailing edge, then a few iterations (23) give a sufficiently converged solution.

The correction δ_m is obtained by a slight variation of (18b, c). Instead of (18c), for which the matrix $\hat{D}_m$ would be needed, we apply a more exact coarse-grid discretization of the system (7) and (8) in the form

$$(24a) \qquad (I_{m\cdot m} - R_{m\leftarrow n} A_{n\cdot n} P_{n\leftarrow m})\delta_m - (R_{m\leftarrow n} A_{n\cdot l})\nu_l^j = R_{m\leftarrow n} \hat{\rho}_n \ ,$$

$$(24b) \qquad (A_{l\cdot n} P_{n\leftarrow m})\delta_m + A_{l\cdot l} \nu_l^j = \hat{\rho}_l \ ,$$

where δ_m is the coarse-grid correction and

$$\hat{\rho}_n = f_n - (I_{n\cdot n} - A_{n\cdot n})\hat{\sigma}_n \ ,$$

$$\hat{\rho}_l = g_n - A_{l\cdot n} \hat{\sigma}_n \ .$$

Let us note that no restriction of ν_l^j to wider strips was needed in (24b). A relatively small system of n/4 + 1 equations (24a, b) was solved by an exact method.

5.2. Method II. The two-grid method, outlined above, has a fast convergence rate when (8) is derived from the impermeability condition in the near wake. However, if the linearized pressure equality condition (10) is used, then iteration (20) diverges, due to the appearance of large coefficients in the diagonal of $A_{l\cdot l}^{-1}$.

For this case a more straightforward method has been used. Let us write the fine-grid system

$$(25a) \qquad (I_{n\cdot n} - A_{n\cdot n})\sigma_n - A_{n\cdot l} \nu_l = f_n \ ,$$

$$(25b) \qquad A_{l\cdot n} \sigma_n + A_{l\cdot l} \nu_l = g_l \ ,$$

corresponding to (7) and (8), in the general form

$$(26) \qquad L\cdot x = b,$$

where L is a (n+l)•(n+l) matrix and x, b are vectors of n+l components (for brevity we shall omit subscripts denoting the matrix and vector sizes). Again we split

$$L = \hat{L} + \tilde{L}$$

where $\hat{L}$ is a sparse part, containing a certain number of the largest coefficients. The system (26) can be solved by the iteration

$$(27) \qquad \hat{L}\, x^{i+1} = b - \tilde{L}\, x^i, \qquad i=1,2,...,$$

with an initial guess x^1. Since $\tilde{L}$ has a small norm, the iteration (27) converges at a fast rate. To accelerate the convergence even further, we apply the following two-grid iteration:

(28a) solve $\hat{L}\,\hat{x} = b - \tilde{L}\,x^i$,

(28b) $\rho := b - L\,\hat{x}$,

(28c) solve $(\hat{L} + PR\tilde{L})\,\delta = \rho$,

(28d) $x^{i+1} := \hat{x} + \delta$,

for i=1,2,..., where the correction δ is calculated on the fine grid, the residual ρ is unrestricted, R and P denote some restriction and prolongation operators acting between $C^{o,n+l}$ and $C^{o,k}$, $k < n+l$. For a general choice of the operator $\hat{L}$, the equation in (28c) cannot be restricted to a coarser grid. In fact, even when $\tilde{L}$ is a smoothing operator, the solution of (28a) and thus also the residual ρ need not be a smooth function (this happens, e.g., near the trailing edge).

For (28c) we use the simple iteration

(29) $\hat{L}\,\delta^i = \rho - PR\tilde{L}\delta^{i-1}$, $i=1,2,\ldots,$

$\delta^o = 0$.

If R is the pointwise restriction, i.e., if the value of $R\tilde{L}\delta^{i-1}$, associated with the group of panels, is chosen to be equal to the value of $\tilde{L}\delta^{i-1}$ on a central panel, then the iteration (29) is much less expensive than (27), since only every fourth row of the first n rows in $\tilde{L}$ is needed for the calculation of $R\tilde{L}\delta^{i-1}$. The operator P extends the result to the other panels in a group. The last l rows, corresponding to the equations (25b), were always included in $\hat{L}$ (hence, $\tilde{L}$ produces only the zero values of the wake components).

The solution of the sparse systems (28a) and (29) of the common form $\hat{L}\,y = v$, can be obtained by the iterative process

(30) $\hat{L}'\,y^j = v - \hat{L}''\,y^{j-1}$, $j=1,2,\ldots,$

where $\hat{L}'$ is a sparse and easily invertible main part of $\hat{L}$ and $\hat{L}''$ is the small remainder. Again, $\hat{L}'$ contains the influence coefficients for the pairs of the mutually most sensitive panels. For the convergence of (30) it was also necessary to include the triads of the influence coefficients for the triads of the surface elements adjacent to the trailing edge (see the bordered detail in Fig. 2). The corresponding triads of equations are solved independently for each segment of the trailing edge, and then the remaining pairs are dealt with. For a thin body a few iterations (30) are sufficient.

The procedure (28a-d) has the following iteration operator

(31) $Q = (\hat{L} + PR\tilde{L})^{-1}\,(I-PR)\,\tilde{L}\,\hat{L}^{-1}\,\tilde{L}$.

The appearance of the product (I-PR) $\tilde{L}$ in the above expression shows that the acceleration of convergence, characteristic of a multigrid method (i.e., increasing when the panel size goes to 0), will be achieved if $\tilde{L}$ is a smoothing operator and the remaining factors in (31) have the uniformly bounded norms. The smoothing properly will be ensured, for example, if every four rows of $\tilde{L}$, corresponding to the four panels in each group, contain only the coefficients due to the influenece of the same panels, separated from the members of a group by a fixed distance, independent of the panel size.

For our calculations we chose $\tilde{L}$ to be the approximation of the compact operator (14). In this case it can also be proved that the norm (I-PR) $\tilde{L}$ tends to zero for the decreasing panel size.

6. Results of the calculations

The calculations were carried out for the flow around the elliptic wing with the axis ratio 1/5 and the NACA 0012 cross-section, at the angle of attack $\alpha = 5^\circ$ (see Fig. 1).

Here, we present some results for the case of $n = 96$ panels and $l = 8$ wake strips. Convergence histories for both cases of the additional condition (8) and the respective multigrid methods of the previous section are shown in the following table.

2-grid iter. no.	impermeability condition (Method I)			pressure condition (Method II)
	$n_D = 7$	$n_D = 15$	$n_D = 23$	$n_L = 23 + l$
1	$0.436_{10}{-1}$	$0.248_{10}{-1}$	$0.228_{10}{-1}$	$0.535_{10}{-4}$
2	$0.793_{10}{-3}$	$0.472_{10}{-3}$	$0.467_{10}{-3}$	$0.851_{10}{-7}$
3	$0.571_{10}{-4}$	$0.978_{10}{-5}$	$0.906_{10}{-5}$	$0.903_{10}{-10}$
4	$0.751_{10}{-5}$	$0.202_{10}{-6}$	$0.176_{10}{-6}$	$0.871_{10}{-13}$
5	$0.883_{10}{-6}$	$0.421_{10}{-8}$	$0.339_{10}{-8}$	

Table. The final value of the maximum residual in (25a) after each two-grid iteration.

For the impermeability condition and Method I the three cases of $n_D = 7$, 15 and 23 non-zero coefficients in each row of the matrix $\tilde{D}_{n \cdot n}$ were examined, corresponding to the ratio $n_D/n \approx 1/12$, 1/6 and 1/4. The number of iterations in (20) and (23) was chosen to be 3 and 2, respectively.

For all the three cases the first three two-grid iterations of Method I gave the practically converged solution.

Let us define the work unit as the amount of the computational work necessary to multiply a matrix of n·n entries by a vector of n components. If we assume that the QR-decomposition for $A_{1 \cdot 1}$ and for the matrix of the system (24a, b) are given in advance, then the total cost of the first three iterations of Method I with the starting guess $\sigma_n^o = 0$ is about 5, 6.2 and 7.5 work units for the respective three choices of n_D. The asymptotic convergence factors are close to 0.12, 0.02 and 0.019, respectively. In practice, it seems best to choose the small values of n_D.

Method II, used in connection with the pressure equality condition (10), gives the practically converged solution after the first two-grid iteration. The calculations have been carried out for $n_L = 23 + 1$ nonzero coefficients in each of the first n rows of the matrix $\hat{L}$. The number of iterations (29) and (30) was chosen to be 3 and 5, respectively. The first two-grid iteration (28a-d), with the starting guess $x^o = 0$ requires about 6.2 work units. Method II is purely iterative, and hence no preliminary matrix decompositions are needed. The asymptotic convergence factor is close to 0.001. Due to its simplicity and generality, Method II seems superior to Method I.

The qualitatively correct pressure coefficient distributions for the case of the impermeability condition, represented in Fig. 3, have been derived from the final results for the velocity potential distribution on the surface.

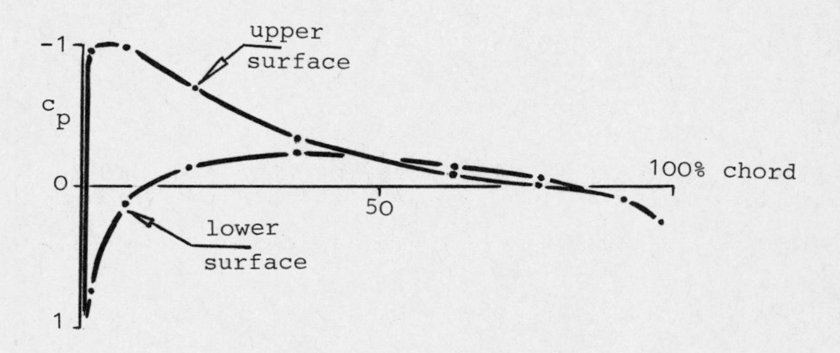

Figure 3. The c_p-distribution in a cross-section of the wing.

For the pressure condition (10) the results were almost identical. The impermeability condition was imposed at the distance equal to 5% of the local chord length downstream of the trailing edge. The pressure equality condition was imposed at the centres of the opposite panels, adjacent to the trailing edge. The wake surface was generated by the trailing edge bisectors.

References

[1] P. M. Anselone, Collectively compact operator approximation theory and applications to integral equations, Prentice Hall, 1971.

[2] W. Hackbusch, Multi-grid methods and applications, Springer, Berlin 1985.

[3] J. L. Hess, The problem of three-dimensional lifting potential flow and its solution by means of surface singularity distribution, Computer Methods in Applied Mechanics and Engineering 4, pp. 283-319, North Holland Publishing Company, 1974.

[4] B. Hunt, The mathematical basis and numerical principles of the boundary integral method for incompressible potential flow over 3-D aerodynamic configurations, Numerical Methods in Fluid Dynamics, B. Hunt, ed., pp. 49-105, Academic Press, 1980.

[5] H. Schippers, Multigrid methods for equations of the 2nd kind wth applications in fluid mechanics, Thesis, Amsterdam 1982.

DEFECT CORRECTION AND HIGHER ORDER SCHEMES FOR THE MULTI GRID SOLUTION OF THE STEADY EULER EQUATIONS

P.W.Hemker

CWI, Centre for Mathematics and Computer Science
P.O.Box 4079, 1009 AB Amsterdam, The Netherlands

ABSTRACT

In this paper we describe 1st and 2nd order finite volume schemes for the solution of the steady Euler equations for inviscid flow. The solution for the first order scheme can be efficiently computed by a FAS multigrid procedure. Second order accurate approximations are obtained by linear interpolation in the flux- or the state space. The corresponding discrete system is solved (up to truncation error) by defect correction iteration. An initial estimate for the 2nd order solution is computed by Richardson extrapolation. Examples of computed approximations are given, with emphasis on the effect for the different possible discontinuities in the solution.

1. INTRODUCTION

As soon as viscosity and heat conduction are neglected, the flow of a gas is described by the Euler equations. In two dimensions these equations are given by

$$\frac{\partial q}{\partial t} + \frac{\partial}{\partial x} f(q) + \frac{\partial}{\partial y} g(q) = 0 , \tag{1.1}$$

with

$$q = \begin{bmatrix} \rho \\ \rho u \\ \rho v \\ \rho e \end{bmatrix}, \quad f = \begin{bmatrix} \rho u \\ \rho u^2 + p \\ \rho u v \\ \rho u H \end{bmatrix}, \quad g = \begin{bmatrix} \rho v \\ \rho v u \\ \rho v^2 + p \\ \rho v H \end{bmatrix}, \tag{1.2}$$

where ρ , u , v , e and p respectively represent density, velocity in x- and y- direction, specific energy and pressure; $H=e+p/\rho$ is the specific enthalpy. The pressure is obtained from the equation of state, which - for a perfect gas - reads

$$p=(\gamma-1)\rho(e-\tfrac{1}{2}(u^2+v^2)),$$

γ is the ratio of specific heats. $q(t,x,y)$ describes the state of the gas as a function of time and space and f and g are the flux in the x- and y- direction. We denote the open domain of definition of (1.1) by Ω^* .

It is well known that solutions of the nonlinear equations (1.1) may develop discontinuities, even if the initial flow ($t=t_0$) is smooth. To allow discontinuous solutions, (1.1) is rewritten in its integral form

$$\frac{\partial}{\partial t}\iint_{\Omega} q \, dx \, dy \; + \; \int_{\partial\Omega}(f.n_x \; + \; g.n_y \;) \, ds \; = \; 0 \, , \qquad \text{for all } \Omega \subset \Omega^* \; ; \tag{1.3}$$

$\partial\Omega$ is the boundary of Ω and (n_x,n_y) is the outward normal vector at the boundary $\partial\Omega$.

The form (1.3) of equation (1.1) shows clearly the character of the system of conservation laws: the increase of q in Ω can be caused only by the inflow of q over $\partial\Omega$. In symbolic form we write (1.1) as

$$q_t \; + \; N(q) = 0. \tag{1.4}$$

In the numerical computations we are only interested in the solution of the steady state Euler equations

$$N(q) = 0. \tag{1.5}$$

The solution of the weak form (1.3) of (1.1) is known to be non-unique and a physically realistic solution (which is the limit of a flow with vanishing viscosity) is known to satisfy the additional entropy condition (cf. [15,16]). Further, the equation (1.1) is hyperbolic, i.e. written in the form

$$\frac{\partial q}{\partial t} + \frac{\partial f}{\partial q} . \frac{\partial q}{\partial x} + \frac{\partial g}{\partial q} . \frac{\partial q}{\partial y} = 0$$

the matrix

$$k_1 \frac{\partial f}{\partial q} \; + \; k_2 \, \frac{\partial g}{\partial q}$$

has real eigenvalues for all (k_1,k_2).

These eigenvalues are $(k_1u+k_2v)\pm c$ and (k_1u+k_2v) (a double eigenvalue); $c \; = \; \sqrt{\gamma p \, / \, \rho}$ is the local speed of sound. The sign of the eigenvalues determines the direction in which the information about the solution is carried along the line (k_1,k_2) as time develops (the direction of the characteristics). It locates the domain of dependence. The entropy condition implies that characteristics do not emerge at a discontinuity in the flow.

2. THE BASIC DISCRETIZATION

In order to discretize eq. (1.1) on a domain with an irregular grid, there are two ways to proceed. First, a mapping can be defined from the physical domain to a computational domain, so that the irregular grid in the physical domain corresponds to a regular grid in the computational domain. By means of this transformation the equation and the boundary conditions are reformulated for the computational domain, where they will contain metric information about the mapping. Now an (arbitrarily accurate) discretization of the transformed equations can be used on the regular grid to solve the original problem.

A second approach (a *finite volume* technique) is to divide the domain of definition in the physical space into a number of disjunct cells $\{\Omega_\alpha\}$ and to require the equation (1.3) to hold on each $\bigcup \Omega_\alpha$. In this way the essential global property of the flow -the conservation character- is easily preserved as long as we take care that for any two neighboring cells Ω_α and Ω_β with $\Gamma_{\alpha\beta} \; = \; \partial\overline{\Omega}_\alpha \bigcup \partial\Omega_\beta$, the same approximation is used for the flow quantities $\int_{\Gamma_{\alpha\beta}} f.n_x \; + \; g.n_y \; ds$, both for the outflow of Ω_α and for the inflow of Ω_β . In that case (1.3) will hold for any Ω which is the union of an arbitrary subset of $\{\Omega_\alpha\}$. In this approach there is no need to transform the equations (1.1) or the boundary conditions.

We found it most convenient to use this finite volume technique and to divide the domain Ω^* in quadrilateral cells Ω_{ij} in a way that is topologically equivalent with a regular division in squares (i.e. $\Omega_{i\pm1,j\pm1}$ are the only possible neighbors of Ω_{ij}).

In order to define subsequent refinements of the irregular mesh and to define a meaningful order of accuracy for our schemes on these non-uniform grids, we introduce a mapping from a "computational domain" divided into regular squares, to the physical domain Ω^*. We assume this mapping to be non-singular (i.e. with a non-vanishing Jacobian J on $\overline{\Omega}^*$) and sufficiently smooth (bounded partial derivatives of J).

The discrete approximation $\tilde{q}(t,x,y)$ to $q(t,x,y)$ is represented by the values q_{ij} for each Ω_{ij}, where q_{ij} represents the mean value of $\tilde{q}$ over Ω_{ij}

$$q_{ij} \approx \frac{\int\int_{\Omega_{ij}} \tilde{q}(x,y)\, dx\, dy}{\text{meas}(\Omega_{ij})}. \tag{2.1}$$

The space discretization method is now completely determined by the way of approximating

$$\int_{\Gamma_{ijk}} (f.n_x + g.n_y)\, ds\,, \qquad k = 1,2,3,4, \tag{2.2}$$

at the four walls of the quadrilateral cell Ω_{ij}. The wall Γ_{ijk} may be either a common boundary with another cell Ω_{ijk} or a part of the boundary $\partial\Omega^*$. In the first case we have to take into account the requirement of conservation of q. To satisfy this requirement we compute the approximation of (2.2) as

$$f^k(q^k_{ij},q^k_{ijk}) \cdot \text{meas}(\Gamma_{ijk}), \tag{2.3}$$

i.e. we approximate $fn_x + gn_y$ by a constant value at Γ_{ijk}, which depends only on q^k_{ij}, a uniform (constant) approximation to $\tilde{q}(t,x,y)$ in Ω_{ij} at the wall Γ_{ijk}, and on q^k_{ijk}, a similar approximation to $\tilde{q}(t,x,y)$ in Ω_{ijk} at Γ_{ijk}. (Notice that $\tilde{q}(t,x,y)$ is not assumed to be continuous over Γ_{ijk} .)

The semi-discretization of the equations (1.4) is now

$$\frac{\partial}{\partial t} q_h \,|_{i,j} = -N_h(q_h)\,|_{i,j} := -\frac{\sum_{k=1,2,3,4} f^k(q^k_{ij},q^k_{ijk})\ \text{meas}(\Gamma_{ijk})}{\text{meas}(\Omega_{ij})}, \tag{2.4}$$

and the steady discrete equations $N_h(q_h) = 0$ are equivalent with

$$\sum_{k=1,2,3,4} f^k(q^k_{ij},q^k_{ijk}) \cdot \text{meas}(\Gamma_{ijk}) = 0, \qquad \text{for all } \Omega_{ij} \subset \Omega^*. \tag{2.5}$$

The approximate flux function $f^k(q^k_{ij},q^k_{ijk})$ depends on the direction (n^k_x,n^k_y), of the side Γ_{ijk}. However, by the rotation invariance of the Euler equations, we may relate $f^k(\,.\,,\,.\,)$ to a local coordinate system (rotated such that it is aligned with Γ_{ijk}). Hence, only a single function $f(\,.\,,\,.\,)$, the *numerical flux function,* is needed to approximate the flux between two cells (cf.[9,10]):

$$f(\,.\,,\,.\,) = f^k(\,.\,,\,.\,) \quad \text{if} \quad n^k_x = 1,\ n^k_y = 0\,. \tag{2.6}$$

In this way the freedom in the approximation of (2.3) is in the choice of a numerical flux function and in the computation of q^k_{ij} and q^k_{ijk} from $\{q_{ij}|\Omega_{ij} \subset \Omega^*\}$. We shall first consider two elementary possibilities for the choice of a numerical flux function. Then we describe the computation of $\{q^k_{ij}\}$ for the first order scheme. In the next section we shall consider second order schemes, generated by other computations of $\{q^k_{ij}\}$ or f^k .

For consistency of the resulting scheme, $f(\,.\,,\,.\,)$ should satisfy $f(q,q) = f(q)$, cf. [7]. A usual representation of $f(\,.\,,\,.\,)$ is given by

$$f(q_0,q_1) = \tfrac{1}{2}f(q_0) + \tfrac{1}{2}f(q_1) - \tfrac{1}{2}\, d(q_0,q_1), \tag{2.7}$$

with $d(q_0,q_1) = \mathcal{O}(\,||q_1 - q_0||\,)$.

The *central difference flux* is defined by $d(q_0,q_1) = 0$. For an *upwind* numerical flux functions we have ([7]),

$$d(q_0,q_1) = \left| \frac{\partial f}{\partial q} \left[\frac{q_0 + q_1}{2} \right] \right| (q_1 - q_0) + o(\| q_1 - q_0 \|). \tag{2.8}$$

For reasons explained in [10], we prefer to use a slight modification of a numerical flux function that was proposed by Osher [19,21]. We take

$$d(q_0,q_1) = \int_{q_0}^{q_1} \left| \frac{\partial f}{\partial q} \right| (w)\, dw , \tag{2.9}$$

where the integration path in the state space follows three sub-paths along the eigenspaces of $\partial f / \partial q$. These sub-paths correspond to the eigenvalues $\lambda_1 = u - c$, $\lambda_2 = \lambda_3 = u$, and $\lambda_4 = u + c$ respectively.

In the case that $\Gamma_{ijk} \subset \partial\Omega^*$, interpolation from the interior of Ω^* yields a value q_{ijk}^{IN} which corresponds to the mean value of $\tilde{q}_{ij}$ at Γ_{ijk} in Ω_{ij}:

$$q_{ijk}^{IN} = \frac{\int_{\Gamma_{ijk}} \tilde{q}(t,x,y)\ ds}{\text{meas}(\Gamma_{ijk})} .$$

For well posed boundary conditions $B(q) = 0$ at Γ_{ijk} , a value q_{ijk}^{OUT} can be determined such that

$$f^k(q_{ijk}^{OUT},q_{ijk}^{IN}) = f^k(q_{ijk}^{OUT}), \tag{2.10a}$$

and

$$B(q_{ijk}^{OUT}) = 0 \tag{2.10b}$$

is satisfied at a point on Γ_{ijk} . In [10] we showed how q_{ijk}^{OUT} is connected with q_{ijk}^{IN} with respect to the outgoing characteristics; the incoming characteristic information is taken from the boundary conditions. This corresponds with the use of Riemann invariants to derive non-reflecting numerical boundary conditions.

In our first order scheme we use a piecewise constant numerical approximation for $\tilde{q}$:

$$q(x,y) = q_{ij} \qquad \text{for} \quad (x,y) \in \Omega_{ij}. \tag{2.11a}$$

This uniform state in Ω_{ij} is assumed for all (i,j), and hence

$$f^k(q_{ij}^k,q_{ijk}^k) = f^k(q_{ij},q_{ijk}) . \tag{2.11b}$$

The flux at Γ_{ijk} now corresponds with the flux at a discontinuity between two uniform states. Such a flux can be computed by solving the Riemann problem of gasdynamics. However, this is a nontrivial nonlinear computation, and we approximate it by (2.7) , (2.9), cf.[10], which is a slight variant of Osher's "approximate Riemann solver".

The order of accuracy of the resulting schemes on the nonuniform mesh is not immediate. It can be proved that in general (at most) second order accuracy can be obtained when q_{ij}^k, q_{ijk}^k are computed properly. We shall give here the principles along which these accuracy results are derived. For the detailed proof see [23].

The relation between equation (1.1) and (2.4) can be described by the following diagram

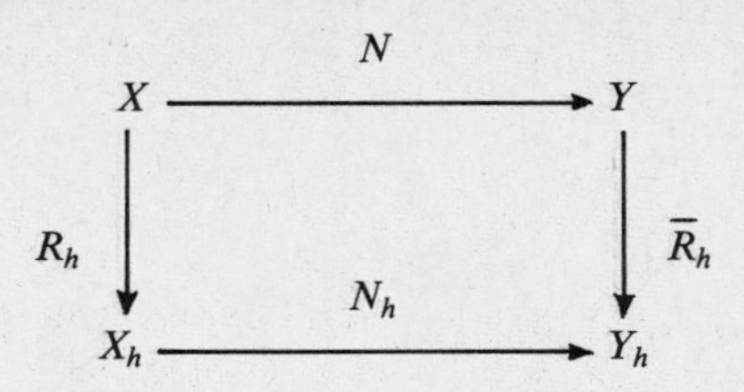

where X (X_h) and Y (Y_h) are the continuous (discrete) spaces of state and change-of-state respectively: $q \in X$, $q_h \in X_h$, $N(q) \in Y$, $N_h(q_h) \in Y_h$.

To compute the order of consistency, we introduce the restrictions R_h and $\bar{R}_h$, as well as the parametrization $h \to 0$ for the mesh refinement.
For a given parameter value $h > 0$, and an arbitrary $q \in X$ we define

$$\bar{q}(\xi,\eta) = \frac{1}{(2h)^2} \cdot \int_{\xi-h}^{\xi+h} \int_{\eta-h}^{\eta+h} q(x(\xi,\eta),y(\xi,\eta)) \;\; d\xi \;\, d\eta \; \cdot \tag{2.12}$$

Here $x(\xi,\eta)$ and $y(\xi,\eta)$ describe the mapping between the computational (ξ,η)-domain and the physical (x,y)-domain, such that

$$(\, x(2hi \pm h, 2hj \pm h) \, , \, y(2hi \pm h, 2hj \pm h) \,) \tag{2.13}$$

denote the vertices of the cell Ω_{ij} in the physical space. We consider the order of consistency for the discretization on the irregular grid, assuming that the mesh refines corresponding to $h \to 0$ in (2.13), and that the mapping $(\xi,\eta) \to (x,y)$ is independent of h. Moreover, we require the Jacobian $J(\xi,\eta) = x_\xi y_\eta - x_\eta y_\xi$ not to vanish and to be smooth enough, i.e. $|J(\xi,\eta)| \geqslant C_1 > 0$, and

$$\left| \left[\frac{\partial}{\partial \xi} \right]^{m_1} \left[\frac{\partial}{\partial \eta} \right]^{m_2} J(\xi,\eta) \right| \;\leqslant\; C_2 \, |J(\xi,\eta)| \; .$$

A mean value of q in Ω_{ij} is now given by $\bar{q}(2hi, 2hj)$. In this way a restriction operator R_h: $q \to q_h$ is defined.
Notice that the assumptions on the Jacobian imply

$$\frac{\iint\limits_{\Omega_{ij}} q \;\; dx \;\, dy}{\iint\limits_{\Omega_{ij}} \;\; dx \;\, dy} \;=\; \bar{q}(2hi,\, 2hj) \; (\, 1 \,+\, \mathcal{O}(h^2)) \cdot \tag{2.14}$$

The restriction $\bar{R}_h$: $Y \to Y_h$ is simply defined by

$$(\bar{R}_h r)_{i,j} \;=\; \frac{1}{(2h)^2} \cdot \iint\limits_{\Omega_{ij}} r \; dx \, dy \; . \tag{2.15}$$

Now the truncation error for $q \in X$ is defined as

$$N_h \, (R_h \, q) \;-\; \bar{R}_h \, N(q) \;=\; \tau_h(q), \tag{2.16}$$

and for a smooth function q we can determine the order of consistency. It can be shown that the order of consistency is 1 if eq.(2.11) holds. We denote this first order semi-discretization (2.4) - (2.11) in symbolic form by

$$(q_h)_t \;+\; N_h^1(q_h) \;=\; 0 \; . \tag{2.17}$$

Note: In the actual computations each discrete equation at the h-level is multiplied by a factor $(2h)^2$. This can be seen as discretization of the integral form (1.3) rather than of the differential form (1.1). The advantage is a simpler implementation.

3. SECOND ORDER SCHEMES

The first order discretization discussed in section 2 has a number of advantages: it is conservative, satisfies an entropy condition, is monotonous and gives a sharp representation of discontinuities (shocks and contact discontinuities), as long as these are aligned with the mesh. Further it allows an efficient solution of the discrete equations by a multigrid method [9]. Disadvantages are: the low order of accuracy (many points are required to find an accurate representation of a smooth solution) and the fact that it is highly diffusive for oblique discontinuities (the discontinuities are smeared out over a large number of cells). For a first order (upwind) scheme these are well known facts (cf. e.g. [8]) and it leads to the search for higher order methods.

A key property of the discretization, that we want to maintain in a 2nd order scheme, is the conservation of q, because it allows discontinuities to be captured as weak solutions of (1.1) and avoids the necessity of a shock fitting technique. Therefore, we consider only schemes that are still based on (2.4), and we select $f^k(q_{ij}^k, q_{ijk}^k)$ that yield a better approximation to (2.2) than (2.11b).

The higher order schemes can be obtained in two different ways. Higher order interpolation (or extrapolation) is used either for the states (i.e. in X_h) or for the fluxes (i.e. in Y_h). The first approach is used e.g. in [1,4,28], the second in [20,24]. In the first case, in (2.5) q_{ij}^k and q_{ijk}^k are obtained by some interpolation from $q_h = \{q_{ij}\}$. In the latter, $f^k(q_{ij}^k, q_{ijk}^k)$ is obtained from $\{ f^k(q_{ij}^k, q_{ijk}^k) \} \cup \{ f^k(q_{ij}^k, q_{ij}^k) \}$

From the point of view of finite volume discretization, a straightforward way to form a more accurate approximation is to replace the 1st order approximation (2.11) by a 2nd order one. Instead of the piecewise constant $\tilde{q}(x,y)$, we may consider a piecewise bilinear function $\tilde{q}(x,y)$ on a set of 2×2 cells (a "superbox"). Such a superbox on the h-level corresponds with a single cell at the 2h-level. Over the boundaries of the superbox $\tilde{q}(x,y)$ can be discontinuous; in the superbox $\tilde{q}(x,y)$ is determined by q_{ij}, $q_{i+1,j}$, $q_{i,j+1}$ and $q_{i+1,j+1}$ as defined by (2.1). Using such a bilinear function, we see that the central difference approximation is used for flux computations inside the superboxes; at superbox boundaries interpolation is made from the left and the right and the approximate Riemann solver is used to compute the flux. We denote the corresponding discrete operator by N_h^S. It is easily shown that this *superbox scheme* is 2nd order accurate in the sense that

$$\overline{R}_{2h,h} \left(N_h^S(R_h q) - \overline{R}_h N(q) \right) = \mathcal{O}(h^2) .$$

Instead of the finite volume superbox scheme, we can also adopt a finite difference approach. In the case of interpolation of states, interpolation from the left (right) can be used to obtain a value q_{ijk}^L (q_{ijk}^R) at the left (right) side of all walls Γ_{ijk}. In the case of interpolation of fluxes, it may be necessary to split flux-differences in positive and negative (right- and left- going) parts.

In both cases the simplest 2nd order schemes are central differencing schemes. Here the interpolation is done irrespective of a particular characteristic direction. Central differencing in the X_h-space yields $f(q_0, q_1) = f((q_0 + q_1)/2)$ for the numerical flux function (2.6). By central differencing in Y_h we obtain $f(q_0, q_1) = \frac{1}{2} f(q_0) + \frac{1}{2} f(q_1)$. In contrast with the first order schemes, the central difference schemes are under-diffusive, which may lead to instabilities. An uncoupling of odd and even points may occur and spurious oscillations may appear in the solution. When these schemes are used alone, an artificial additional diffusion (dissipation) term is added to stabilize the solution [11,22].

To improve the stability behavior, both for the X_h- and for the Y_h-interpolation, it is better to take into account the domain of dependence of the solution (the direction of the characteristics) and to distinguish between interpolation from the left and from the right of a cell wall. For simplicity of notation we shall exemplify this only for the 1-D case. Generalization to 2-D is straightforward.
In 1-D , eq.(2.4) reduces to

$$N_h(q_h)_i = \frac{f_{i+\frac{1}{2}} - f_{i-\frac{1}{2}}}{x_{i+\frac{1}{2}} - x_{i-\frac{1}{2}}} , \tag{3.1}$$

where $f_{i+\frac{1}{2}} = f(q^L_{i+\frac{1}{2}}, q^R_{i+\frac{1}{2}})$.

Interpolation in X_h

At a cell wall $x_{i+1/2}$ we distinguish between the interpolated values from the left, $q^L_{i+1/2}$, and from the right, $q^R_{i+1/2}$. We define $\Delta q_{i+1/2} = q_{i+1} - q_i$ and find the 2nd order upwind interpolated values

$$q^L_{i+\frac{1}{2}} = q_i + \frac{1}{2}\, \Delta q_{i-\frac{1}{2}}\,, \tag{3.2}$$

$$q^R_{i+\frac{1}{2}} = q_{i+1} - \frac{1}{2}\, \Delta q_{i+\frac{3}{2}}\,.$$

The stability properties of these one-sided approximations are better than for central approximations, but monotonicity is not preserved (see section 5). The usual way to force the monotonicity is to introduce a limiting function ϕ, $0 \leqslant \phi \leqslant 2$ and to interpolate by

$$q^L_{i+\frac{1}{2}} = q_i + \frac{1}{2}\, \phi^L_{i+\frac{1}{2}}\, \Delta q_{i-\frac{1}{2}} \tag{3.3}$$

$$q^R_{i-\frac{1}{2}} = q_i - \frac{1}{2}\, \phi^R_{i-\frac{1}{2}}\, \Delta q_{i+\frac{1}{2}}\,,$$

where ϕ^L and ϕ^R are chosen depending on $\{\Delta q_{i\pm\frac{1}{2}}\}$ such that $q^L_{i-1/2}$ lies between q_{i-1} and q_i, and $q^R_{i+1/2}$ between q_i and q_{i+1}, (cf. [1,26]).

Van Leer [28] also introduces a linear combination of the one-sided and central interpolation. Parametrized by κ we obtain

$$q^L_{i+\frac{1}{2}} = q_i + \frac{1}{4}\left[(1-\kappa)\Delta q_{i-\frac{1}{2}} + (1+\kappa)\Delta q_{i+\frac{1}{2}}\right], \tag{3.4}$$

$$q^R_{i-\frac{1}{2}} = q_i - \frac{1}{4}\left[(1-\kappa)\Delta q_{i+\frac{1}{2}} + (1+\kappa)\Delta q_{i-\frac{1}{2}}\right].$$

This formula contains: ($\kappa = -1$) the one-sided 2nd order scheme, ($\kappa = \frac{1}{3}$) a "3rd order" upwind biased scheme, and ($\kappa = 1$) the central difference scheme. (Notice that the "3rd order" scheme is 3rd order consistent in a 1-D situation; in 2-D the scheme is still 2nd order accurate.) We use (2.4) - (3.4) for the construction of a 2nd order discretization of (1.4)

$$(q_h)_t + N^2_h(q_h) = 0\,. \tag{3.5}$$

In 1-D the superbox scheme corresponds to the use of $\kappa = +1$ for odd i, and $\kappa = -1$ for even i.

The interpolation (3.4) is well defined in the interior cells of the domain. In the cells near the boundary $\partial\Omega^*$, one of the values $\Delta q_{i\pm 1/2}$ is not defined, by the absence of a value q_i corresponding to a point outside Ω^*. Here a different approximation should be used. In our computations we set $\Delta q_{i+1/2} = \Delta q_{i-1/2}$ at the cell Ω_i near the boundary. This corresponds with the "superbox" approximation for these cells. For the superbox scheme and for the scheme (3.4), with different values of κ, we show some results in section 5.

Interpolation in Y_h

To take into account the domain of dependence (the direction of the characteristics), we here distinguish between flux differences in the positive and in the negative direction. We define

$$\Delta f^+_{i+\frac{1}{2}} = -f(q_i, q_{i+1}) + f(q_{i+1}), \tag{3.6}$$

$$\Delta f^-_{i+\frac{1}{2}} = +f(q_i, q_{i+1}) - f(q_i).$$

It is easily seen that

$$f(q_{i+1}) - f(q_i) = \Delta f^+_{i+\frac{1}{2}} + \Delta f^-_{i+\frac{1}{2}}\,,$$

and

$$f(q_i,q_{i+1}) - f(q_{i-1},q_i) = \Delta f^+_{i-\frac{1}{2}} + \Delta f^-_{i+\frac{1}{2}} \ .$$

Further, the numerical flux has been constructed such that

$$\Delta f^+_{i+\frac{1}{2}} \ / \ \Delta q_{i+\frac{1}{2}} \geqslant 0 \ , \tag{3.7}$$

$$\Delta f^-_{i+\frac{1}{2}} \ / \ \Delta q_{i+\frac{1}{2}} \leqslant 0.$$

For vectors f and q we mean by (3.7) that the matrices of partial derivatives have real non-negative (non-positive) eigenvalues. Hence, $\Delta f^+_{i+\frac{1}{2}}$ (or $\Delta f^-_{i+\frac{1}{2}}$) corresponds to information going to the right (left).

A 2nd order upwind scheme is now constructed as

$$f^k_{i+\frac{1}{2}} := f^k(q_i,q_{i+1}) + \frac{1}{2}\,\Delta f^+_{i-\frac{1}{2}} - \frac{1}{2}\Delta f^-_{i+\frac{3}{2}} \tag{3.8}$$

Notice that with this notation central differencing is written as

$$f^k_{i+\frac{1}{2}} := f^k(q_i,q_{i+1}) + \frac{1}{2}\,\Delta f^+_{i+\frac{1}{2}} - \frac{1}{2}\Delta f^-_{i+\frac{1}{2}} \ . \tag{3.9}$$

Also here a linear combination of (3.8) and (3.9) is easily realized and flux limiting functions can be introduced to maintain monotonicity of the solution as for (3.3) [20].

4. DEFECT CORRECTION ITERATION

The 2nd order space discretization of the timedependent equations (1.4) yields a semi-discretization (3.5). The usual way to find the solution of the steady state equations

$$N^2_h(q_h) = 0 \ , \tag{4.1}$$

is to take an initial guess and to solve (3.5) for $t \to \infty$, i.e. to compute $q_h(t)$ until initial disturbances have sufficiently died out. The advantage is that, starting with a physically meaningful situation, we may expect that a meaningful steady state will be reached, even when unicity of the steady equations is not guaranteed. The drawback is that many timesteps may be necessary before the solution has sufficiently converged. For the acceleration of the convergence, many devices have been developed such as local time stepping, residual smoothing, implicit residual averaging or enthalpy damping [22].

Multigrid is also used as an acceleration device [14,18,22]. Here discretizations (3.5) are given on a sequence of grids. The coarse grids are used to move low frequency disturbances rapidly out of the domain Ω^* by large timesteps, whereas high frequency disturbances should be locally damped on the fine grids, e.g. by a sufficiently dissipative timestepping procedure.

We take another approach [9,10,12,13,17], and consider directly the steady state equations. By the stability of the first order discretization, a relatively simple relaxation method (Collective Symmetric Gauss Seidel iteration, i.e. a SGS relaxation where the 4 variables corresponding to a single cell Ω_{ij} are relaxed collectively) is able to reduce the high frequency error components efficiently, and -therefore- a FAS-algorithm with this relaxation is well suited to solve the discrete first order equations.

Although no explicit artificial viscosity is added to the scheme, a suitable amount of "numerical diffusivity" is automatically introduced by the upwind discretization. As $h \to 0$, this "artificial diffusion" vanishes and the sequence of discretizations converges to the Euler equations as the limit of an equation with vanishing viscosity.

Another advantage of the introduction of this "artificial viscosity" just by the use of the upwind scheme is that the coarser discretizations, including their larger amount of "numerical viscosity", are now Galerkin approximations to the corresponding finer grid discretizations. Hence coarse and fine discretizations are relatively consistent. (For a discussion of related problems when multigrid is applied to the convection diffusion equation and various amounts of artificial viscosity are used on the different grids cf [30].)

When we try to solve the 2nd order discretization (4.1) in the same manner as we do the first order equations, we may expect difficulties for two reasons. First, the set of 4 equations to be solved at each cell Ω_{ij} in the Collective SGS relaxation is much more complex . The set up of these equations would increase the amount of computational work considerably. Secondly, the nonlinear equations (4.1) are less stable. The 2nd order discretizations are less diffusive, and in the case of central differences clearly "anti-diffusive". This may lead not only to non-monotonous solutions, but it also can cause a Gauss Seidel relaxation not to reduce sufficiently the rapidly varying error components.

A local mode analysis of smoothing properties of GS for 1st and 2nd order upwind Euler discretizations can be found in [12]. There, the flux splitting upwind scheme of Steger and Warming[25] is analyzed, whereas we apply Osher's scheme [19,21]. Numerical evidence that convergence for the relaxation process of a 2nd order upwind procedure is slower than for a 1st order scheme, is also found in [17,29]. Here van Leer's flux splitting scheme [27] was used.

To obtain 2nd order accurate solutions, we do not try to solve the system $N_h^2(q_h) = 0$ as such. We use the first order operator N_h^1 to find the higher order accurate approximation in a defect correction iteration:

$$N_h^1(q_h^{(1)}) = 0\,, \tag{4.2a}$$

$$N_h^1(q_h^{(i+1)}) = N_h^1(q_h^{(i)}) - N_h^2(q_h^{(i)}). \tag{4.2b}$$

For an introduction to the defect correction principle see [2]. It is well known [6] that -if the problem is smooth enough- the accuracy of $q_h^{(i)}$ is of order 2 for $i \geqslant 2$. If the solution is not smooth (higher order derivatives are dominating) there is no clear reason to expect the solution of (4.1) to be more accurate than the solution of (4.2a). Nevertheless, in section 5 evidence is given that a few defect correction steps may improve the solution considerably.

In fact we may use $q_h^{(i+1)} - q_h^{(i)}$ as an error indicator. In the smooth parts of the solution $q_h^{(1)} - q_h^{(1+i)} = \mathcal{O}(h)$, $q_h^{(2)} - q_h^{(2+i)} = \mathcal{O}(h^2)$; where these differences are larger, e.g. $\mathcal{O}(1)$, the solution is not smooth (relative to the the grid used). Then grid adaptation is to be considered rather than the choice of a higher order method, if a more accurate solution is wanted.

In a multigrid environment, where solutions on more grids are available, we should -of course- also consider other approaches to compute higher order solutions, such as
(1) Richardson extrapolation,
(2) τ-extrapolation, or
(3) Brandt's double discretization.

The two extrapolation methods can be well used to find a more accurate solution if the solution is smooth indeed. Then no additional difference scheme (4.1) is required . A drawback is that these methods rely on the existence of an asymptotic expansion of the (truncation) error for $h \rightarrow 0$, and -globally- no a-priori information about the validity of this assumption is available. Another disadvantage is that the accurate solution (for Richardson extrapolation) or the estimate for the truncation error (τ-extrapolation) is obtained at the one-but-finest level and no high resolution of local phenomena is obtained. Whereas we want not only a high order of accuracy, but also an accurate representation of possible discontinuities, we use Richardson extrapolation (only) as a possibility to find a higher order initial estimate for the iteration process (4.2b).

Since the evaluation of $N_h^2(q_h)$ is hardly more expensive than the evaluation of $N_h^1(q_h)$, the costs to compute the defect in (4.2b) is of the same order as the evaluation of the relative truncation error $\tau_{2h,h}(q_h) = N_{2h}^1(R_{2h,h}q_h) - R_{2h,h}N_h^1(q_h)$. This makes us to prefer (4.2b) to τ-extrapolation.

Having both a 1st and a 2nd order discrete operator at our disposal, Brandt's double discretization [3] seems another efficient way to find a 2nd order accurate solution. However, we have bad experience in applying it to the Euler equations. In particular when solving (contact) discontinuities. Using the Collective SGS relaxation and a 2nd order scheme based on (3.4), we experienced serious problems in the computation of the numerical fluxes (2.11b), caused by virtual cavitation of the flow. Our explanation is the following. In Brandt's double discretization each iteration cycle consists of a smoothing step towards the solution of $N_h^1(q_h) = r_h^1$, and a coarse grid correction step towards the solution of $N_h^2(q_h) = r_h^2$. If a discontinuity in the solution is present, the differences between the results after the first and the second half-step may be considerable. In our case these differences resulted in such large differences in values for q_{ij}^k and q_{ijk}^k, that the numerical flux $f^k(q_{ij}^k, q_{ijk}^k)$ could not properly be evaluated. (The solution of the Riemann problem with the two states q_{ij}^k and q_{ijk}^k shows cavitation.) The responsibility for this problem lies in part on the type of relaxation used: for the non-elliptic Euler equations CSGS-relaxation is not a pure local smoothing procedure. However, we did not succeed in finding a local smoothing procedure that was satisfactory for the Euler equations. E.g. experimentation with a damped collective Jacobi relaxation was not successful.

The Full Multi Grid algorithm

We aim at the approximate solution q_h of the Euler equations for a given mesh and we assume that also L coarser meshes exist. We denote the level of refinement by m and the approximate solution at level m by $q_{(m)} = q_{2^{(L-m)}h}$. The coarser grids, $m<L$, are not only used for the realization of FAS-iteration steps as described in [9,10], but also for the construction of the initial estimate for the iteration process. The algorithm used to obtain the initial estimate and further iterands in the defect correction process is as follows:

(0) start with an approximation for $q_{(0)}$;
(1) **for** $m := 0\ (1)\ L-1$ **do**
(.) **begin**
(1.a) **for** i := 1 (1) k_m **do** FAS ($N_{(m)}^1\, q_{(m)} = 0$);
(1.b) $q_{(m+1)} := P_{m+1,m}^1\, q_{(m)}$;
(.) **end;**
(.) $m := L$;
(2) **for** i := 1 (1) k_L **do** FAS ($N_{(m)}^1\, q_{(m)} = 0$);
(3) $q_{(m)} := q_{(m)} + P_{m,m-1}^S\,(R_{m-1,m}^0\, q_{(m)} - q_{(m-1)})$;
(4) **for** d := 1 (1) dcps **do**
(.) **begin**
(4.a) $r_{(m)} := N_{(m)}^1(q_{(m)}) - N_{(m)}^2(q_{(m)})$;
(4.b) **for** i := 1 (1) kd **do** FAS ($N_{(m)}^1\, q_{(m)} = r_{(m)}$);
(.) **end;**

Step (1) is an FMG process to obtain a 1st order accurate initial estimate at level L. The prolongation $P_{m+1,m}^1$ is a linear interpolation procedure and, hence, accurate enough to retain the 1st order accuracy on the finer mesh. Asymptotically, the discretization error for $q_{(m)}$ is bounded by $C\,h_{(m)} = \mathcal{O}(2^{L-m}h)$ for $h_{(L)} = h \to 0$. Now a simple analysis shows that, for a fixed $k_m = k$ at all levels, the iteration error at level m is $\approx Ch_{(m)}\, s^k / (1-2s^k)$, where s is an upper bound for the FAS-convergence factor. Therefore, to obtain a 1st order accurate solution, for iteration (1.a) it is not necessary to reduce the iteration error in $q_{(m)}$ by a factor much smaller than $s^k \approx 1/3$. This means that a single FAS step as described in [9,10] may be sufficient. Not being sure about the validity of the asymptotic assumption, we set $k_m = 2$, $m = 1,2,\ldots,L$. Step (2) is the FAS-iteration to obtain the solution to $N_h^1(q_h) = 0$ up to truncation error accuracy.

Step (3) is a Richardson extrapolation step to find a 2nd order initial estimate for q_h. The prolongation $P_{m,m-1}^S$ and the restriction $R_{m-1,m}^0$ are piecewise bilinear interpolation over superboxes and averaging over cells, respectively, so that $R_{m-1,m}^0 P_{m,m-1}^S = I_{m-1}$ is the identity, and $P_{m,m-1}^S R_{m-1,m}^0$ is a projection operator. With the asymptotic expansion for the error in q_h as

$$q_h = R_h \hat{q} + h^p R_h\, e + \mathcal{O}(h^{p+1})\,, \tag{4.3}$$

where $\hat{q}$ is the exact solution, we obtain for p=1 the 2nd order extrapolation

$$R_{2h}\, \hat{q} = 2\, R_{2h,h} q_h - q_{2h} + \mathcal{O}(h^2)\,. \tag{4.4}$$

We find the extrapolated value of q_h in (3) as the sum of (4.4) and $(I_m - P^S_{m,m-1} R^0_{m-1,m}) q_h \in \mathrm{Ker}(R_{2h})$. We notice that formally the approximation of $q_{(L)}$ after stage (3) is still $\mathcal{O}(h)$, unless $q_{(L-1)}$ is an $\mathcal{O}(h^2)$ approximation, and stage (2) can reduce the (smooth) error component $R_h e$ by a factor $\mathcal{O}(h)$. Nevertheless, we see in practice that already for small values of k_m, $m = 1,2,...,L$, the Richardson extrapolation can reduce the error significantly.
Step (4) is the defect correction iteration (4.2b). If the defect correction iteration starts with a 1st order initial approximation, for 2nd order accuracy it is sufficient to take dcps = 1. This necessitates an improvement of the error by a factor $\mathcal{O}(h)$ in the iteration (4.b), i.e. we need kd $= \mathcal{O}(\log(h))$. However, since the FAS iteration step is the expensive part of the computation in (4), for most purposes we take kd = 1 and a sufficiently large number for dcps.

5. NUMERICAL RESULTS

To see the effect of the various different 2nd order schemes and their combination with (a few steps) in the defect correction iteration (4.2), we consider three model problems. We take (1) a smooth subsonic flow through a channel with a curved wall, (2) an oblique shock, and (3) an oblique contact discontinuity. The three problems are all defined on a rectangular domain. The first problem may clearly show the 2nd order accuracy. The other two problems contain the two kinds of discontinuities that may appear in Eulerian flow. In the shock, the characteristics converge and there is a natural mechanism to steepen a smeared shock [5]. In the contact discontinuity the characteristics are parallel and no such mechanism exists. This kind of discontinuity is more like discontinuities that may appear in the solution of the linear convection diffusion equation [8].

We first give a precise description of the 3 problems and then comment on the various numerical results obtained.
Problem 1 The smooth problem.
The domain Ω^* is $(-1,1)\times(0,1)$; the coarsest mesh (m = 0) contains 4×2 square cells. $y = -1$ is the inflow boundary, with boundary conditions $\rho = 1.0$, $u = 0.75$, $v = 0.0$; $y = 1$ is the outflow boundary: $p = 1/\gamma$; $x = 0$ and $x = 1$ are solid walls: at $x = 1$ we take $v = 0$, and at $x = 0$ we use a slender body approximation for a curved boundary: $v/u = 0.02 \cdot \sin(\pi x)$.
The initial approximation is uniform flow in Ω^* with $\rho = 1.0$, $u = 0.75$, $v = 0.0$ and $p = 1.0$.
Problem 2 The oblique shock
The domain Ω^* is $(0,4)\times(0,1)$; the coarsest mesh contains 6×2 cells. The exact solution has 3 subregions with uniform states as given in figure 5.1.

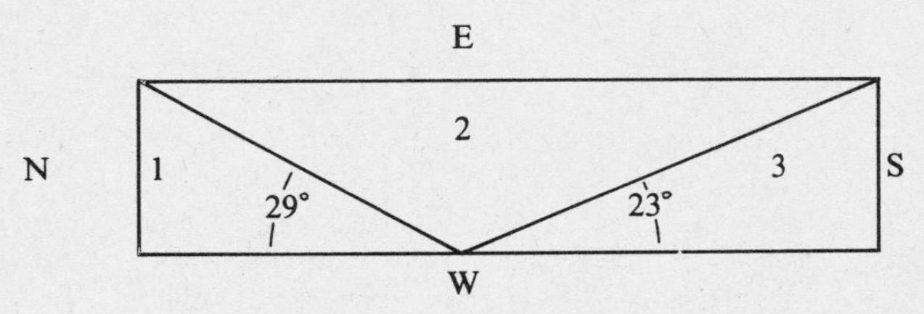

Figure 5.1

The states are resp.:
state 1 : $u = 2.9$, $v = 0.0$, $c = 1.0$, $p = 1.0$;
state 2 : $u = 2.6$, $v = -0.5$, $c = 1.1$, $p = 2.1$;
state 3 : $u = 2.4$, $v = 0.0$, $c = 1.2$, $p = 4.0$.
The boundary conditions are: at N supersonic inflow, at W a solid wall, and at E and S the boundary conditions are overspecified (i.e. for all variables Dirichlet boundary conditions are given, but

these are partly neglected by the difference scheme).

Problem 3 The contact discontinuity

Here $\Omega^* = (0,1)\times(0,1)$, the coarsest mesh is 2×2 cells. The exact solution of the problem has a discontinuity at $x+y = 1$. In both parts of the domain the solution has a uniform state: for $x+y<1$ we take $p = 1.0$, $u = 0.3$, $v = -0.3$ and $c = 0.6$; for $x+y>1$ we have $p = 1.0$, $u = 0.6$, $v = -0.6$ and $c = 1.0$. At the outflow boundaries $p = 1.0$ is given and at the inflow boundaries we give the correct values for u, v and the entropy. The initial estimate has a uniform state over all Ω^* which has mean values between the two uniform states that define the exact solution.

In the figures 5.2 we show for problem 1 the pressure at the curved wall $y = 0$. The results are obtained by the algorithm (0)-(4) described in section 4. In figure 5.2.a the 1st order solution (i.e. the solution after stage (2)) is given for $L = 3,4,5$ and in figure 5.2.b-c we show the second order solutions (at stage (4)) obtained from scheme (3.4) with $\kappa = -1$, after $d = 0$ and $d = 1$ defect correction steps ($k_m = k_L = 4$, kd $= 1$). I.e. fig. 5.2.b shows the solution before and fig. 5.2.c the solution after the first defect correction step. More defect correction steps ($d>1$), or the use of $\kappa = 1/3$ or the "superbox" scheme all yield very similar pressure profiles.

In figure 5.2.a we see clearly 1st order convergence for the 1st order scheme; 5.2.b and 5.2.c show more accurate solutions. Under assumption of the asymptotic expansion

$$q_h(x,y) = \hat{q}(x,y) + h^p e(x,y) + \mathcal{O}(h^{p+1}),$$

the order of convergence p is derived from the solutions for $L = 3,4,5$, computed as described above. The same computations were made for $\kappa = -1$, $\kappa = 1/3$ and for the "superbox" scheme, both with and without the Richardson extrapolation (i.e. stage (3) of the FMG algorithm). The results are shown in Table 5.1. They seem to confirm the hypothesis of the validity of the asymptotic expansion (4.3) with $p = 1$.

	with Richardson extrapolation			without Richardson extrapolation		
	$\kappa = -1$	$\kappa = 1/3$	SB	$\kappa = -1$	$\kappa = 1/3$	SB
d = 0	2.08	2.08	2.08	1.00	1.00	1.00
d = 1	2.20	1.88	2.23	1.64	1.78	1.50
d = 2	2.11	1.93	1.81	2.18	1.83	1.50
d = 3	1.88	2.01	1.96	1.88	2.13	2.02
d = 4	2.15	1.93	1.96	2.10	1.99	1.95
d = 5	1.92	1.92	1.92	1.98	1.93	1.92

Table 5.1. The measured (mean) order of convergence (at cell corners, boundaries excluded). The second order schemes are (3.4) with $\kappa = -1$ and $\kappa = 1/3$, and (SB) the "superbox" scheme.

For problem 2 we show results in the figures 5.3. For the level $L = 4$ we show the 1st order solution, the solution obtained after Richardson extrapolation and the solution after 1 and 3 defect correction steps.

In the figures 5.4 we show the same results for problem 3. For the problems 2 and 3 results are shown only for the scheme (3.4) with $\kappa = -1$. From the figures 5.3 and 5.4 it is clear that not only a higher order of accuracy is obtained; we also find a better resolution of skew discontinuities.

ACKNOWLEDGEMENT

I want to acknowledge the cooperation with B. Koren, S. Spekreijse and P.M. de Zeeuw in this Euler research. The investigations were supported in part by the Netherlands Technology Foundation (STW).

REFERENCES

[1] Anderson, W.K., Thomas, J.L., and Van Leer, B., "A comparison of finite volume flux vector splittings for the Euler equations" AIAA Paper No. 85-0122.

[2] Böhmer, K., Hemker, P. & Stetter, H., "The Defect Correction Approach." Computing Suppl. 5 (1984) 1-32.

[3] Brandt, A., "Guide to Multigrid Development." In: Multigrid Methods (W. Hackbusch and U. Trottenberg eds), Lecture Notes in Mathematics 960, pp.220-312, Springer Verlag 1982.

[4] Colella, P. and Woodward, P.R., "The Piecewise Parabolic Method (PPM) for Gas Dynamical Simulations" J. Comp. Phys 52 (1984) 174-201.

[5] Davis, S.F., "A rotationally biased upwind difference scheme for the Euler equations." J. Comp. Phys. 57 (1984) 65-92.

[6] Hackbusch, W., "Bemerkungen zur iterierten Defektkorrektur und zu ihrer Kombination mit Mehrgitterverfahren." Rev. Roum. Math. Pures Appl. 26 (1981) 1319-1329.

[7] Harten, A., Lax, P.D. & Van Leer, B., "On upstream differencing and Godunov-type schemes for hyperbolic conservation laws." SIAM Review 25 (1983) 35-61.

[8] Hemker, P.W., "Mixed defect correction iteration for the solution of a singular perturbation problem." Comp. Suppl. 5 (1984) 123-145.

[9] Hemker, P.W. & Spekreijse, S.P., "Multigrid solution of the Steady Euler Equations." In: Advances in Multi-Grid Methods (D.Braess, W.Hackbusch and U.Trottenberg eds) Proceedings Oberwolfach Meeting, Dec. 1984, Notes on Numerical Fluid Dynamics, Vol.11, Vieweg, Braunschweig, 1985.

[10] Hemker, P.W. & Spekreijse, S.P., "Multiple Grid and Osher's Scheme for the Efficient Solution of the the Steady Euler Equations." (Report NM-8507, CWI, Amsterdam, 1985.) To appear in Applied Numerical Mathematics.

[11] Jameson, A., "Numerical Solution of the Euler Equations for Compressible Inviscid Fluids." In: Procs 6th International Conference on Computational Methods in Applied Science and Engineering, Versailles, France, Dec. 1983.

[12] Jespersen, D.C. "Design and implementation of a multigrid code for the steady Euler equations." Appl. Math. and Computat. 13 (1983) 357-374.

[13] Jespersen, D.C. "Recent developments in multigrid methods for the steady Euler equations." Lecture Notes, March 12-16, 1984, von Karman Inst., Rhode-St.Genese, Belgium.

[14] G.M. Johnson, "Multiple grid convergence acceleration of viscous and inviscid flow computations." Appl. Math. and Computat. 13 (1983) 375-398.

[15] Lax, P.D., "Hyperbolic systems of conservation laws and the mathematical theory of shock waves." Regional conference series in applied mathematics 11. SIAM Publication, 1973

[16] Lax, P.D., "Shock waves and entropy" In: Contributions to Nonlinear Functional Analysis (E.H. Zarantonello ed.) Acad. Press, New York, 1971.

[17] Mulder, W.A. "Multigrid Relaxation for the Euler equations." To appear in: J. Comp. Phys. 1985.

[18] Ni Ron-Ho , "A multiple grid scheme for solving the Euler equations." AIAA Journal 20 (1982) 1565-1571.

[19] Osher, S. "Numerical solution of singular perturbation problems and hyperbolic systems of conservation laws", In: Analytical and Numerical Approaches to Asymptotic problems in Analysis, O. Axelsson, L.S. Frank and A. van der Sluis eds.), North Holland Publ. Comp., 1981.

[20] Osher, S. & Chakravarthy, S., "High resolution schemes and the entropy condition." SIAM J. Numer. Anal. 21 (1984) 955-984.

[21] Osher, S & Solomon, F., "Upwind difference schemes for hyperbolic systems of conservation laws." Math. Comp. 38 (1982) 339-374.

[22] Schmidt, W. and Jameson, A., "Euler solvers as an analysis tool for aircraft aerodynamics." In: Advances in Computational Transonics (W.G.Habashi (ed.) Pineridge Press, Swansea.

[23] Spekreijse, S. "Second order accurate upwind solutions of the 2D steady state Euler equations by the use of a defect correction method." CWI-report, in preparation, 1985.

[24] Steger, J.L., "A preliminary study of relaxation methods for the inviscid conservative gasdynamics equations using flux splitting." Nasa Contractor Report 3415 (1981).

[25] Steger, J.L. & Warming, R.F., "Flux vector splitting of the inviscid gasdynamics equations with applications to finite difference methods." J. Comp. Phys. 40 (1981) 263-293.

[26] Sweby, P.K. "High resolution schemes using flux limiters for hyperbolic conservation laws", SIAM J.Numer.Anal. 21 (1984) 995-1011.

[27] Van Leer, B., "Flux-vector splitting for the Euler equations." In: Procs. 8th Intern. Conf. on numerical methods in fluid dynamics, Aachen, June, 1982. Lecture Notes in Physics 170, Springer Verlag.

[28] Van Leer, B., "Upwind difference methods for aerodynamic problems governed by the Euler equations" Report 84-23, Dept. Math. & Inf., Delft Univ. Techn., 1984.

[29] Van Leer, B. and Mulder, W.A., "Relaxation methods for hyperbolic conservation laws." In: Dynamics of Gas in a rotating Galaxy (W.A. Mulder, thesis, Leiden Univ., 1985).

[30] de Zeeuw, P.M. and van Asselt, E.J., "The convergence rate of multi-level algorithms applied to the convection diffusion equation." SIAM J.S.S.C. 6 (1985) 492-503.

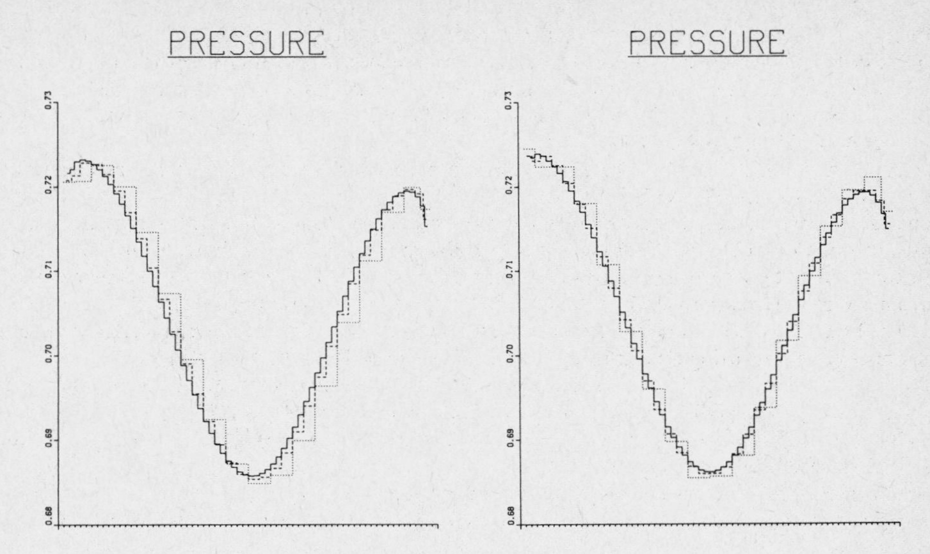

Figure 5.2.a Problem 1. The 1st order solution.

Figure 5.2.b Problem 1.
The 2nd order solution [(3.4) with $\kappa = -1$], dcps = 0.

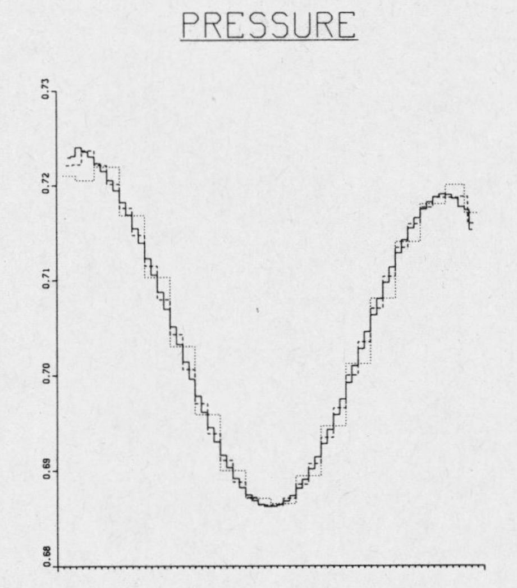

Figure 5.2.c Problem 1. The 2nd order solution [(3.4) with $\kappa = -1$], dcps = 1.

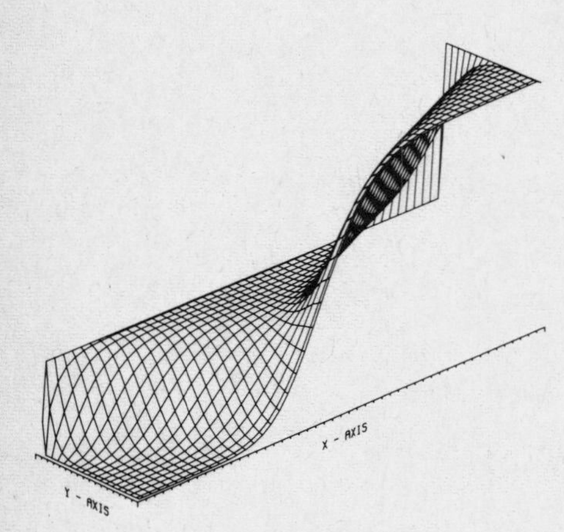

Figure 5.3.a Problem 2. The 1st order solution.

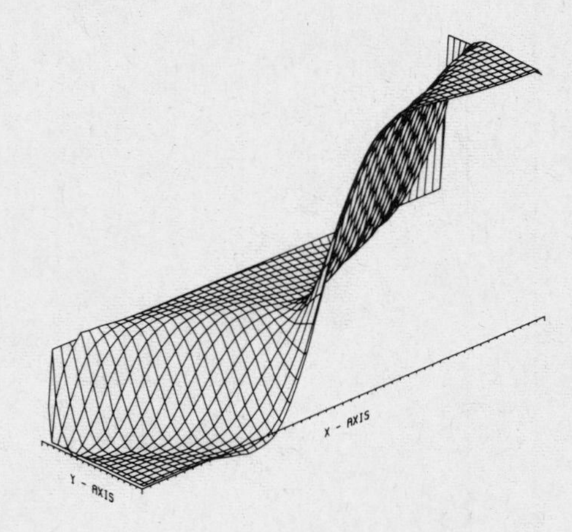

Figure 5.3.b Problem 2.
The 2nd order solution after Richardson extrapolation.

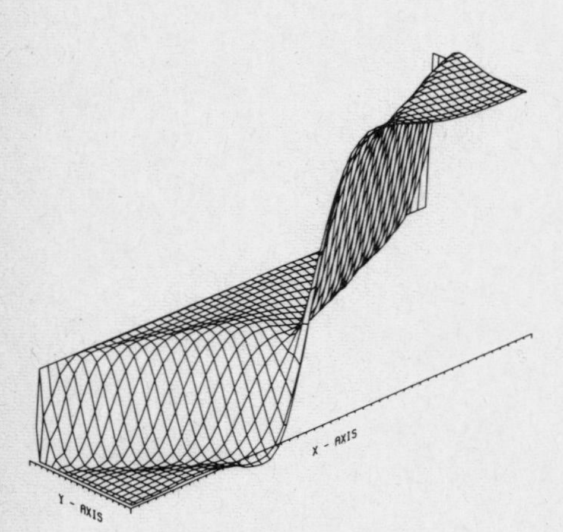

Figure 5.3.c Problem 2. After 1 DCP step

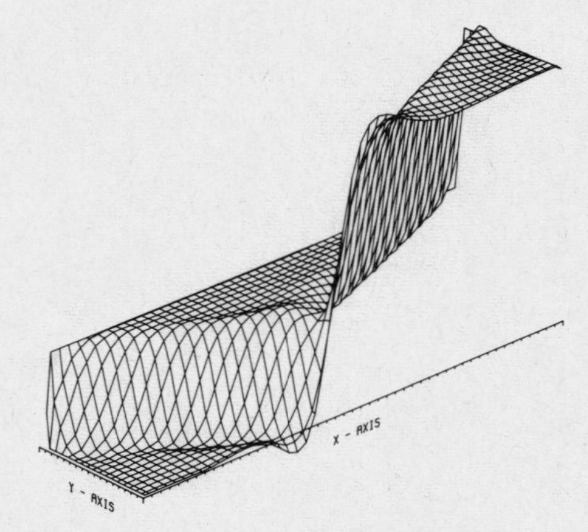

Figure 5.3.d Problem 2. After 3 DCP steps

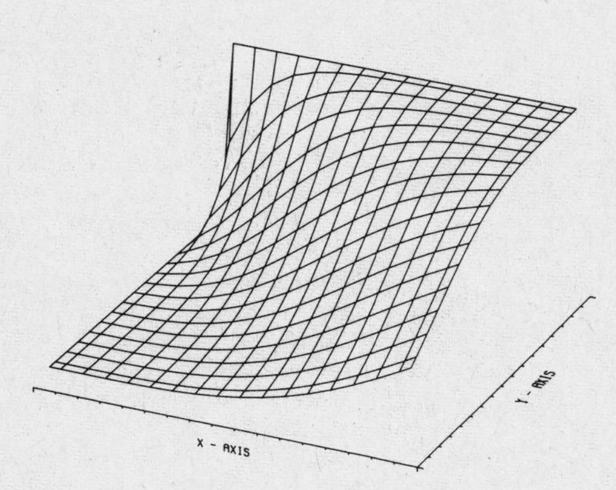

Figure 5.4.a Problem 3. The 1st order solution.

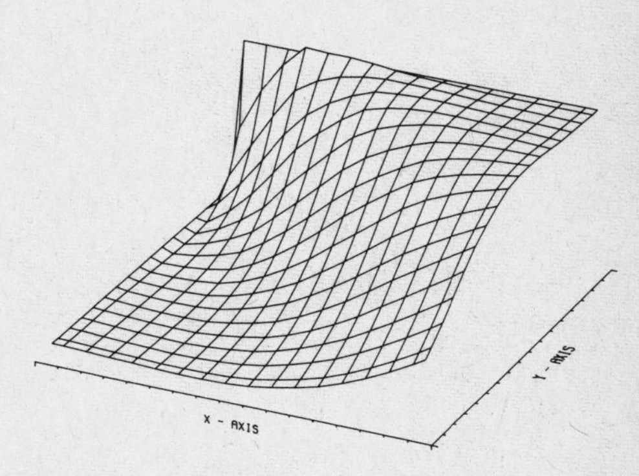

Figure 5.4.b Problem 3.
The 2nd order solution after Richardson extrapolation.

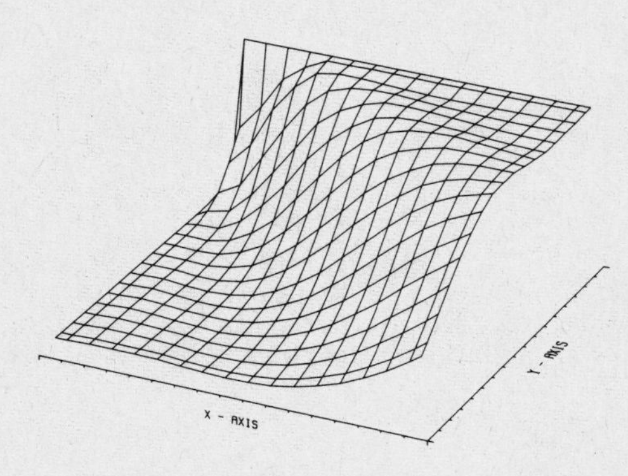

Figure 5.4.c Problem 3. After 1 DCP step

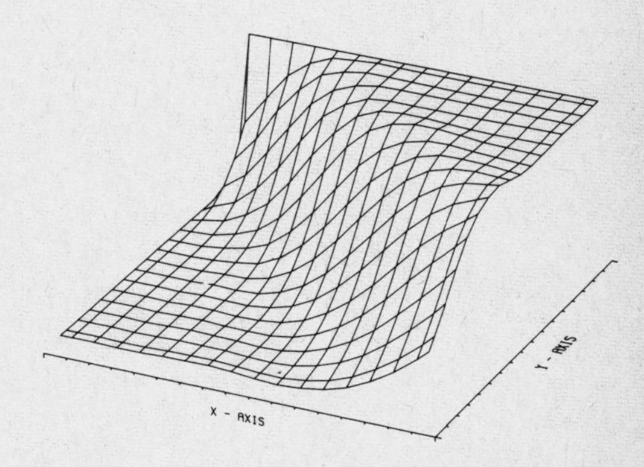

Figure 5.4.d Problem 3. After 3 DCP steps

MULTIGRID ALGORITHMS
FOR COMPRESSIBLE FLOW CALCULATIONS

by

Antony Jameson
Princeton University
Princeton, N.J. 08544

1. Introduction

During the last two decades computational methods have transformed the science of aerodynamics. Following the introduction of panel methods for subsonic flow in the sixties [1,2], and major advances in the simulation of transonic flow by the potential flow approximation in the seventies [3-6], the eighties have seen rapid developments in methods for solving the Euler and Navier Stokes equations [7-11]. Multigrid techniques have penetrated this rapidly burgeoning field, with some notable successes [12-21], but they have not yet advanced to the point of gaining general acceptance, nor is their appliction to non-elliptic problems securely anchored on a firm theoretical foundation. Multigrid time stepping schemes, which are the subject of this paper, have emerged as a promising way to extend multigrid concepts to the treatment of problems governed by hyperbolic equations. The paper falls into two main parts. The first part (Sections 2 and 3) reviews methods of solving the Euler equations of compressible flow, and discusses the trade-offs which underly the choice of schemes for space and time discretization. The second part (Sections 4 and 5) discusses multigrid time stepping schemes, and presents a new method of analyzing the stability of these schemes.

The major considerations in the design of effective methods for the computation of aerodynamic flows are the capability to treat flows over complex geometrical shapes, proper representation of shock waves and contact discontinuities, treatment of viscous effects, and computational efficiency. In practice the viscosity of air is so low that viscous effects are largely confined to thin boundary layers adjacent to the surface. These can only be resolved by the introduction of tightly bunched meshes. At Reynolds numbers typical of full scale flight, of the order of thirty million, the flow in the boundary layer also becomes turbulent, and is unsteady on small scales. The representation of these effects can become prohibitively expensive, and poses a challenge for the future. This paper is restricted to the treatment of inviscid flow. This is already a sufficiently testing problem, and good inviscid methods are needed as a platform for the development of sound viscous methods. The immediate application of the present work is the calculation of steady inviscid transonic flow, and some results are presented for a swept wing. Transonic flow is important because it is the principal operating regime of both commercial and military aircraft. Similar trade-offs, however, apply to a broader class of problems, and the multigrid time stepping technique which I describe should prove useful whenever there is a need to find the steady state solution of a system which is governed by a hyperbolic equation.

The underlying idea is to integrate the time dependent Euler equations until they reach a steady state. In this work the discretization is performed in two stages. The problem is first reduced to a set of ordinary differential equations by subdividing the domain into polygonal or polyhedral cells, and writing the conservation laws in integral form for each cell. The resulting semi-discrete model can then be integrated in time by a variety of discrete time stepping schemes, either implicit or explicit. If one assumes that the optimal time step increases with the space interval, then one can anticipate a faster rate of convergence on a coarser grid. This motivates the concept of time stepping on multiple grids. The cells of the fine mesh can be amalgmated into larger cells which form a coarser mesh. In each coarse mesh cell the conservation laws are then represented by summing the flux balances of its constituent fine mesh cells, with the result that the evolution on the coarse mesh is driven by the disequilibrium of the fine mesh equations. Finally the corrections on the coarse mesh are interpolated back to the fine mesh. This process can be repeated through a sequence of meshes in each of which the mesh spacing is doubled from the previous mesh. If the time step is also doubled each time the process passes to a coarser mesh, then a four level multigrid cycle consisting of one step on each mesh represents a total advance

$$\Delta t + 2\Delta t + 4\Delta t + 8\Delta t = 15\Delta t,$$

where Δt is the step on the fine mesh.

The potential for acceleration through the use of large time steps on the coarse grids is apparent. Its realization in practice is not so easy, and is contingent on ensuring the stability of the composite process, and preventing too much attrition of the convergence rate from the errors introduced by interpolation from coarser to finer grids. The results presented in Section 6, however, indeed confirm the power of the method. In fact, three dimensional solutions of the Euler equations can be obtained in 10-20 multigrid cycles.

2. Space Discretization of the Euler Equations

Let p, ρ, u, v, w, E and H denote the pressure, density, Cartesian velocity components, total energy and total enthalpy. For a perfect gas

$$E = \frac{p}{(\gamma-1)\rho} + \frac{1}{2}(u^2 + v^2 + w^2) \quad , \quad H = E + p/\rho$$

where γ is the ratio of specific heats. The Euler equations for flow of a crompressible inviscid fluid can be written in integral form as

$$\frac{\partial}{\partial t} \iiint_{\Omega} w d\Omega + \iint_{\delta\Omega} \underline{F} \cdot d\underline{S} = 0 \tag{2.1}$$

for a domain Ω with boundary $\delta\Omega$ and directed surface elements $d\underline{S}$. Here w represents the conserved quantity and $\underline{F}$ is the corresponding flux. For mass conservation

$$w = \rho \ , \ \underline{F} = (\rho u, \rho v, \rho w)$$

For conservation of momentum in the x direction

$$w = \rho u, \ \underline{F} = (\rho u^2 + p \ , \ \rho uv, \ \rho uw)$$

with similar definitions for the y and z directions, and for energy conservation

$$w = \rho E, \ \underline{F} = (\rho Hu, \ \rho Hv, \ \rho Hw)$$

If we divide the domain into a large number of small subdomains, we can use equation (2.1) to estimate the average rate of change of w in each subdomain. This is an effective method to obtain discrete approximations to equation (2.1) which preserve its conservation form. In general the subdomains could be arbitrary, but it is convenient to use either distorted cubic or tetrahedral cells. Alternative discretizations may be obtained by storing sample values of the flow variables at either the cell centers or the cell corners. These variations are illustrated in Figure 1 for a two dimensional case.

Figures 1(a) and 1(b) show cell centered schemes on rectilinear and triangular meshes [7,18]. In either case equation (1) is written for the cell labelled 0 as

$$\frac{d}{dt}(Vw) + Q = 0 \tag{2.2}$$

where V is the cell volume and Q is the net flux out of the cell. This can be approximated as

$$Q = \sum_k \underline{F}_{0k} \cdot \underline{S}_{0k} \tag{2.3}$$

where the sum is over the faces of cell 0, $\underline{S}_{0k}$ is the directed area of the face separating cell 0 from cell k, and the flux $\underline{F}_{0k}$ is evaluated by taking the average of its value in cell 0 and cell k.

$$\underline{F}_{0k} = \frac{1}{2}(\underline{F}_0 + \underline{F}_k) \tag{2.4}$$

An alternative averaging procedure is to multiply the average value of the convected quantity, ρ_{0k} in the case of the continuity equation, for example, by the transport vector

$$Q_{0k} = \frac{1}{2}\left(\underline{q}_0 + \underline{q}_k\right) \cdot \underline{S}_{0k} \qquad (2.4^*)$$

obtained by taking the inner product of the mean of the velocity vector $\underline{q}$ with the directed face area.

Figures 1(c) and 1(d) show corresponding schemes on rectilinear and triangular meshes in which the flow variables are stored at the vertices. We can now form a control volume for each vertex by taking the union of the cells meeting at that vertex. Equation (2.1) then takes the form

$$\frac{d}{dt}\left(\sum_k V_k\right) w + \sum_k Q_k = 0 \qquad (2.5)$$

where V_k and Q_k are the cell volume and flux balance for the kth cell in the control volume. The flux balance for a given cell is now approximated as

$$Q = \sum_\ell \underline{F}_\ell \cdot \underline{S}_\ell \qquad (2.6)$$

where $\underline{S}_\ell$ is the directed area of the ℓth face, and $\underline{F}_\ell$ is an estimate of the mean flux vector across that face. Fluxes across internal faces cancel when the sum $\sum_k Q_k$ is taken in equation (2.5), so that only the external faces of the control volume contribute to its flux balance. The flux balance at each vertex can be evaluated directly by summing either the flux balances of its constituent cells or the contributions of its faces. Alternatively the flux balance at every vertex can be evaluated indirectly by a loop over the faces. In this case the flux $\underline{F} \cdot \underline{S}$ across a given face is accumulated into the flux balance of those control volumes which contain that face as an external face. On a tetrahedral mesh each face is shared by exactly two control volumes (centered at the outer vertices of the two tetrahedra which have that face as a common base, as illustrated in Figure 2), and this is a very efficient procedure [11].

In the two dimensional case the mean flux across an edge can be conveniently approximated as the average of the values at its two end points,

$$\overline{\underline{F}}_{12} = \frac{1}{2}\left(\underline{F}_1 + \underline{F}_2\right)$$

in Figure 1(c) or 1(d), for example. The sum ΣQ_k in equation (2.5), which then amounts to a trapezoidal integration rule around the boundary of the control area, should remain fairly accurate even when the mesh is irregular. This is an advantage of the vertex formulation over the cell centered formulation, in which the midpoint of the line joining the sample values does not necessarily coincide with the midpoint of the corresponding edge, with a consequent reduction of accuracy on a distorted or kinked mesh (see Figure 3).

Storage of the solution at the vertices has a similar advantage when a tetrahedral mesh is used in a three dimensional calculation. The use of a simple average of the three corner values of each triangular face

$$\overline{\underline{F}} = \frac{1}{3}\left(\underline{F}_1 + \underline{F}_2 + \underline{F}_3\right)$$

is a natural choice, which is consistent with the assumption that $\underline{F}$ varies linearly over the face. It can be shown that the resulting scheme is essentially equivalent to the use of a Galerkin method with piecewise linear basis fuctions [11]. If a rectilinear mesh is used in a three dimensional problem, the difficulty arises that the four corners of a face are not necessarily coplanar. The use of a simple average of the four corner values is a natural choice on a smooth mesh, but in order to preserve accuracy on an irregular mesh it may be necessary to use more complex integration formulas based on local mappings.

The approximations (2.2) or (2.5) need to be augmented by artificial dissipative terms for two reasons. First there is the possibility of undamped oscillatory modes. For example, when either a cell centered or a vertex formulation is used to represent a conservation law on a rectilinear mesh, a mode with values ± 1 alternately at odd and even points leads to a numerically evaluated flux balance of zero in every interior control volume. Although the boundary conditions may suppress such a mode in the steady state solution, the absence of damping at interior points may have an adverse effect on the rate of convergence to the steady state.

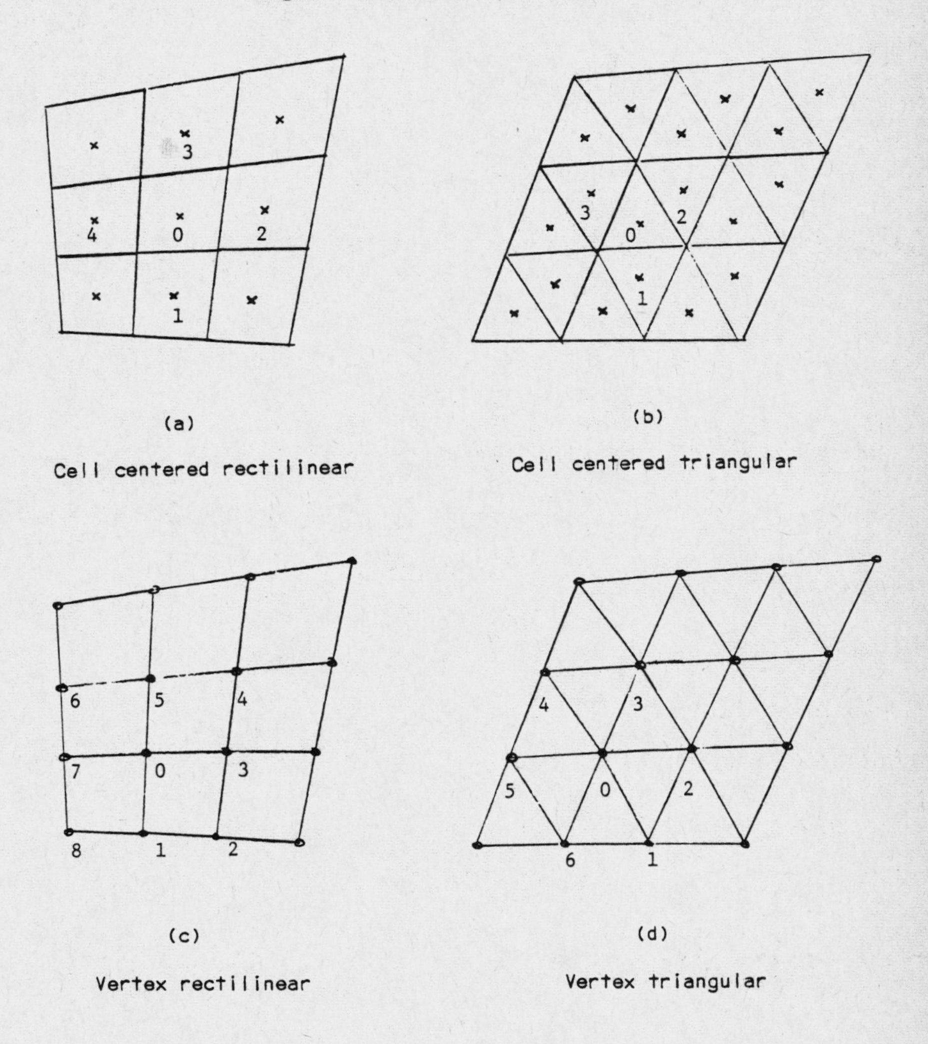

Figure 1

Alternative discretization schemes

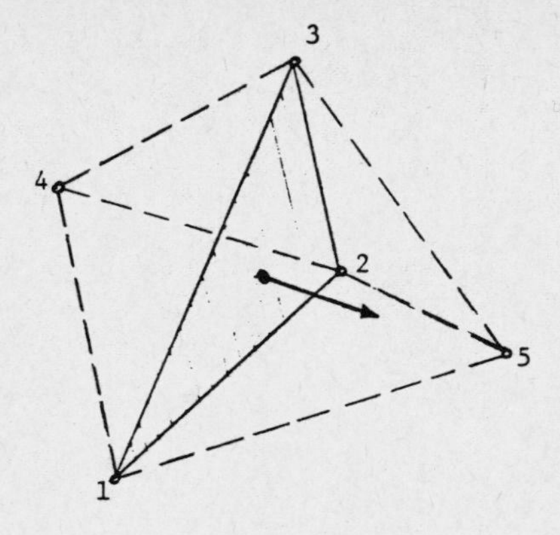

Figure 2

Vertex scheme on a tetrahedral mesh.
Flux through face 1 2 3 influences
balance in control volumes centered at 4 and 5.

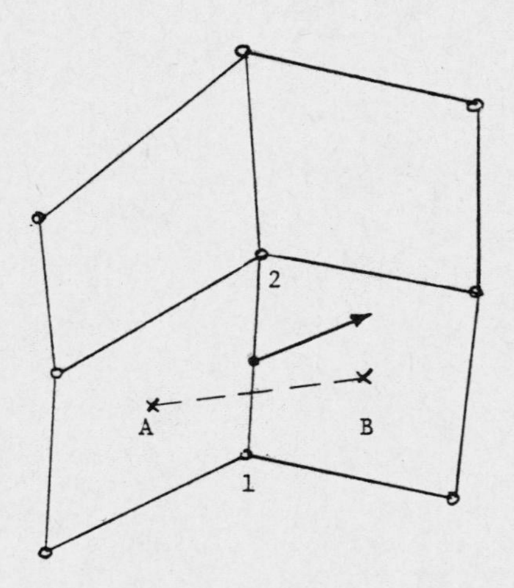

Figure 3

Comparison of discretization schemes on a kinked mesh.
Evaluation of $\underline{F}$ at P by averaging $\underline{F_1}$ and $\underline{F_2}$
is more accurate than averaging $\underline{F_A}$ and $\underline{F_B}$.

The second reason for introducing dissipative terms is to allow the clean capture of shock waves and contact discontinuities without undesirable oscillations. Following the pioneering work of Godunov [22], a variety of dissipative and upwind schemes designed to have good shock capturing properties have been developed during the past decade [23-33]. The one dimensional scalar conservation law

$$\frac{\partial u}{\partial t} + \frac{\partial}{\partial x} f(u) = 0 \tag{2.7}$$

provides a useful model for the analysis of these schemes. The total variation

$$TV = \int_{-\infty}^{\infty} \left| \frac{\partial u}{\partial x} \right| dx$$

of a solution of (2.7) does not increase, provided that any discontinuity appearing in the solution satisfies an entropy condition [34]. The concept of total variation diminishing (TVD) difference schemes, introduced by Harten [28], provides a unifying framework for the study of shock capturing methods. These are schemes with the property that the total variation of the discrete solution

$$TV = \sum_{-\infty}^{\infty} |v_j - v_{j-1}|$$

cannot increase. The general conditions for a multipoint one dimensional scheme to be TVD have been stated and proved by Jameson and Lax [35]. For a semi-discrete scheme expresses in the form

$$\frac{d}{dt} v_j = \sum_{j=-Q}^{Q-1} c_q(j)(v_{j-q} - v_{j-q-1}) \tag{2.8}$$

these conditions are

$$c_{-1}(j-1) \geq c_{-2}(j-2) \ldots \geq c_{-j}(j-Q) \geq 0 \tag{2.9a}$$

and

$$-c_0(j) \geq -c_1(j+1) \ldots \geq -c_{Q-1}(j+Q-1) \geq 0 \tag{2.9b}$$

Specialized to a three point scheme these conditions imply the scheme

$$\frac{d}{dt} v_j = c_{j+1/2}(v_{j+1} - v_j) - c_{j-1/2}(v_j - v_{j-1})$$

is TVD if $c_{j+1/2} \geq 0$, $c_{j-1/2} \geq 0$.

A conservative semi-discrete approximation to equation (2.7) can be derived by subdividing the line into cells. Then the evolution of the value v_j in the jth cell is given by

$$\Delta x \frac{d}{dt} v_j + h_{j+1/2} - h_{j-1/2} = 0 \tag{2.10}$$

where $h_{j+1/2}$ is the estimate of the flux between cells j and j + 1. Conditions (2.9) are satisfied by the upwind scheme

$$h_{j+1/2} = \begin{cases} f(v_j) & \text{if } a_{j+1/2} \geq 0 \\ f(v_{j+1}) & \text{if } a_{j+1/2} < 0 \end{cases} \tag{2.11}$$

where $a_{j+1/2}$ is a numerical estimate of the wave speed $a = \partial f/\partial u$,

$$a_{j+1/2} = \begin{cases} \dfrac{f_{j+1} - f_j}{v_{j+1} - v_j} & \text{if } v_{j+1} \neq v_j \\ \dfrac{\partial f}{\partial v}\Big|_{v=v_j} & \text{if } v_{j+1} = v_j \end{cases} \tag{2.12}$$

More generally, if one sets

$$h_{j+1/2} = \frac{1}{2}(f_{j+1} + f_j) + \alpha_{j+1/2}(v_{j+1} - v_j) \tag{2.13}$$

where $\alpha_{j+1/2}$ is a dissipative coefficient, the scheme is TVD if

$$\alpha_{j+1/2} \geq \frac{1}{2}|\ a_{j+1/2}\ | \tag{2.14}$$

since one can write

$$h_{j+1/2} = f_j + \frac{1}{2}(f_{j+1} - f_j) - \alpha_{j+1/2}(v_{j+1} - v_j)$$

$$= f_j + (\frac{1}{2} a_{j+1/2} - \alpha_{j+1/2})(v_{j+1} - v_j)$$

and

$$h_{j-1/2} = f_j - \frac{1}{2}(f_j - f_{j-1}) - \alpha_{j-1/2}(v_j - v_{j-1})$$

$$= f_j - (\frac{1}{2} a_{j-1/2} + \alpha_{j-1/2})(v_j - v_{j-1})$$

Thus TVD schemes in general require the use of a dissipative coefficient with a magnitude of at least half the wave speed, while the minimum sufficient value

$$\alpha_{j+1/2} = \frac{1}{2}|\ a_{j+1/2}\ |$$

produces the upwind scheme.

TVD schemes preserve the monotonicity of an initially monotone profile, because the total variation would increase if the profile ceased to be monotone. Consequently they prevent the formation of spurious oscillations. In this simple form, however, they are at best first order acccurate. Higher order schemes can be constructed by using multipoint extrapolation formulas to estimate $h_{j+1/2}$, or by adding higher order dissipative terms. In either case flux limiters are then needed to control the signs of the coefficients $c_q(j)$ [28-33].
A convenient way of applying these ideas to a system of equations was proposed by Roe [26]. Let $A_{j+1/2}$ be a matrix with the property that the flux difference satisfies the relation

$$f(w_{j+1}) - f(w_j) = A_{j+1/2}(w_{j+1} - w_j) \tag{2.15}$$

Roe gives a method of constructing of such a matrix, which is a numerical approximation to the Jacobian matrix $\partial f/\partial w$. Its eigenvalues λ_ℓ are thus numerical estimates of the wave speeds associated with the system. Now decompose the difference $w_{j+1} - w_j$ as a sum of the eigenvectors r_ℓ of $A_{j+1/2}$,

$$w_{j+1} - w_j = \Sigma\ \alpha_\ell\ r_\ell \tag{2.16}$$

Then

$$f_{j+1} - f_j = \Sigma\ \lambda_\ell\ \alpha_\ell\ r_\ell \tag{2.17}$$

and the desired dissipative term can be constructed as

$$\Sigma\ \mu_\ell\ \alpha_\ell\ r_\ell \tag{2.18}$$

where

$$\mu_\ell \geq \frac{1}{2}|\ \lambda_\ell\ |$$

This method amounts to constructing separate dissipative terms for the characteristic variables defined by the eigenvectors of $A_{j+1/2}$. It is closely related to the concept of flux splitting first introduced by Steger and Warming [25], in which the flux vector itself is split into components corresponding to the wave speeds, and backward differencing is used for the part propagating forwards, while forward differencing is used for the part propagating backwards. Alternative methods of flux splitting which lead to excellent shock capturing schemes have been proposed by Osher [27] and VanLeer [31].

These concepts can been applied to two and three dimensional problems by separately applying the one dimensional construction in each coordinate direction. There is no theoretical basis for this, but it generally leads to good results in practice. The cell centered finite volume formulation is readily adapted to this kind of construction. A first order upwind scheme can be constructed by splitting the flux across each face into components corresponding to forward and backward propagation, and then evaluating each component by taking values from the cell on the upwind side of the face.

An alternative approach is as follows. Consider a two dimensional scalar conservation law of the form

$$\frac{\partial v}{\partial t} + \frac{\partial}{\partial x} f(v) + \frac{\partial}{\partial y} g(v) = 0 \tag{2.20}$$

The mesh may be either rectilinear or triangular, as sketched in Figure 1. Assume that the evolution equation at the mesh point 0 depends on contributions from the nearest neighbors, numbered as in the figure. Suppose that this is expressed in the form

$$\frac{dv_0}{dt} = \sum_k c_k(v_k - v_0) \tag{2.21}$$

where the sum is over the neighbors. Then we reguire all the coefficients to be nonnegative

$$c_k \geq 0, \quad k = 1,2... \tag{2.22}$$

This condition on the signs of the coefficients, which is a direct generalization of the conditions for a one dimensional three point scheme to be TVD, assures that a maximum cannot increase. Finite volume approximations to equation (2.20) can be reduced to the form (2.21) by making use of the fact that the sums $\sum_k \Delta x$ and $\sum_k \Delta y$ taken around the perimeter of the control area are zero, so that a multiple of $f(v_0)$ or $g(v_0)$ can be subtracted from the flux. Consider, for example, a formulation in which v is stored at the vertices of a rectilinar mesh, as in Figure 1(c). Then equation (2.20) is replaced by

$$S\frac{dv_0}{dt} + \frac{1}{2}\sum_k \{(f_k + f_{k-1})(y_k - y_{k-1}) - (g_k + g_{k-1})(x_k - x_{k-1})\} = 0 \tag{2.23}$$

where k ranges from 1 to 8. This can be rearranged as

$$S\frac{dv_0}{dt} + \sum_k \{f(v_k)\Delta y_k - g(v_k)\Delta x_k\} = 0$$

where

$$\Delta x_k = \frac{1}{2}(x_{k-1} - x_{k-1}), \quad \Delta y_k = \frac{1}{2}(y_{k+1} - y_{k-1}),$$

and this is equivalent to

$$S\frac{dv_0}{dt} + \sum_k \{(f(v_k) - f(v_0))\Delta y_k + (g(v_k) - g(v_0))\Delta x_k\} = 0 \tag{2.24}$$

Define the coefficient a_{k0} as

$$a_{k0} = \begin{cases} \dfrac{(f_k-f_0)\Delta y_k - (g_k-g_0)\Delta x_k}{v_k - v_0}, & v_k \neq v_0 \\ \left(\dfrac{\partial f}{\partial v}\,\Delta y_k - \dfrac{\partial g}{\partial v}\,\Delta x_k\right)\Big|_{v = v_0}, & v_k = v_0 \end{cases} \tag{2.25}$$

Then equation (2.24) reduces to

$$S\,\frac{dv_0}{dt} + \sum_k a_{k0}(v_k - v_0) = 0.$$

To produce a scheme satisfying the sign condition (2.22), add a dissipative term on the right hand side of the form

$$\sum_k \alpha_{k0}\,(v_k - v_0)$$

where the coefficients α_{k0} satisfy the condition

$$\alpha_{k0} \geq |\,a_{k0}\,| \tag{2.26}$$

The definition (2.25) and condition (2.26) correspond to the definition (2.12) and condition (2.14) in the one dimensional case. The extension to a system can be carried out with the aid of Roe's construction. Now a_{k0} is replaced by the corresponding matrix A_{k0} sucht that

$$A_{k0}(w_k - w_0) = (f_k - f_0)\,\Delta y_k - (g_k - g_0)\,\Delta x_k$$

Then $w_k - w_0$ is expanded as a sum of the eigenvectors of A_{k0}, and a contribution to the dissipative term is formed by multiplying each eigenvector by a positive coefficient with a magnitude not less than that of the corresponding eigenvalue.

The use of flux splitting allows precise matching of the dissipative terms to introduce the minimum amount of dissipation needed to prevent oscillations. This in turn reduces the thickness of the numerical shock layer to the minimum attainable, one or two cells for a normal shock. In practice, however, it turns out that shock waves can be quite cleanly captured without flux splitting by using adaptive coefficients. The dissipation then has a low background level which is increased in the neighborhood of shock waves to a peak value proportional to the maximum local wave speed. The second difference of the pressure has been found to be an effective measure for this purpose. The dissipative terms are constructed in a similar manner for each dependent variable by introducing dissipative fluxes which preserve the conservation form. For a three dimensional rectilinear mesh the added terms have the form

$$d_{i+1/2,j,k} - d_{i-1/2,j,k} + d_{i,j+1/2,k} - d_{i,j-1/2,k} + d_{i,j,k+1/2} - d_{i,j,k-1/2} \tag{2.27}$$

These fluxes are constructed by blending first and third differences of the dependent variables. For example, the dissipative flux in the i direction for the mass equation is

$$d_{i+1/2,j,k} = R(\varepsilon^{(2)} - \varepsilon^{(4)}\,\delta_x^2)(\rho_{i+1,j,k} - \rho_{i,j,k}) \tag{2.28}$$

where δ_x^2 is the second difference operator, $\varepsilon^{(2)}$ and $\varepsilon^{(4)}$ are the adaptive coefficients, and R is a scaling factor proportional to an estimate of the maximum local wave speed. For an explicit scheme the local time step limit Δt^* is a measure of the time it takes for the fastest wave to cross a mesh interval, and R can accordingly by made proportional to $1/\Delta t^*$. The coefficient $\varepsilon^{(4)}$ provides the background dissipation in smooth parts of the flow, and can be used to improve the capability of the scheme to damp high frequency modes. Shock capturing is controlled by the coefficient $\varepsilon^{(2)}$, which is made proportional to the normalized second difference of the pressure

$$v_{i,j,k} = \left| \frac{p_{i+1,j,k} - 2p_{i,j,k} + p_{i-1,j,k}}{p_{i+1,j,k} + 2p_{i,j,k} + p_{i-1,j,k}} \right|$$

in the adjacent cells.

Schemes constructed along these lines combine the advantages of simplicity and economy of computation, at the expense of an increase in thickness of the numerical shock layer to three or four cells. They have also proved robust in calculations over a wide range of Mach Numbers (extending up to 20 in recent studies [36]).

3. Time Stepping Schemes

The discretization procedures of Section 2 lead to set of coupled ordinary differential equations, which can be written in the form

$$\frac{dw}{dt} + R(w) = 0 \qquad (3.1)$$

where w is the vector of the flow variables at the mesh points, and R(w) is the vector of the residuals, consisting of the flux balances defined by equations (2.2) or (2.5), together with the added dissipative terms. These are to be integrated to a steady state. Since the unsteady problem is used only as a vehicle for reaching the steady state, alternative iterative schemes might be contemplated. Two possibilities in particular are the least squares method, which has been successfully employed to solve a variety of nonlinear problems by Glowinski and his co-workers [6], and the Newton iteration, which has recently been used to solve the two dimensional Euler equations by Giles [37]. The strategy of the present work, however, is to rely upon the simplest possible method, and to attempt to obtain a fast rate of convergence by the of multiple grids.

Since the objective is simply to reach the steady state and details of the transient solution are immaterial, the time stepping scheme may be designed solely to maximize the rate of convergence without having to meet any constraints imposed by the need to achieve a specified level of accuracy, provided that it does not interfere with the definition of the residual R(w). Figure 4 indicates some of the principal time stepping schemes which might be considered. The first major choice is whether to use an explicit or an implicit scheme.

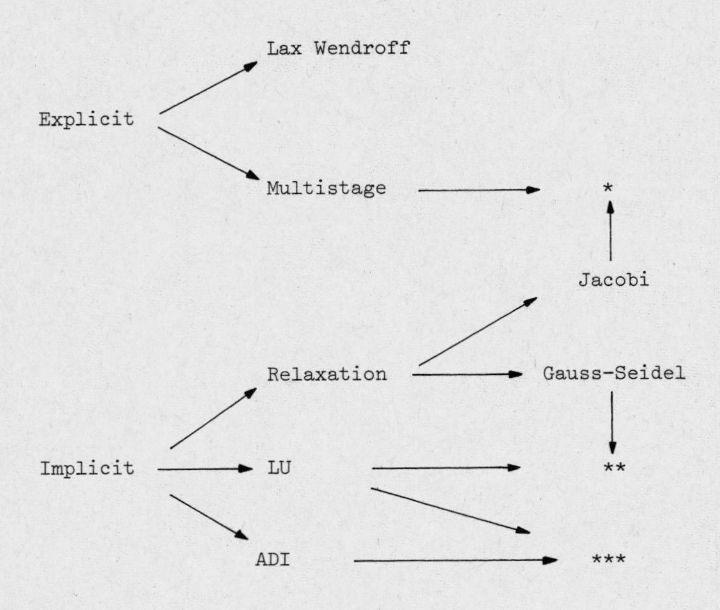

Figure 4

Time stepping schemes

* facilitates vector and parallel processing

A number of popular explicit schemes are based on the Lax Wendroff formulation, in which the change δw during a time step is calculated from the first two terms of the Taylor series

$$\delta w = \Delta t \frac{\partial w}{\partial t} + \frac{\Delta t^2}{2} \frac{\partial^2 w}{\partial t^2}$$

In a two dimensional case for which $R(w)$ approximates $\frac{\partial}{\partial x} f(w) + \frac{\partial}{\partial y} g(w)$, the second derivative is then estimated by substituting

$$\frac{\partial^2 w}{\partial t^2} = - \frac{\partial}{\partial t} R(w) = - \frac{\partial R}{\partial w} \frac{\partial w}{\partial t}$$

$$= (\frac{\partial}{\partial x} A + \frac{\partial}{\partial y} B)\ R(w)$$

where A and B are the Jacobian matrices

$$A = \frac{\partial f}{\partial w}, \quad B = \frac{\partial y}{\partial w}$$

In a variation which has been successfully used to drive multigrid calculations by Ni [16], and Hall [20], the flow variables are stored at the cell vertices. The correction at a vertex is then calculated from the average of the residuals in the four neighbouring cells, augmented by differences in the x and y directions of the residuals multiplied by the Jacobian matrices. Accordingly

$$\delta w = - \{\mu_x \mu_y - \frac{\Delta 1}{2} (\frac{1}{\Delta x} \mu_y \delta_x A + \frac{1}{\Delta y} \mu_x \delta_y B)\}\ \Delta t\ Q(w) \qquad (3.2)$$

where μ and δ denote averaging and difference operators, and $Q(w)$ is the flux balance in one cell, calculated in the manner described in Section 2. Ni views this is a rule for distributing a correction calculated at the center of each cell unequally to its four corners. As it stands, the distribution rule is consistent with a steady state solution $Q(w) = 0$, but it does not damp a high frequency mode with alternate signs at odd and even points. In shock capturing applications it has proved necessary to augment the correction with a dissipative term $\Delta t\ D(w)$. The steady state solution then depends on the Time step Δt because the second term in the expansion is multiplied by Δt^2. Consequently the time stepping procedure cannot be regarded as an iterative scheme independent of the equations to be solved.

The same is true of a number of two step Lax Wendroff schemes in which $\Delta t \frac{\partial w}{\partial t} + \frac{\Delta t^2}{2} \frac{\partial^2 w}{\partial t^2}$ is replaced by an estimate of $\frac{\partial w}{\partial t}$ at the time $t + \frac{\Delta t}{2}$. In the widely used MacCormack scheme, the predictor and corrector steps are

$$(1) \qquad w^* = w^n - \Delta t (D_x^+ f^n + D_y^+ g^n)$$

and

$$(2) \qquad w^{n+1} = w^n - \frac{\Delta t}{2} (D_x^+ f^n + D_y^+ g^n) - \frac{\Delta t}{2} (D_x^- f^* + D_y^- g^*)$$

where D_x^+, D_x^-, D_y^+ and D_y^- are forward and backward difference operators approximating $\partial/\partial x$ and $\partial/\partial y$. Here the use of different approximations for $\partial f/\partial x + \partial g/\partial y$ in the two stages leads to a dependence of the steady state solution on Δt.

If one regards equation (3.1) as a set of ordinary differential equations in which $R(w)$ has a fixed form then the steady state solution is unambiguously $R(w) = 0$. Explicit schemes which might be considered include linear multistep methods such as the leap frog and Adams-Bashforth schemes, and one step multistage methods such as the classical Runge-Kutta schemes. The one step multistage schemes have the advantages that they require no special start up procedure, and that they can readily be tailored to give a desired stability region. They have proved extremely effective in practice as a method of solving the Euler equations [7-8,11].

Let w^n be the result after n steps. The general form of an m stage scheme is

$$\begin{aligned}
w^{(0)} &= w^n \\
w^{(1)} &= w^{(0)} - \alpha_1 \ \Delta t \ R^{(0)} \\
&\dots \\
w^{(m-1)} &= w^{(0)} - \alpha_{m-1} \ \Delta t \ R^{(m-2)} \\
w^{(m)} &= w^{(0)} - \Delta t \ R^{(m-1)} \\
w^{n+1} &= w^{(m)}
\end{aligned} \tag{3.3}$$

The residual in the q+1st stage is evaluated as

$$R^{(q)} = \sum_{r=0}^{q} \beta_{qr} R(w^{(r)}) \tag{3.4}$$

where

$$\sum_{r=0}^{q} \beta_{qr} = 1$$

In the simplest case

$$R^{(q)} = R(w^{(q)})$$

It is then known how to choose the coefficients α_q to maximize the stability interval along the imaginary axis, and consequently the time step [38]. Since only the steady state solution is needed, it pays, however, to separate the residual R(w) into its convective and dissipative parts Q(w) and D(w). Then the residual in the (q+1)st stage is evaluated as

$$R^{(q)} = \sum_{r=0}^{q} \{\beta_{qr} Q(w^{(r)}) - \gamma_{qr} D(w^{(r)})\} \tag{3.4*}$$

where

$$\sum_{r=0}^{q} \beta_{qr} = 1 \ , \quad \sum_{r=0}^{q} \gamma_{qr} = 1$$

Blended multi-stage schemes of this type, which have been analyzed in reference 39, can be tailored to give large stability intervals along both the imaginary and negative real axes.

Instead of recalculating the residuals at each stage it is possible to estimate the change in the residual by the first term of a Taylor expansion

$$R(w + \delta w) = R(w) + \frac{\partial R}{\partial w} \delta w + \dots$$

If R(w) approximates $\frac{\partial}{\partial x} f(w) + \frac{\partial}{\partial y} g(w)$ and the dissipative terms are frozen, this leads to the following reformulation of the multi-stage scheme:

$$\begin{aligned}
\delta w^{(1)} &= -\alpha_1 \ \Delta t \ R(w^{(0)}) \\
\delta w^{(2)} &= -\alpha_2 \ \Delta t (R(w^{(0)}) + (D_x A + D_y B) \delta w^{(1)}) \\
\delta w^{(3)} &= -\alpha_3 \ \Delta t (R(w^{(0)}) + (D_x A + D_y B) \delta w^{(2)}) \\
&\dots
\end{aligned}$$

This formulation in which the current estimate of the correction is repeatedly multiplied by the Jacobians to produce an improved estimate, resembles the Lax Wendroff scheme. It differs from the Lax Wendroff scheme, however, in applying the difference operator $D_x A + D_y B$ to the actual δw_j calculated at the mesh points x_j in the previous stage. Accordingly it has the stability properties of the multi-stage scheme.

The properties of multi-stage schemes can be further enhanced by residual averaging [8]. Here the residual at a mesh point is replaced by a weighted average of neighboring residuals. The average is calculated implicitly. In a one dimensional case $R(w)$ is replaced by $R(\overline{w})$, where at the jth mesh point

$$- \varepsilon \overline{R}_{j-1} + (1+2\varepsilon)\overline{R}_j - \varepsilon \overline{R}_{j+1} = R_j$$

It can easily be shown that the scheme can be stabilized for an arbitrarily large time step by choosing a sufficiently large value for ε. In a nondissipative one dimensional case one needs

$$\varepsilon \geq \frac{1}{4}\left(\left(\frac{\Delta t}{\Delta t^*}\right)^2 - 1\right)$$

where Δt^* is the maximum stable time step of the basic scheme, and Δt is the actual time step. The method can be extended to three dimensions by using smoothing in product form

$$(1 - \varepsilon_x \delta_x^2)(1 - \varepsilon_y \delta_y^2)(1 - \varepsilon_z \delta_z^2)\, \overline{R} = R \tag{3.5}$$

where δ_x^2, δ_y^2 and δ_z^2 are second difference operators in the coordinate directions, and ε_x, ε_y and ε_z are the corresponding smoothing coefficients. Residual averaging can also be used on triangular meshes [11,19]. The implicit equations are then solved by a Jacobi iteration.

One can anticipate that implicit schemes will yield convergence in a smaller number of time steps, since the time step is no longer constrained by a stability limit. This will only pay, however, if the decrease in the number of time steps outweighs the increase in the computational effort per time step consequent upon the need to solve coupled equations. The prototype implicit scheme can be formulated by estimating $\partial w/\partial t$ at $t + \mu\Delta t$ as a linear combination of $R(w^n)$ and $R(w^{n+1})$. The resulting equation

$$w^{n+1} = w^n - \Delta t\{(1-\mu)\, R(w^n) + \mu R(w^{n+1})\} \tag{3.6}$$

can be linearized as

$$(I + \mu\Delta t \frac{\partial R}{\partial w})\,\delta w + \Delta t\, R(w^n) = 0 \tag{3.7}$$

Equation (3.7) reduces to the Newton iteration of one sets $\mu = 1$ and lets $\Delta t \to \infty$. In a three dimensional case with an N×N×N mesh its bandwidth is of order N2. Direct inversion requires a number of operations proportional to the number of unknowns multiplied by the square of the bandwidth, that is O(N7). This is prohibitive, and forces recourse to either an approximate factorization method or an iterative solution method.

The main possibilities for approximate factorization are the alternating direction and LU decompositon methods. The alternating direction method, which may be traced back to the work of Gourlay and Mitchell [40], was given an elegant formulation for nonlinear problems by Beam and Warming [41]. In a two dimensional case equation (3.7) is replaced by

$$(I + \mu\Delta t D_x A)(I + \mu\Delta t D_y B)\,\delta w + \Delta t\, R(w) = 0$$

where D_x and D_y are difference operators approximating $\partial/\partial x$ and $\partial/\partial y$, and A and B are the Jacobian matrices. This may be solved in two steps:

$$(1) \quad (I + \mu\Delta t D_x A)\delta w^* = - \Delta t\ R(w)$$

$$(2) \quad (I + \mu\Delta t D_y B)\delta w^*$$

Each step requires block tridiagonal inversions, and may be performed in $O(N^2)$ operations on an $N \times N$ mesh. The algorithm is amenable to vectorization by simultaneous solution of the tridiagonal equations along parallel coordinate lines. The method has been refined to a high level of efficiency by Pulliam and Steger [9], and Yee has extended it to incorporate a TVD scheme [32]. Its main disadvantage is that its extension to three dimensions is inherently unstable according a Von Neumann analysis.

The idea of the LU decompostion method [42] is to replace the operator in equation (3.3) by the product of lower and upper block triangular factors L and U,

$$LU\ \delta w + \Delta t\ R(w) = 0 \tag{3.9}$$

Two factors are used independent of the number of dimensions, and the inversion of each can be accomplished by inversion of its diagonal blocks. The method can be conveniently illustrated by considering a one dimensional example. Let the Jacobian matrix $A = \partial f/\partial w$ be split as

$$A = A^+ + A^-$$

where the eigenvalues of A^+ and A^- are positive and negative respectively. Then we can take

$$L \equiv I + \mu\Delta t\ D_x^- A^+ \ ,\ U \equiv I + \mu\Delta t\ D_x^- A^t \tag{3.10}$$

where D_x^+ and D_x^- denote forward and backward difference operators approximating ∂/∂_x. The reason for splitting A is to ensure the diagonal dominance of L and U, independent of Δt. Otherwise stable inversion of both factors will only be possible for a limited range of Δt. A crude choice is

$$A^\pm = \frac{1}{2}\ (A \pm \rho I)$$

where ρ is at least equal to the spectral radius of A. If flux splitting is used in the calculation of the residual, it is natural to use the corresponding splitting for L and U. An interesting variation is to combine an alternating direction scheme with LU decomposition in the different coordinate directions [43,44].

If one chooses to adopt the iterative solution technique, the principal alternatives are variants of the Gauss-Seidel and Jacobi methods. These may be applied to either the nonlinear equation (3.6) or the linearized equation (3.7). A Jacobi method of solving (3.6) can be formulated by regarding it as an equation

$$w - w^{(0)} + \mu\Delta t\ R(w) + (1-\mu)\ \Delta t\ R(w^{(0)}) = 0$$

to be solved for w. Here $w^{(0)}$ is a fixed value obtained as the result of the previous time step. Now using bracketed superscripts to denote the iterations, we have

$$w^{(0)} = w^n$$

$$w^{(1)} = w^{(0)} + \Delta t\ R(w^{(0)})$$

and for $k > 1$

$$w^{(k+1)} = w^{(k)} + \sigma_{k+1}\{(w^{(k)} - w^{(0)} + \mu\Delta t\ R(w^{(k)}) + (1-\mu)\Delta t\ R(w^{(0)})\}$$

where the parameters σ_{k+1} can be chosen to optimize convergence. Finally, if we stop after m iterations,

$$w^{n+1} = w^{(m)}$$

We can express $w^{(k+1)}$ as

$$w^{(k+1)} = w^{(0)} + (1 + \sigma_{k+1})(w^{(k)} - w^{(0)}) + \sigma_{k+1}\{(\mu\Delta t\ R(w^{(k)}) + (1-\mu)\Delta t\ R(w^{(0)})\}$$

Since

$$w^{(1)} - w^{(0)} = \sigma_1\ \Delta t\ R(w^{(0)})$$

it follows that for all k we can express $w^{(k)} - w^{(0)}$ as a linear combination of $R(w^{(j)})$, $j < k$. Thus this scheme is a variant of the multi-stage time stepping scheme described by equations (3.3) and (3.4). It has the advantage that it permits simultaneous or overlapped calculation of the corrections at every mesh point, and is readily amenable to parallel and vector processing.

A symmetric Gauss-Seidel scheme has been successfully employed in several recent works [10,21,45]. Consider the case of a flux split scheme in one dimension, for which

$$R(w) = D_x^+ f^-(w) + D_x^- f^+(w)$$

where the flux is split so that the Jacobian matrices

$$A^+ = \frac{\partial f^+}{\partial w} \text{ and } A^- = \frac{\partial f^-}{\partial w}$$

have positive and negative eigenvalues respectively. Now equation (3.7) becomes

$$\{I + \mu\Delta t\ (D_x^+ A^- + D_x^- A^+)\}\ \delta w + \Delta t\ R(w) = 0.$$

At the jth mesh point this is

$$\{I + \alpha(A_j^+ - A_j^-)\}\ \delta w_j + \alpha\ A_{j+1}^-\ \delta w_{j+1} - \alpha\ A_{j-1}^+\ \delta w_{j-1} + \Delta t R_j = 0$$

where

$$\alpha = \mu\frac{\Delta t}{\Delta x}$$

Set $\delta w_j^{(0)} = 0$. A two sweep symmetric Gauss-Seidel scheme is then

(1) $\{I + \alpha(A_j^+ - A_j^-)\}\ \delta w_j^{(1)} - \alpha\ A_{j-1}^+\ \delta w_{j-1}^{(1)} + \Delta t R_j = 0$

(2) $\{I + \alpha(A_j^+ - A_j^-)\}\ \delta w_j^{(2)} + \alpha\ A_{j+1}^-\ \delta w_{j+1}^{(2)} - \alpha\ A_{j-1}^+\ \delta w_{j-1}^{(2)} + \Delta t R_j = 0$

Subtracting (1) from (2) we find that

$$\{I + \alpha(A_j^+ - A_j^-\}\ \delta w_j^{(2)} + \alpha\ A_{j+1}^-\ \delta w_{j+1}^{(2)} = \{I + \alpha(A_j^+ - A_j^-)\}\delta w_j^{(1)}$$

Define the lower triangular, upper triangular and diagonal operators L, U and D as

$$L \equiv I - \alpha\ A^- + \mu t\ D_x^-\ A^+$$

$$U \equiv I + \alpha\ A^+ + \mu t\ D_x^+\ A^-$$

$$D \equiv I + \alpha(A^+ - A^-)$$

It follows that the scheme can be written as

$$L\ D^{-1} U\ \delta w = -\ \Delta t\ R(w)$$

Commonly the iteration is terminated after one double sweep. The scheme is then a variation of an LU implicit scheme.

Some of these interconnections are illustrated in Figure (4). Schemes in three main classes appear to be the most appealing:

1) Varations of multi-stage time stepping, including the application of a Jacobi iterative method to the implicit scheme, (indicated by a single asterisk).

2) Variations of LU decomposition, including the application of a Gauss-Seidel iterative method to the implicit scheme (indicated by a double asterisk).

3) Alternating direction schemes, including schemes in which an LU decomposition is separately used in each coordinate direction (indicated by a triple asterisk).

Schemes of all three classes have been successfully used in conjunction with multigrid techniques [17-19,21,46-48]. The optimal choice may finally depend on the computer architecture. One might anticipate that the Gauss-Seidel method of iteration could yield a faster rate of convergence than a Jacobi method, and it appears to be a particularly natural choice in conjunction with a flux split scheme which yields diagonal dominance. The efficiency of this approach has been confirmed in the recent work of Hemker and Spekreijse [21]. This class of schemes, however, restricts the use of vector or parallel processing. Multistage time stepping, or Jacobi iteration of the implicit scheme, allow maximal use of vector or parallel processing. The alternating direction formulation removes any restriction on the time step (at least in the two dimensional case), while permitting vectorization along coordinate lines. The ADI-LU scheme is an interesting compromise.

Viewed in the broader context of Runge-Kutta methods for solving ordinary differential equations, the coefficients of a multi-stage scheme can be tailored to optimize the stability region without any requirement of diagonal dominance. As has been noted by Hall, multigrid time stepping methods also expand the domain of dependence of the discrete scheme in a way that corresponds to signal propagation of the physical system. This allows a large effective time step to be attained by a multigrid cycle without the need to introduce an implicit time stepping scheme. The results presented in Section 6 confirm that rapid convergence can indeed be obtained by explicit multi-stage methods in conjunction with a multigrid scheme.

4. Multigrid Time Stepping Schemes

The discrete equations (3.1) describe the local evolution of the system in the neighborhood of each mesh point. The underlying idea of a multigrid time stepping scheme is to transfer some of the task of tracking the evolution of the system to a sequence of successively coarser meshes. This has two advantages. First, the computational effort per time step is reduced on a coarser mesh. Second, the use of larger control volumes on the coarser grids tracks the evolution on a larger scale, with the consequence that global equilibrium can be more rapidly attained. In the case of an explicit time stepping scheme, this manifests itself through the possibility of using successively large time steps as the process passes to the coarser grids, without violating the stability bound.

Suppose that successively coarser auxiliary grids are introduced, with the grids numbered from 1 to m, where grid 1 is the original mesh. Then after one or more time steps on grid 1 one passes to grid 2. Again, after one or more steps one passes to grid 3, and so on until grid m is reached. For $k > 1$, the evolution on grid k is driven by a weighted average of the residuals calculated on grid k-1, so that each mesh simulates the evolution that would have occurred on the next finer mesh. When the coarsest grid has been reached, changes in the solution calculated on each mesh are consecutively interpolated back to the next finer mesh. Time steps may also be included between the interpolation steps on the way back up to grid 1. In practice it has been found that an effective multigrid strategy is to use a simple saw tooth cycle, with one time step on each grid on the way down to the coarsest grid, and no Euler calculation between the interpolation steps on the way up.

In general one can conceive of a multigrid scheme using a sequence of independently generated coarser meshes which are not associated with each other in any structured way. Here attention will be restricted to the case in which coarser meshes are generated by eliminating alternate points in each coordinate direction. Accordingly each cell on grid k coincides either exactly or approximately with a group of four cells on grid k-1 in the two dimensional case, or eight cells in the three dimensional case. This allows the formulation of simple rules for the transfer of data between grids.

In order to give a precise description of the multigrid scheme it is convenient to use subscripts to indicate the grid. Several transfer operations need to be defined. First the solution vector on grid k must be initialized as

$$w_k^{(0)} = T_{k,k-1}\, w_{k-1}$$

where w_{k-1} is the current value on grid k-1, and $T_{k,k-1}$ is a transfer operator. Next it is necessary to transfer a residual forcing function such that the solution on grid k is driven by the residuals calculated on grid k-1. This can be accomplished by setting

$$P_k = Q_{k,k-1}\, R_{k-1}(w_{k-1}) - R_k(w_k^{(0)})$$

where $Q_{k,k-1}$ is another transfer operator. Then $R_k(wk)$ is replaced by $R_k(w_k) + P_k$ in the time stepping scheme. For example, the multi-stage scheme defined by equation (3.3) is reformulated as

$$w_k^{(1)} = w_k^{(0)} - \alpha_1 \Delta t_k\, (R_k^{(0)} + P_k)$$

...

$$w_k^{(q+1)} = w_k^{(0)} - \alpha_{q+1}\, \Delta t_k\, (R_k^{(q)} + P_k)$$

...

The result $w_k^{(m)}$ then provides the initial data for grid k+1. Finally the accumulated correction on grid k has to be transferred back to grid k-1. Let w_k^+ be the final value of w_k resulting from both the correction calculated in the time step on grid k and the correction transferred from grid k+1. Then one sets

$$w_{k-1}^+ = w_{k-1} + I_{k-1,k}(w_k^+ - w_k^{(0)})$$

where w_{k-1} is the solution on grid k-1 after the time step on grid k-1 and before the transfer from grid k, and $I_{k-1,k}$ is an interpolation operator.

In the case of a cell centered scheme the solution transfer operator $T_{k,k-1}$ is defined by the rule

$$T_{k,k-1}\, w_{k-1} = (\Sigma\, V_{k-1}\, w_{k-1})/V_k$$

where the sum is over the constituent cells on grid k-1, and V is the cell area or volume. This rule conserves mass, momentum and energy. The residual transferred to grid k is the sum of the residuals in the constituent cells

$$Q_{k,k-1}\, R_{k-1} = \Sigma\, R_{k-1}$$

The corrections are transferred up using either bilinear or trilinear interpolation for the operator $I_{k-1,k}$.

When the flow variables are stored at the cell vertices the solution transfer rule is simply to set $w_k^{(0)}$ to w_{k-1} at the coincident mesh point in grid k-1. The residual transfer rule is a weighted sum over the 9 nearest points in two dimensions, or the 27 nearest points in three dimensions. The corresponding transfer operator $Q_{k,k-1}$ can be expressed as a product of summation operators in the coordinate directions. Let μ_x denote an averaging operator in the x direction:

$$(\mu_x R)_{i+1/2,j,k} = \frac{1}{2}(R_{i,j,k} + R_{i+1,j,k})$$

and

$$(\mu_x^2 R)_{i,j,k} = \frac{1}{4} R_{i-1,j,k} + \frac{1}{2} R_{i,j,k} + \frac{1}{4} R_{i+1,j,k}$$

Then in the three dimensional case

$$Q_{k,k-1} \equiv 8\, \mu_x^2\, \mu_y^2\, \mu_z^2$$

The interpolation operator $I_{k-1,k}$ transfers the corrections at coincident mesh points, and fills in the corrections at intermediate points by bilinear or trilinear interpolation.

In this formulation the residuals on each mesh should be re-evaluated after the time step to provide a proper estimate of the current value $R_k(w_k^+)$ for transfer to the next mesh k+1 in the sequence. Just as the multistage time stepping scheme can be modified to eliminate the recalculation of the residuals by substituting a one term Taylor expansion for $R(w+\delta w)$, so can the multigrid scheme be modified by a similar substitution to allow the unmodified residuals to be passed to the coarser mesh. This requires the collection operator $Q_{k,k-1}$ to be constructed so that $Q_{k,k-1}\, R_{k-1}(w_{k-1})$ approximates a weighted average of the residuals $R_{k-1}(w_{k-1} + \delta w_{k-1})$. If $R(w)$ approximation $\partial/\partial x\, f(w) + \partial/\partial y\, g(w)$, and the change in the dissipative term is ignored, $Q_{k,k-1}$ should then be a nonsymmetric operator approximating a multiple of $I + \Delta t_k\, (D_x A + D_y B)$, where A and B are the Jacobian matrices. Hall uses a procedure of this type in his formulation of a multigrid scheme with Lax Wendroff time stepping [20].

5. Analysis of Multigrid Time Stepping Schemes

The analysis of multigrid schemes is complicated by the nonuniformity of the process. If a mesh point is common to two meshes then corrections can be directly transferred from the coarse to the fine mesh. On the other hand the correction at a point of the fine mesh which is not contained in the coarse mesh has to be interpolated from the corrections at neighboring points. It is proposed here to circumvent this difficulty by modeling the multigrid process as a combination of two processes. The first is a uniform process in which every mesh point is treated in the same way, and the second is a nonlinear filtering scheme which eliminates the data from alternate points. For the sake of simplicity the analysis will be restricted to a one dimensional model. It also proceeds on the assumption that each coarser mesh is produced by eliminating alternate points of the finer mesh, so that there exists a set of points which are common to all the meshes.

Figure 5(a) illustrates the data flow of a two level scheme in which grid 1 is the finer mesh and grid 2 is the coarser mesh. Suppose that the calculation is simulating an equation of the form

$$\frac{du_j}{dt} = R_j(u) \tag{5.1}$$

where u_j is the dependent variable at mesh point j of grid 1, and $R(u_j)$ is the residual. Here it will be convenient to use bracketed superscripts to indicate the grid level, and to reserve the use of subscripts for the indication of the location of the mesh point in the fine grid. Suppose that the points 0,2,4... are common to both meshes, while the points 1,3,5... are eliminated in grid 2. A simple multigrid scheme can be described as follows. On grid 1 u_j is updated by a correction

$$\delta u_j^{(1)} = -\Delta t^{(1)} f(R_j(u)) \qquad (5.2)$$

where the function f depends on the time stepping scheme. On grid 2 corrections are calculated as

$$\delta u_j^{(2)} = -\Delta t^{(2)} f(R_j^{(2)}) , \; j = 1,3,5... \qquad (5.3)$$

where the residual $R_j^{(2)}$ is calculated by accumulating the residuals at the nearest neighbors after first allowing for the correction introduced on grid 1. For example,

$$R_j^{(2)} = \varepsilon R_{j-1}^{+} + (1-2\varepsilon) R_j^{+} + \varepsilon R_{j+1}^{+} \qquad (5.4)$$

where

$$R_j^{+} = R_j(u + \delta u^{(1)}) \qquad (5.5)$$

Then on interpolating the corrections on grid 2 back to grid 1, the total correction of the complete multigrid scheme is

$$\delta u_j = \delta u_j^{(1)} + \delta u_j^{(2)} , \; j \text{ even}$$

$$\delta u_j = \delta u_j^{(1)} + \frac{1}{2} (\delta u_{j-1}^{(2)} + \delta u_{j+1}^{(2)}), \; j \text{ odd}$$

This process can be broken down into two stages as illustrated in Figure 5.1(b). First the corrections $\delta u_j^{(2)}$ are calculated for all points of grid 1 by formulas (5.3)-(5.5) for j both even and odd. In effect the two level process is now calculated uniformly on the original fine grid. In the second stage $\delta u_j^{(2)}$ is then replaced by

$$\overline{\delta u}_j^{(2)} = \delta u_j^{(2)} \quad , \; j \text{ even}$$

$$\overline{\delta u}_j^{(2)} = \frac{1}{2} (\delta u_{j-1}^{(2)} + \delta u_{j+1}^{(2)}), \; j \text{ odd.}$$

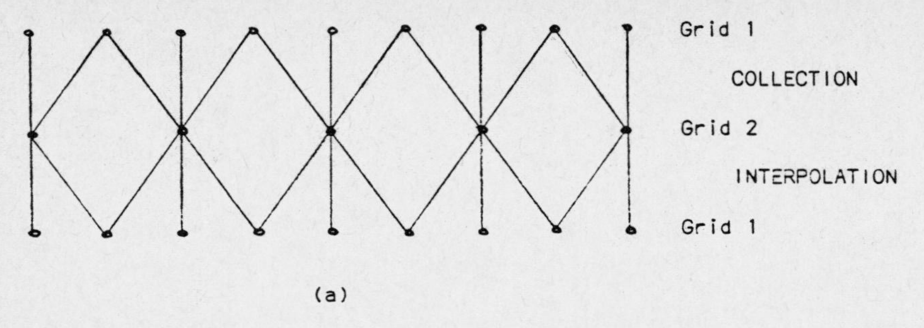

(a)

Multigrid scheme

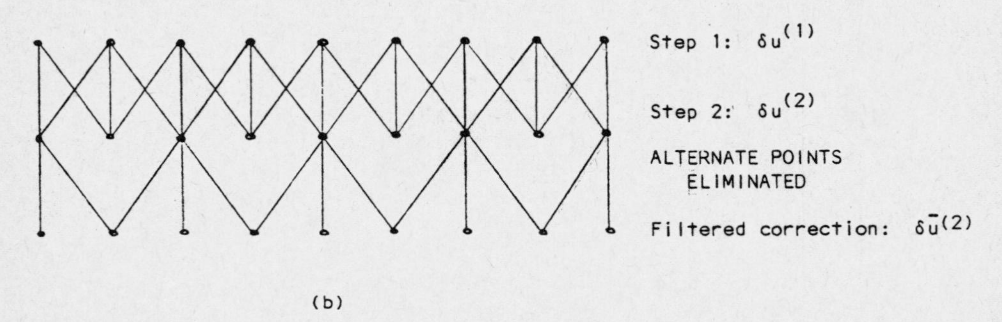

(b)

Uniform scheme with nonlinear filter

Figure 5

Data flow of multigrid and uniform schemes

This nonlinear filtering process eliminates the need to calculate $\delta u_j^{(2)}$ at the odd points, allowing these calculations to be shifted to a coarser grid. It introduces an additional error

$$e_j = 0 \quad , \ j \text{ even}$$

$$e_j = \frac{1}{2}\,(\delta u_{j-1}^{(2)} - 2\delta u_j^{(2)} + \delta u_{j+1}^{(2)}), \quad j \text{ odd}$$

Assuming the mesh to be uniform, this can be written as

$$e_j = \frac{1}{4}\,(\delta u_{j-1}^{(2)} - 2\delta u_j^{(2)} + \delta u_{j+1}^{(2)})(1 - \cos\frac{\pi}{\Delta x}x_j) \qquad (5.6)$$

where Δx is the mesh interval of the fine mesh, and $x_j = j\Delta x$ are its mesh points. Thus the filter introduces additional errors in the form of a carrier wave at the mesh frequency $\pi/\Delta x$ of the fine mesh, modulated by the second difference of the corrections $\delta u_j^{(2)}$ which would be calculated in the second stage of the uniform scheme. If we make the usual assumptions of linearity and periodicity, the multilevel uniform scheme can be analyzed by the Fourier method. If the multilevel uniform scheme is unstable, we can anticipate that the corresponding multigrid scheme will be unsound. Because of the injection of additional errors at various mesh frequencies by the interpolation process of the multigrid scheme, a reasonable criterion is to require the multilevel uniform scheme to have a substantial stability margin at the mesh frequencies of all the meshes above the coarsest mesh in the sequence.

The following paragraphs address the question of the stability of the multilevel uniform scheme. The analysis is carried out for an initial value problem on an infinite interval governed by an equation of the form

$$\frac{\partial v}{\partial t} + Av = 0 \qquad (5.7)$$

where A is a linear differential operator in one space dimension. The operator A may contain a forcing term, so that v is not zero when the system reaches a steady state. Let the vector u with elements u_j represent the discrete solution. The residual is

$$R = Pu \qquad (5.8)$$

where P is a difference operator approximating $\Delta t\ A$. In the case of a pth order accurate scheme, if P is applied to the values $v_j = v(x_j)$ of the exact solution, then

$$Pv = \Delta t(Av + O(\Delta x)^p)$$

Using supercripts to denote the time steps,

$$u^{n+1} = u^n + \delta u$$

where the correction δu depends on the residual through the action of a time stepping operator F, corresponding to equation (5.2). For example, if we use the multi-stage scheme

$$u^{(0)} = u^n$$

$$u^{(1)} = u^{(0)} - \alpha_1\ Pu^{(0)}$$

$$u^{(2)} = u^{(0)} - \alpha_2\ Pu^{(1)}$$

$$u^{(3)} = u^{(0)} - \alpha_3\ Pu^{(2)}$$

$$u^{n+1} = u^{(3)}$$

we find that

$$u^{(3)} = u^{(0)} - \alpha_3(I - \alpha_2\ P + \alpha_2\ \alpha_1\ P^2)\ Pu^{(0)}$$

Consequently

$$F = \alpha_3\ (I - \alpha_2 P + \alpha_2 \alpha_1 P^2)$$

For the Crank Nicolson scheme

$$u^{n+1} = u^n - \frac{1}{2}\ (Pu^{n+1} + Pu^n)$$

we obtain

$$F = (I + \frac{1}{2}\ P)^{-1}$$

If we set

$$\hat{u}(\xi) = \Delta x \sum_{-\infty}^{\infty} u_j \; e^{-i\xi x_j/\Delta x}$$

then the Fourier transform of the residual (5.8) is $\hat{P}\hat{u}$ where $\hat{P}(\xi)$ is the Fourier symbol of the difference operator. Suppose, for example, that

$$A \equiv a \frac{\partial}{\partial x}$$

and that we use a central difference scheme with added dissipative terms. Then

$$(Pu)_j = \frac{\lambda}{2}(u_{j+1} - u_{j-1}) - \lambda\mu_2(u_{j+1} - 2u_j + u_{j-1}) \tag{5.9}$$

$$+\lambda\mu_4(u_{j+2} - 4u_{j+1} + 6u_j - 4u_{j-1} + u_{j-2})$$

where λ is the Courant number,

$$\lambda = a\frac{\Delta t}{\Delta x}$$

and μ_2 and μ_4 are dissipative coefficients. Also

$$\hat{P}(\xi) = \lambda\, i \sin\xi + 2\lambda\mu_2(1 - \cos\xi) + 4\lambda\mu_4(1 - \cos\xi)^2 \tag{5.9*}$$

Similarly if $\hat{F}(\xi)$ is the Fourier symbol of the time stepping operator, then

$$\delta\hat{u}(\xi) = -\hat{F}(\xi)\,\hat{P}(\xi)\,\hat{u}^n(\xi)$$

and

$$\hat{u}^{n+1}(\xi) = g(\xi)\,\hat{u}^n(\xi) \tag{5.10}$$

where $g(\xi)$ is the amplification factor

$$g(\xi) = 1 - \hat{F}(\xi)\,\hat{P}(\xi)$$

Suppose that we have a nested set of grids with successively doubled mesh intervals. It is now convenient to revert to denoting the grids by subscripts 1,2,3... (Since the individual elements of the solution vector do not appear in the analysis this leads to no confusion). Consider a multigrid time stepping scheme in which time steps are taken on successive grids sequentially down to the coarsest grid, and the cycle is then repeated. In order to produce the same final steady state as a scheme using only the fine grid, the evolution on every grid except grid 1 should driven by the residuals calculated on the next finer grid. Let R_1^+ be the residual on grid 1 after the change δu_1, and let R_2 be the residual calculated on grid 2. Also let Q_{21} be the operator transferring residuals from grid 1 to grid 2, so that $Q_{21}\,R_1$ is a weighted sum of fine grid residuals corresponding to the coarse grid residual R_2. Then on grid 2 replace R_2 by

$$\overline{R}_2 = R_2 + S_2$$

where

$$S_2 = Q_{21}\,R_1^+ - R_2$$

and on grid 3 replace R_3 by

$$\overline{R}_3 = R_3 + S_3$$

where

$$S_3 = Q_{32}\, R_2 - \overline{R}_3^{+} - R_3$$

$$= Q_{32}(Q_{21}\, R_1^{+} + R_2^{+} - R_2) - R_3$$

With a single stage time stepping scheme δu_2 is determined by substituting the corresponding fine grid residual $Q_{21}R_1^{+}$ for R_2, but R_2 needs to be calculated because $R_2^{+} - R_2$ appears in S_3. With a multi-stage time stepping scheme R_2 would be recalculated several times while S_2 would be frozen at its initial value on grid 2. If we examine the action of m stage scheme on one of the coarser grids, we have

$$u_k^{(0)} = u_{k-1}^{+}$$

$$u_k^{(1)} = u_k^{(0)} - \alpha_1\, Q_{k,k-1}\, R_{k-1}^{+}$$

$$u_k^{(2)} = u_k^{(0)} - \alpha_2(R_k^{(1)} + Q_{k,k-1}\, R_{k-1}^{+} - R_k^{(0)})$$

$$\ldots$$

$$u_k^{(m)} = u_k^{(0)} - (R_k^{(m-1)} + Q_{k,k-1}\, R_{k-1}^{+} - R_k^{(0)})$$

$$u_k^{+} = u_k^{(m)}$$

Here in the second stage

$$R_k^{(1)} - R_k^{(0)} = P_k(u_k^{(1)} - u_k^{(0)})$$

$$= -\,\alpha_1 P_k\, Q_{k,k-1}\, R_{k-1}^{+}$$

whence

$$u_k^{(2)} - u_k^{(0)} = -\,\alpha_2\,(I - \alpha_1\, P_k)\, Q_{k,k-1}\, R_{k-1}^{+}$$

Following through the remaining stages, we find that

$$\delta u_k = u_k^{(m)} - u_k^{(0)} = -\,F_k\, Q_{k,k-1}\, R_{k-1}^{+} \qquad (5.12)$$

where F_k is the time stepping operator on grid k as it would appear for a single grid.

Now consider the evolution of all quantities in the multigrid process, assuming that it is uniformly applied at every mesh point of grid 1. Suppose that the collection operators Q_{21}, Q_{32} all have the same generic form. On the fine grid denote this by Q, with corresponding Fourier symbol $\hat{Q}(\xi)$. For example, if

$$(QR)_j = \frac{1}{2}\, R_{j-1} + R_j + \frac{1}{2}\, R_{j+1} \qquad (5.13)$$

then

$$\hat{Q}(\xi) = 1 + \cos\xi \qquad (5.13^*)$$

On grid 1 denote the Fourier symbols of the residual and time stepping operators by

$$p_1 = \hat{P}(\xi),\; f_1 = \hat{F}(\xi) \qquad (5.14a)$$

and the symbol of the first collection operator by

$$q_{21} = \hat{Q}(\xi) \tag{5.14b}$$

For a system of equations these symbols will be matrices. On the subsequent levels the corresponding symbols are

$$p_k = \hat{P}(2^{k-1}\xi), \ f_k = \hat{F}(2^{k-1}\xi) \tag{5.14c}$$

and

$$q_{k,k-1} = \hat{Q}(2^{k-1}\xi) \tag{5.14d}$$

Now on the first grid

$$\delta\hat{u}_1 = - f_1 r_1$$

where r_1 is the Fourier transform of the residual

$$r_1 = p_1 \hat{u}_1$$

On subsequent grids it follows from equation (5.12) that

$$\delta\hat{u}_k = - f_k r_k$$

where

$$r_k = q_{k,k-1} \, r^{+}_{k-1}$$

Since the system is linear

$$r^{+}_{k-1} = r_{k-1} + p_{k-1} \, \delta\hat{u}_{k-1}$$

(but in general r^{+}_{k-1} is not equal to $p_{k-1} \, u^{+}_{k-1}$ when $k > 2$). Substituting for $\delta\hat{u}_{k-1}$, we find that

$$r_k = q_{k,k-1}(I - p_{k-1} \, f_{k-1})r_{k-1} \tag{5.15}$$

Finally for an m level scheme

$$\hat{u}^{+}_m = \hat{u}_1 - \sum_{k=1}^{m} f_k \, r_k \tag{5.16}$$

Equations (5.14-5.16) define the stability of the complete multilevel scheme. The final formula may be evaluated directly as a sum in which each new term is obtained recursively from the previous term, or as a nested product by the loop

$$Z_m = f_m \ ;$$

for $k = m - 1$ to 1

$$Z_k = f_k + Z_{k+1} \, q_{k+1,k}(I - p_k \, f_k)$$

and

$$\hat{u}_m = (1 - Z_1 p_1)\hat{u}_1$$

If the operators F and P commute, then equation (5.15) may be simplified by the substitution

$$I - p_k f_k = I - f_k p_k = g_k$$

where g_k is the amplification factor of the basic time stepping scheme applied on level k. This will be the case for any scheme applied to a scalar equation, and for typical multi-stage schemes applied to a system of equations.

In the special case that

$$Q_{k,k-1} P_{k-1} = P_k \ ,$$

for example, if at the jth mesh point

$$R_j = \frac{\lambda}{2} (u_{j+1} - u_{j+1}), \ (QR)_j = R_{j-1} + R_{j+1}$$

equation (5.16) reduces to

$$\hat{u}_m^+ = g_m g_{m-1} \cdots g_1 \hat{u}_1$$

In general it does not. This result can be proved by noting that

$$r_2 = q_{21} r_1^+ = q_{21} p_1 \hat{u}_1^+ = p_2 \hat{u}_1^+ = p_2 \hat{u}_2$$

and

$$r_2^+ = p_2 \hat{u}_2 + p_2 \delta\hat{u}_2 = p_2 \hat{u}_2^+$$

Then

$$r_3 = q_{32} r_2^+ = q_{32} p_2 \hat{u}_2^+ = p_3 \hat{u}_2^+ = p_3 \hat{u}_3$$

and so on. Consequently it follows that

$$\hat{u}_k^+ = (I - f_k q_{k,k-1}, p_{k-1}) \hat{u}_{k-1}^+ = g_k \hat{u}_{k-1}^+$$

Formulas (5.14)-(5.16) can easily be evaluated for any particular choices of residual operator, time stepping operator and collection operator with the aid of a computer program. Figures 6 and 7 show typical results for the dissipative central difference scheme (5.9), with the collection operator (5.13). Both results are for blended multi-stage time stepping schemes of the class defined by equations (3.3) and (3.4*). Figure 6 shows the amplification factor of a three stage scheme in which the dissipative terms are evaluated once. The Courant number is 1.5 and the coefficients are

$$\begin{aligned} &\alpha_1 = .6, && \alpha_2 = .6 \\ &\beta_{qq} = 1, && \beta_{qr} = 0, \ q > r \\ &\gamma_{q0} = 1, && \gamma_{qr} = 0, \ r > 0 \end{aligned} \tag{5.17}$$

As the number of levels is increased the stability curve defined by the amplification factor is compressed to the left, retaining a large margin of stability at all high frequencies. Thus the scheme should be resistant to the injection of interpolation errors. Figure 7 schows the amplification factor of a five stage scheme in which the dissipative terms are evaluated twice. In this case the coefficients are

$$\alpha_1 = 1/4, \quad \alpha_2 = 1/6, \quad \alpha_3 = 3/8, \quad \alpha_4 = 1/2$$

$$\beta_{qq} = 1, \quad \beta_{qr} = 0, \quad q > r \qquad (5.18)$$

$$\gamma_{00} = 1, \quad \gamma_{q1} = 1, \quad \gamma_{qr} = 0, \quad r \neq 1$$

Residual averaging is also included with a coefficient of .75, and the Courant number is 7.5. Although the stability curve exhibits a bump, there is still a substantial margin of safety , and this scheme has proved very effective in practice [39].

The formulas of this section can be modified to allow for alternative multigrid strategies, including more complicated V and W cycles. Nor is it necessary to use the same time stepping and residual operators on every grid. It may pay, for example, to use a simplified lower order scheme on the coarse grids. This method of analysis, in which the multigrid process is regarded as a multilevel uniform process on a single grid, subject to the injection of additional interpolation errors, is also easily extended to two and three dimensional problems.

6. Some Results for an Explicit Multi-stage Scheme

This section presents some results for a simple mutigrid method in which an explicit multi-stage scheme was used for time stepping. The application is the calculation of three dimensional transonic flow past a swept wing. The vertex formulation described by equations (2.5) was used for the discretization of the Euler equations. A five stage time stepping scheme with the coefficients defined by equations (5.18) was used in conjunction with a simple saw tooth multigrid cycle. Implicit residual averaging as defined by equation (3.5) was also used.

The mesh was of C type in streamwise vertical planes, generated by the introduction of sheared parabolic coordinates. This was accomplished by a two stage mapping procedure. The first stage introduces parabolic coordinates by the transformation

$$(\bar{X} + i\bar{Y})^2 = \{x - x_o(z) + i\,(y - y_o)\}/t(z)$$
$$\bar{Z} = z$$

where z is the spanwise coordinate, $t(z)$ is a scaling factor which can be used to control the number of cells covering the wing, and $x_o(z)$ and $y_o(z)$ are the coordinates of a singular line lying just inside the leading edge. The effect of this transformation is to unwrap the wing to a shallow bump $Y = S(X,Z)$. The second stage is a shearing transformation

$$X = X, \; Y = Y - S(X,Z), \; Z = Z$$

which maps the wing to the coordinate surface $Y = 0$. The mesh is then constructed by the reverse sequence of mappings from a rectangular grid in the X,Y,Z coordinate system. Meshes of this type contain badly distorted cells in the neighborhood of the singular line where it passes into the flowfield beyond the wing tip. These cells, which have a very high aspect ratio and a triangular cross section, present a severe test of the robustness of the multigrid scheme.

Figure 8 shows a typical result for the well known ONERA M6 wing at a Mach number of .840 and an angle of attack of 3.06 degrees*. The mesh contained 96 cells in the chordwise direction, 16 cells in the direction normal to the wing, and 16 cells in the spanwise direction, and the calculation was performed in two stages. A result was first obtained on a 48×8×8 mesh using three levels in the multigrid scheme. This was then used to provide the initial state for the calculation on the 96×16×16 mesh in which four levels were used in the multigrid scheme. Table 1 shows the rate of

*Calculated on a Cray 1 computer at Grumman: I am indebted to G. Volpe for his assistance in optimizing the computer program to run on the Cray and preparing the graphic display of the result.

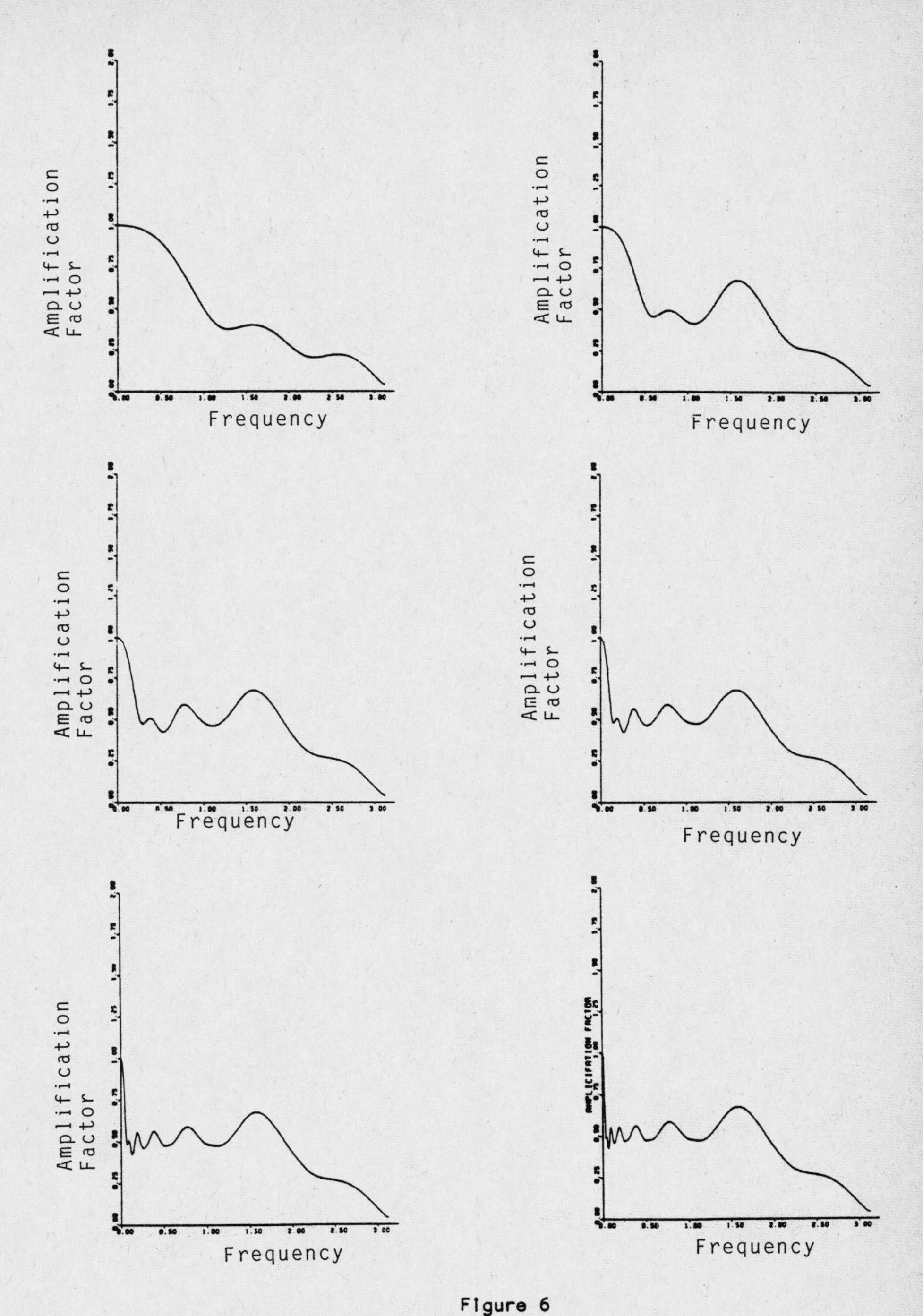

Figure 6

Amplification Diagrams for a 3 Stage Scheme for 1-6 Grid Levels

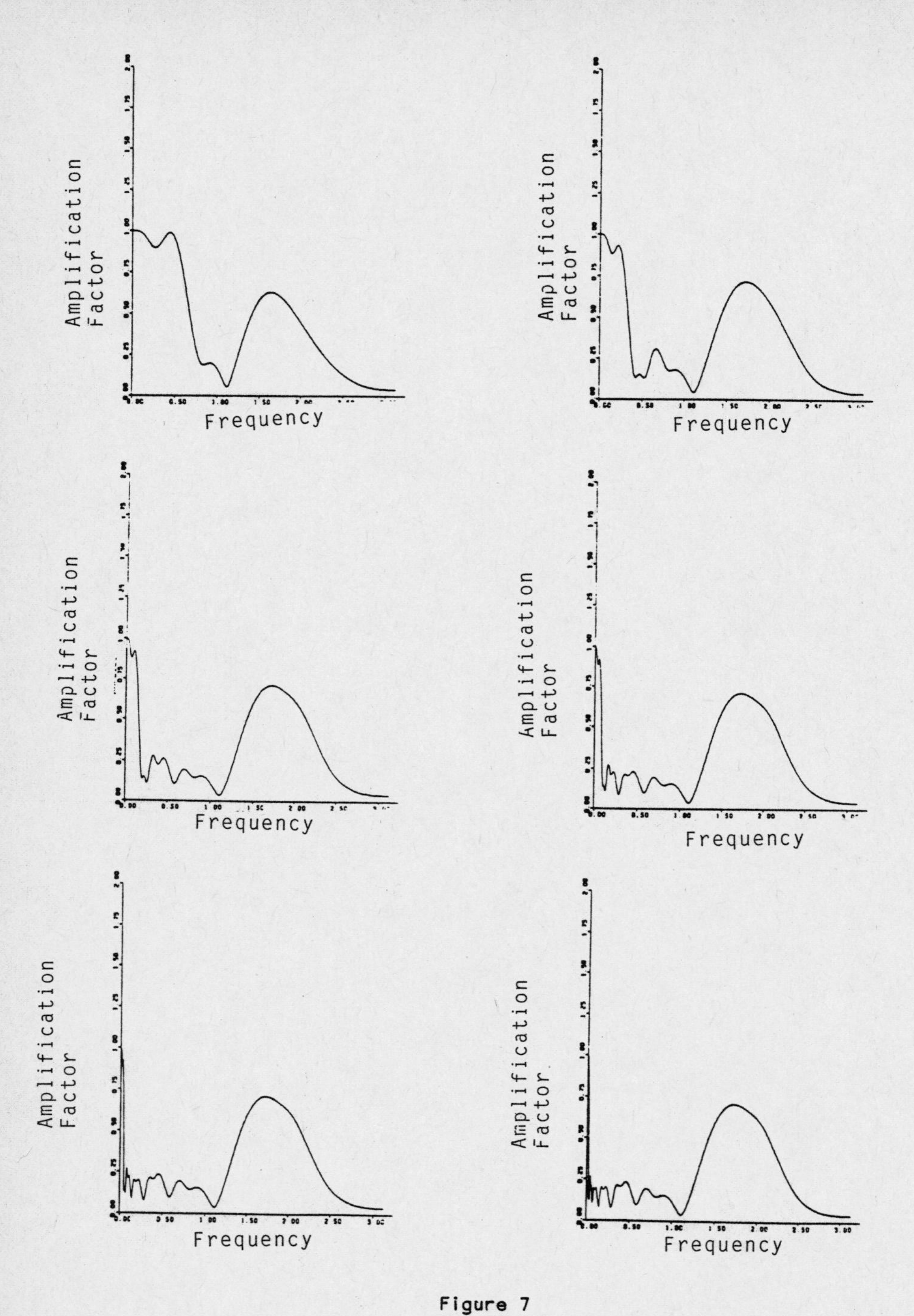

Figure 7

Amplification Diagrams for a 5 Stage Scheme with 2 Evaluations of the Dissipative Terms and Residual Averaging for 1-6 Grid Levels

convergence over 100 multigrid cycles on the 96×16×16 mesh, measured by the average rate of change of density, together with the development of the lift and drag coefficients CL and CD. It can be seen that these are converged to four figures within 20 cycles. Table 2 shows the result of a similar calculation using a sequence of three meshes containing 32×8×8. 64×16×16 amd 128×32×32 cells respectively. Three levels were used in the multigrid scheme on the first mesh, four on the second, and five on the third. After 10 cycles on the 32×8×8 mesh, 10 cycles on the 64×16×16 mesh and 5 cycles on the 128×32×32 mesh, the calculated force coefficients were CL = .3145, and CD = .0167. These are barely different from the final converged values CL = .3144 and CD = .0164. The discretization errors, which may be estimated by comparing fully converged results on the sequence of three meshes, are in fact substantially larger than these differences, confirming that convergence well within the discretization error can be obtained in 5-10 cycles.

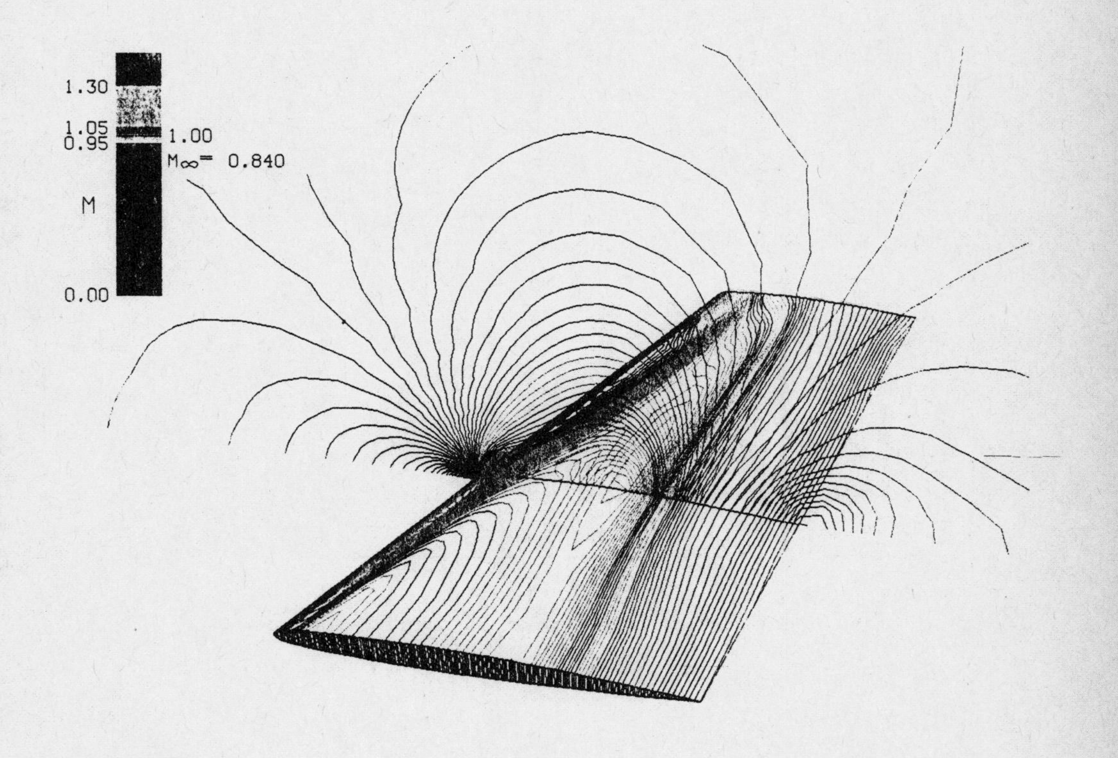

Figure 8

Constant pressure contours of flow over the ONERA M6 wing

Table 1

Calculation of the flow past the ONERA M6 wing at Mach .840, and 3.06° angle of attack on a 96×16×16 mesh.

Cycle	Average dρ/dt	CL	CD
1	$.916\ 10^{-1}$		
10	$.158\ 10^{-2}$	.3110	.0205
20	$.243\ 10^{-3}$	.3118	.0203
30	$.245\ 10^{-4}$	.3118	.0203
40	$.353\ 10^{-5}$	.3118	.0203
50	$.528\ 10^{-6}$	.3118	.0203
60	$.772\ 10^{-7}$	.3118	.0203
70	$.124\ 10^{-8}$	.3118	.0203
80	$.241\ 10^{-9}$	.3118	.0203
90	$.363\ 10^{-9}$	.3118	.0203
100	$.528\ 10^{-10}$	.3118	.0203

Average reduction of dρ/dt per multigrid cycle: .807.

Table 2

Result for the ONERA M6 Wing with a sequence of 3 meshes.

	CL	CD
Result after 10 cycles on 32×8×8 mesh	.2956	.0373
Result after 10 cycles on 64×16×16 mesh	.3167	.0263
Result after 5 cycles on 128×32×32 mesh	.3145	.0167
Final Converged result on 128×32×32 mesh	.3144	.0164

In assessing these results it should be noted that the computational effort of one step of the 5 stage scheme is substantially greater then that of a Lax Wendroff scheme, but appreciably less than that required by an alternating direction or LU decomposition scheme. Measured by a work unit consisting of the computational effort of one time step on the fine grid, the work required for one multigrid cycle with five levels is

$$1 + \frac{1}{8} + \frac{1}{64} + \frac{1}{512} + \frac{1}{4096}$$

plus the work required for additional residual calculations, which is of the order of 25 percent. Using a single processor of a Cray XMP computer, the time required for a multigrid cycle on a 96×16×16 mesh is about 1.3 seconds, and a complete solution on such a mesh can be obtained in about 15 seconds. This is fast enough that interactive analysis of alternative wing designs at the preliminary design stage is now within the realm of possibility.

7. Conclusion

Multigrid techniques for the Euler equations are by now solidly established, and a variety of repidly convergent methods have been demonstrated. The concept of a multigrid time stepping scheme provides an alternative framework for the analysis of these methods. In contrast to the more classical view of the multigrid process based upon assumptions of ellipticity, this concept emphasizes the role of the coarse grids in increasing the speed at which disturbances can be propagated through the domain. It leads rather naturally to the method of analysis proposed in Section 5, which may prove useful for screening alternative multigrid strategies, and identifying those which are most promising.

While the successes which have been achieved to date are enough to indicate the potential of multigrid methods, much work remains to be done. Several particularly important topics of investigation may be singled out. First, the extreme geometrical complexity of the configurations which need to be treated in many engineering applications may well dictate the use of patched and unstructured meshes. The use of an unstructured tetrahedral mesh appears, for example, to be one of the more promising ways to calculate the flow past a complete aircraft [11]. If multigrid methods are to be more widely used, I believe, therefore, that it will be necessary to develop effective methods for unstructured meshes. Second, accurate simulations of real flows must include the effects of viscosity and turbulence, and will accordingly require the treatment of the Reynolds averaged Navier Stokes equations. The need to use meshes with very high aspect ratio cells in the boundary layer region accentuates the difficulties in obtaining rapid convergence. While some acceleration has been demonstrated with multigrid techniques, the speed of convergence still falls far short of the rates achieved in Euler calculations. A third direction of improvement which needs to be pursued is the integration of multigrid solution strategies with procedures for automatic grid refinement. Results which have already been obtained in two dimensional calculations clearly show the potential advantages of such an approach, which could be the key to better resolution of both shock waves and boundary layers [49,50].

The realization of these improvements will bring us closer to the ultimate goal of accurate and economical prediction of flows over complete configurations. Computation methods may then finally fulfill their proper role as a reliable guide for the design of aeroplanes, cars, and any other devices whose performance significantly depends on aerodynamic efficiency.

References

1. Hess, J.L. and Smith, A.M.O., "Calculation of Non-Lifting Potential Flow About Arbitrary Three-Dimensional Bodies", Douglas Aircraft Report, ES 40622, 1962

2. Rubbert, P.E. and Saaris, G.R., "A General Three Dimensional Potential Flow Method Applied to V/STOL Aerodynamics", SAE Paper 680304, 1968.

3. Murman, E.M. and Cole, J.D., "Calculation of Plane Steady Transonic Flows", AIAA Journal, Vol. 9, 1971, pp. 114-121.

4. Jameson, Antony, "Iterative Solution of Transonic Flows Over Airfoils and Wings, Including Flows at Mach 1", Comm. Pure. Appl. Math, Vol. 27, 1974, pp. 283-309.

5. Jameson, Antony and Caughey, D.A., "A Finite Volume Method for Transonic Potential Flow Calculations", Proc. AIAA 3rd Computational Fluid Dynamics Conference, Albuquerque, 1977, pp. 35-54.

6. Bristeau, M.O., Pironneau, O., Glowinski, R., Periaux, J., Perrier, P., and Poirier, G., "On the Numerical Solution of Nonlinear Problems in Fluid Dynamics by Least Squares and Finite Element Methods (II). Application to Transonic Flow Simulations", Proc. 3rd International Conference on Finite Elements in Nonlinear Mechanics, FENOMECH 84, Stuttgart, 1984, edited by J. St. Doltsinis, North Holland, 1985, pp. 363-394.

7. Jameson, A., Schmidt, W., and Turkel, E., "Numerical Solution of the Euler Equations by Finite Volume Methods Using Runge-Kutta Time Stepping Schemes", AIAA Paper 81-1259, AIAA 14th Fluid Dynamics and Plasma Dynamics Conference, Palo Alto, 1981.

8. Jameson, Antony, and Baker, Timothy J., "Solution of the Euler Equations for Complex Configurations", Proc. AIAA 6th Computational Fluid Dynamics Coference, Danvers, 1983, pp. 293-302.

9. Pulliam, T.J., and Steger, J.L., "Recent Improvements in Efficiency, Accuracy and Convergence for Implicit Approximate Factorization Algorithms", AIAA Paper 85-0360, AIAA 23rd Aerospace Sciences Meeting, Reno, January 1985.

10. MacCormack, R.W., "Current Status of Numerical Solutions of the Navier-Stokes Equations", AIAA Paper 85-0032, AIAA 23rd Aerospace Sciences Meeting, Reno, January 1985.

11. Jameson, A., Baker, T.J., and Weatherill, N.P., "Calculation of Inviscid Transonic Flow Over a Complete Aircraft", AIAA Paper 86-0103, AIAA 24th Aerospace Sciences Meeting, Reno, January 1986.

12. Fedorenko, R.P., "The Speed of Convergence of One Iterative Process", USSR Comp. Math. and Math. Phys., Vol. 4, 1964, pp. 227-235.

13. South, J.C. and Brandt, A., "Application of a Multi-Level Grid Method to Transonic Flow Calculations", Proc. of Workshop on Transonic Flow Problems in Turbomachinery, Monterey, 1976, edited by T.C. Adamson and M.F. Platzer, Hemisphere, 1977, pp. 180-206.

14. Jameson, Antony, "Acceleration of Transonic Potential Flow Calculations on Arbitrary Meshes by the Multiple Grid Method", Proc. AIAA 4th Computational Fluid Dynamics Conference, Williamsburg, 1979, pp. 122-146.

15. Caughey, D.A., "Multigrid Calculation of Three-Dimensional Transonic Potential Flows", AIAA Paper 83-0374, AIAA 21st Aerospace Sciences Meeting, Reno, January 1983.

16. Ni, Ron Ho., "A Multiple Grid Scheme for Solving the Euler Equations", AIAA Journal, Vol. 20, 1982, pp. 1565-1571.

17. Jameson, A., "Solution of the Euler Equations by a Multigrid Method", Applied Math. and Computation, Vol. 13, 1983, pp. 327-356.

18. Jameson, A., and Schmidt, W., "Recent Developments in Numerical Methods for Transonic Flows", Proc. 3rd International Conference on Finite Elements in Nonlinear Mechanics, FENOMECH 84, Stuttgart, 1984, edited by J.St. Doltsinis, North-Holland ,1985, pp. 467-493.

19. Jameson, A., and Mavriplis, D., "Finite Volume Solution of the Two Dimensional Euler Equations on a Regular Triangular Mesh", AIAA Paper 85-0435, AIAA 23rd Aerospace Sciences Meeting, Reno, January 1985.

20. Hall, M.G., "Cell Vertex Multigrid Schemes for Solution of the Euler Equations", IMA Conference on Numerical Methods for Fluid Dynamics", Reading, April 1985.

21. Hemker, P.W., and Spekreijse, S.P., "Multigrid Solution of the Steady Euler Equations", Proc. Oberwolfach Meeting on Multigrid Methods, December 1984.

22. Godunov, S.K., "A Difference Method for the Numerical Calculation of Discontinous Solutions of Hydrodynamic Equations", Mat. Sbornik, 47, 1959, pp. 271-306, translated as JPRS 7225 by U.S. Dept. of Commerce, 1960.

23. Boris, J.P., and Book, D.L., "Flux Corrected Transport. 1. SHASTA, A Fluid Transport Algorithm that Works", J. Comp. Phys. Vol. 11, 1973, pp. 38-69.

24. Van Leer, B., "Towards the Ultimate Conservative Difference Scheme. II, Monotonicity and Conservation Combined in a Second Order Scheme," J. Comp. Phys. Vol. 14, 1974, pp. 361-370.

25. Steger, J.L., and Warming, R.F., "Flux Vector Splitting of the Inviscid Gas Dynamics Equations with Applications to Finite Difference Methods," J. Comp. Phys., Vol. 40, 1981, pp. 263-293.

26. Roe, P.L., "Approximate Riemann Solvers, Parameter Vectors, and Difference Schemes", J. Comp. Phys., Vol. 43, 1981, pp. 357-372.

27. Osher, S., and Solomon, F., "Upwind Difference Schemes for Hyperbolic Systems of Conservation Laws", Math. Comp., Vol. 38, 1982, pp. 339-374.

28. Harten, A., "High Resolution Schemes for Hyperbolic Conservation Laws", J. Comp. Phys., Vol. 49, 1983, pp. 357-393.

29. Osher, Stanley, and Chakravarthy, Sukumar, "High Resolution Schemes and the Entropy Condition", SIAM J. Num. Analysis, Vol. 21, 1984, pp. 955-984.

30. Sweby, P.K., "High Resolution Schemes Using Flux Limiters for Hyperbolic Conservation Laws", SIAM J. Num. Analysis, Vol. 21, 1984, pp. 995-1011.

31. Anderson, B.K., Thomas, J.L., and Van Leer, B., "A Comparison of Flux Vector Splittings for the Euler Equations", AIAA Paper 85-0122, AIAA 23rd Aerospace Sciences Meeting, Reno, January, 1984.

32. Yee, H.C., "On Symmetric and Upwind TVD Schemes", Proc. 6th GAMM Conference on Numerical Methods in Fluid Mechanics, Göttingen, September 1985.

33. Jameson, A., "A Nonoscillatory Shock Capturing Scheme Using Flux Limited Dissipation", Lectures in Applied Mathematics, Vol. 22, Part 1, Large Scale Computations in Fluid Mechanics, edited by B.E. Engquist, S. Osher and R.C.J. Sommerville, AMS, 1985, pp. 345-370.

34. Lax, P.D., "Hyperbolic Systems of Conservation Laws and the Mathematical Theory of Shock Waves", SIAM Regional Series on Applied Mathematics, Vol. 11, 1973.

35. Jameson, A., and Lax, P.D., "Conditions for the Construction of Multi-Point Total Variation Diminishing Difference Schemes", Princeton University Report MAE 1650, April 1984.

36. Yoon, S., Private Communication.

37. Gilles, M., Drela, M., and Thompkins, W.T., "Newton Solution of Direct and Inverse Transonic Euler Equations", AIAA Paper 85-1530, Proc, AIAA 7th Computational Fluid Dynamics Conference, Cincinnati, 1985, pp. 394-402.

38. Kinmark, I.P.E., "One Step Integration Methods with Large Stability Limits for Hyperbolic Partial Differential Equations", Advances in Computer Methods for Partial Differential Equations, V, edited by R. Vichnevetsky and R.S. Stepleman, IMACS, 1984, pp. 345-349.

39. Jameson, A., "Transonic Flow Calculations for Aircraft", Lecture Notes in Mathematics, Vol. 1127, Numerical Methods in Fluid Dynamics, edited by F. Brezzi, Springer Verlag, 1985, pp. 156-242.

40. Gourlay, A.R., and Mitchell, A.R., "A Stable Implicit Difference Scheme for Hyperbolic Systems in Two Space Variables", Numer. Math., Vol. 8, 1966, pp. 367-375.

41. Beam, R.W., and Warming, R.F., "An Implicit Finite Difference Algorithm for Hyperbolic Systems in Conservation Form", J. Comp. Phys., Vol. 23, 1976, pp. 87-110.

42. Jameson, A., and Turkel E., "Implicit Schemes and LU Decompositions", Math. Comp. Vol. 37, 1981, pp. 385-397.

43. Obayashi, S., and Kuwakara, K., "LU Factorization of an Implicit Scheme for the Compressible Navier Stokes Equations", AIAA Paper 84-1670, AIAA 17th Fluid Dynamics and Plasma Dynamics Conference, Snowmass, June 1984.

44. Obayashi, S., Matsukima, K., Fujii, K., and Kuwakara, K., "Improvements in Efficiency and Reliability for Navier-Stokes Computations Using the LU-ADI Factorization Algorithm", AIAA Paper 86-0338, AIAA 24th Aerospace Sciences Meeting, Reno, January 1986.

45. Chakravarthy, S.R., "Relaxation Methods for Unfactored Implicit Upwind Schemes", AIAA Paper 84-0165, AIAA 23rd Aerospace Sciences Meeting, Reno, January 1984.

46. Jameson, A., and Yoon, S., "Multigrid Solution of the Euler Equations Using Implicit Schemes", AIAA Paper 85-0293, AIAA 23rd Aerospace Sciences Meeting, Reno, January, 1985.

47. Jameson, A., and Yoon, S., "LU Implicit Schemes with Multiple Grids for the Euler Equations", AIAA Paper 86-0105, AIAA 24th Aerospace Sciences Meeting, Reno, January, 1986.

48. Anderson, W.K., Thomas, J.L., and Whitfield, D.L., "Multigrid Acceleration of the Flux Split Euler Equations", AIAA Paper 86-0274, AIAA 24th Aerospace Sciences Meeting, Reno, January 1986.

49. Berger, M., and Jameson, A., "Automatic Adaptive Grid Refinement for the Euler Equations", AIAA Journal, Vol. 23, 1985, pp. 561-568.

50. Dannenhoffer, J.F., and Baron, J.R., "Robust Grid Adaption for Complex Transonic Flows", AIAA Paper 86-0495, AIAA 24th Aerospace Sciences Meeting, Reno, January 1986.

BUS COUPLED SYSTEMS FOR MULTIGRID ALGORITHMS

O. Kolp and H. Mierendorff
Gesellschaft fuer Mathematik und Datenverarbeitung mbH
Schloss Birlinghoven
D-5205 Sankt Augustin 1, F. R. Germany

Abstract

Speedup and efficiency of some simple parallel multigrid algorithms for a class of bus coupled systems are investigated. We consider some basic multigrid methods (V-cycle, W-cycle) with regular grid generation and without local refinements. Our bus coupled systems consist of many independent processors each with its own local memory. A typical example for our abstract bus concept is a ring bus. The investigation of such systems is restricted to hierarchical orthogonal systems. Simple orthogonal bus systems, tree structures and mixed types are included in our general model. It can be shown that all systems are of identical suitability if the tasks are sufficiently large. The smaller however the degree of parallelism of an algorithm is, the clearer are the differences in the performance of the various systems. We can classify the most powerful systems and systems with lower performance but better technical properties. Complexity investigations enabled us to evaluate the different systems. These investigations are complemented by simulations based on the different parallel algorithms.

In general, the order of the speedup depends only on a few parameters, such as the dimension of the problem, the cycle type and the dimension of the system. The constant factors in our asymptotical expressions for the speedup depend on many parameters, especially on those of the processors and buses. We investigate these relations by simulation of some typical examples. The simulation also clarifies under which circumstances the asymptotical rules are useful for the description of system behavior.

1. Introduction

Parallel computers allow a considerable reduction of absolute computing time for most problems if compared with processing on single processors of the same type. For large problems it is of special importance to know the speedup to be thus obtained if using the fastest known algorithms. Multigrid methods are among the fastest methods for solving partial differential equations. The speedup obtainable by parallelizing these methods has already been investigated from various points of view. Brandt mentions in [Br81] basic possibilities of parallelizing the method and some especially suitable computer structures. In [CS83] Chan and Schreiber discuss a problem-adapted computer architecture. In all these architectures the order of speedup is exclusively determined by the properties of the corresponding multigrid method and by the number of used processors. If we use a computer architecture whose connection structure is not in accordance with problem structure, the communication between processors is of decisive importance to the obtainable speedup. Basic considerations on this issue are contained, for example, in [GR84]. Kolp and Mierendorff investigate in [KM85] some classes of connection structures where the number of connection elements and that of processors are equal with respect to the order.

Here we consider connection elements which are able to connect relatively many processors. Our system, however, includes considerably less connection elements than processors. Buses are typical tools of implementing such connection elements. In this context the essential assumption about these buses is that at a given time only one processor can send data via the bus while all other processors may act as receivers. Bus concepts which allow a dynamic decomposition of a bus into independently working parts are not included in our consideration.

In the sections 2 to 4 we begin with describing our abstract transport model, we precisely define the classes of the bus systems to be considered and present an abstract skeleton of computational and transport work suitable for the following investigations.

Let $A(P)$ be the time (measured in a suitable unit cost measure) which is required for a parallel algorithm on P processors. By

$$S(P) = A(1)/A(P) \quad \text{and} \quad E(P) = S(P)/P$$

we denote as usual the speedup resp. the efficiency of the parallel algorithm on P processors in contrast to the sequential algorithm on a processor of identical type. If $E(P) \geq const > 0$ for all problem sizes N, we call an algorithm efficient. This definition distinguishes the efficient algorithms especially for P increasing with N.

In sections 5 to 7 we first discuss the optimal speedup to be obtained on our systems. It turns out that the optimal speedup is only restricted by the properties of the used cycle type, but the corresponding algorithms are very inefficient and require unrealistically large systems. For this reason the question for the optimal speedup on our systems is not reasonable. Therefore, we investigate efficient algorithms being best with respect to the order of the speedup and prove that a simple projection algorithm will meet the requirements for all cycle types. By projection algorithm we here denote an element of a class of algorithms for which the domain of the differential equation is decomposed into subdomains to be one-to-one assigned to the processors. Each processor will then calculate all values for the grid points of its subdomain. The optimality of the distinguished projection algorithm is to be understood in relation to the set of algorithms fulfilling the following assumptions:

(I) Each computation occurring in the sequential algorithm is carried out exactly once in the parallel version (no computational redundancy).

(II) All processors involved in the computations for a level of the multigrid method must calculate numbers of grid points that are equal with respect to the order.

The question whether these assumptions are required must be left open. Presumably, at least (II) is not necessary for the result. Finally, we consider some simpler partitions of the problems and the modifications of complexity results to be expected.

In sections 8 and 9 we apply the results to an example, namely the SUPRENUM-1 architecture [GM85]. We consider the actual model behavior for some realistic parameter values. For large systems we find its accordance with our complexity results.

2. Transport Model for Bus Coupled Systems

For the transport of data between the local memories of the processors we use the bus connection represented by the diagram in fig. 1. At a given time only one of the connected processors shall be able to send data via the bus. All other processors may only receive data at that time. We will require only transports going to one target processor. We always imagine the transport in form of data packets of a varying number x of data elements.

For enabling the transport of a data packet, a computational work a_1x+a_2 is required in the involved processors. a_1 is the work required for transporting a data element in the processor (i.e. in its memory) to a location to be accessed by the bus. a_2 is the work required for processing a packet (e.g. for one SEND and RECEIVE). In case of suitable organization of hardware and software one can obtain $a_1=0$. a_2 may correspond in real systems to some hundred or thousand instructions.

In the bus a data packet requires the time a_3x+a_4. a_3 corresponds to the transport time for a data element and a_4 to the time required for a packet.

In asymptotical considerations the order of total transport work required can be denoted by $\Theta(x)$. In transporting many packets via a bus system the work corresponding to the term a_1x+a_2 has in any case to be added to the computational work. This work can be distributed at best to all participating processors. The work corresponding to the term a_3x+a_4 can only be distributed to all buses. The problem of balancing must thus be solved simultaneously for the processors and for the buses.

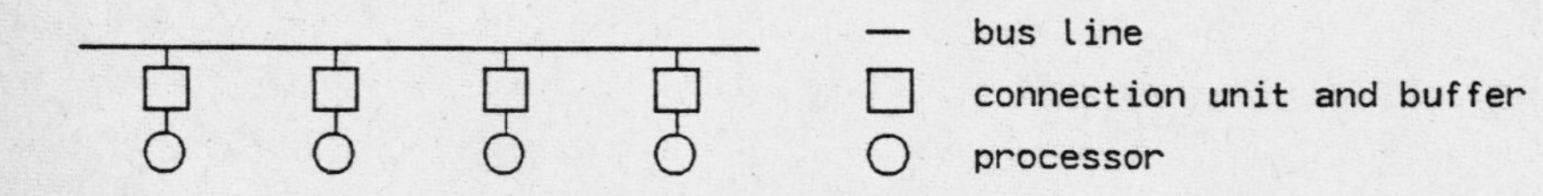

Fig. 1: Model of bus connection

The transport work in the bus system can in the best case be provided in parallel to the computational work in the processors. In the worst case this is done successively. For identifying the optimum order of total work required in complexity investigations this will be of no importance so that we can always start from the pessimistic model. For simulating concrete systems we have to distinguish worst case and best case. In this paper we restrict the illustration to the worst case.

3. The Structure of Hierarchical Orthogonal Systems

As computer structures we consider a set of bus systems that we call hierarchical orthogonal bus systems. The structure consists of m levels that are indexed from the highest to the lowest level by i=1,...,m. In each level there are orthogonal systems of buses (fig. 2). Coupling elements located at the cross points of the bus lines provide connecting points to the next lower level. Let all subsystems of a level have identical form and size. The coupling elements connect each bus of a subsystem in the lower level with the higher level buses belonging to the corresponding cross point. This assumption can be somewhat reduced which is however of no importance to our significant examples. We assume that all orthogonal subsystems are hypercubic. Let b_i be the dimension in the i-th level and let p_i be the ${}^a\log$ of the edge length (number of connecting points). Furthermore let $a \geq 2$ be an integer that will be defined in detail below. The number of all connecting points of the i-th level is then

$$P_i = \prod_{j=1}^{i} a^{b_j p_j} \quad . \tag{3.1}$$

The number of buses of the i-th level is

$$B_i = P_{i-1} b_i a^{(b_i-1)p_i} = b_i a^{\sum_{j=1}^{i} b_j p_j - p_i} \quad . \tag{3.2}$$

To simplify we set

$$b = b_1,\ p = p_1,\ p_i = \beta_i p_{i-1} + \gamma_i,\ \beta = \sum_{i=2}^{m} b_i \prod_{j=2}^{i} \beta_j \text{ and } \gamma = \sum_{i=2}^{m} b_i \sum_{k=2}^{i} \gamma_k \prod_{j=k+1}^{i} \beta_j \ .$$

In the last (m-th) level exactly one processor is located at each coupling element. The number of processors in the system is therefore

$$P = \prod_{i=1}^{m} a^{b_i p_i} = a^{\sum b_i p_i} = a^{(b+\beta)p+\gamma} = \Theta(a^{(b+\beta)p}) \quad \text{for large p.} \tag{3.3}$$

Examples for such structures are all multiprocessor systems without shared memory, but with a bus (m=1, b=1) or purely hierarchical systems ($m \geq 1$; $b_i=1$ for i=1,...,m). The SUPRENUM-1 structure (m=2, $b_1=2$, $b_2=1$) represented by fig. 3 is also included.

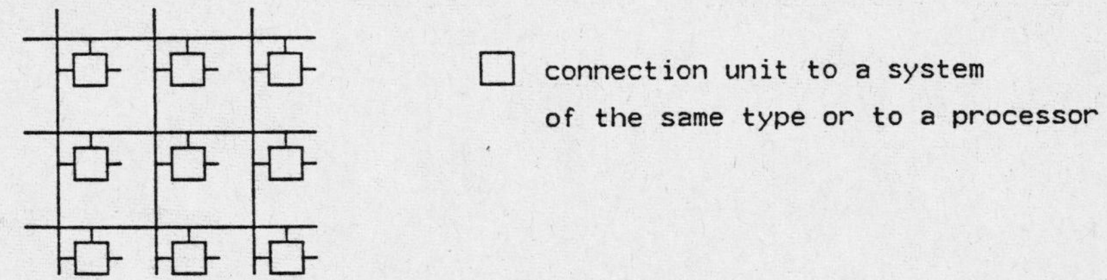

Fig. 2: Orthogonal bus system of dimension b=2

4. An Abstract Skeleton for Simple Multigrid Methods

Multigrid methods are described, for example, in Stüben and Trottenberg [ST82] and are assumed as known here. We denote a general method, which comprises V-cycle and W-cycle, as in [CS83], by Basic Multigrid (BMG). We consider BMG with regular grid generation and without local refinements. Since we intend to study here the decomposition of these algorithms and their mapping onto specific computer architectures, it is sufficient to schematize the considered algorithms to a skeleton and to use only this in any further step.

Let a problem be defined by suitable boundary conditions on a d-dimensional hypercube R as domain and an appropriate partial differential equation. In multigrid methods and after discretization on a set $\{G_1,...,G_n\}$ of orthogonal point grids in R, being in a certain interrelation, discrete equation systems are assigned to this problem. For formal reasons we define G_0 as the grid that only consists of the vertices of R. G_i (i=1,...,n) will then result from G_{i-1} by a-partition of the edges of the meshes of G_{i-1}, where a is a power of 2, a≥2 and in the standard case a=2. Let 'rel' be the effort required for each inner grid point for the relaxation operator of a multigrid method. We here confine ourselves to point relaxation and for simulation we shall always use odd-even relaxation with the advantages involved. Accordingly let the effort required for each point for interpolation and restriction be 'int' and 'res' respectively. In some cases this assumption constitutes a rough schematization. As long as the ratios of the time effort of the operators are however bounded, this will be of no importance to our asymptotic consideration. In the numerical evaluation of our model we shall assume simple elliptic differential equations and simple operators (that need at most the 2d next neighbors of a point), hence with good approximation rel=int=res.

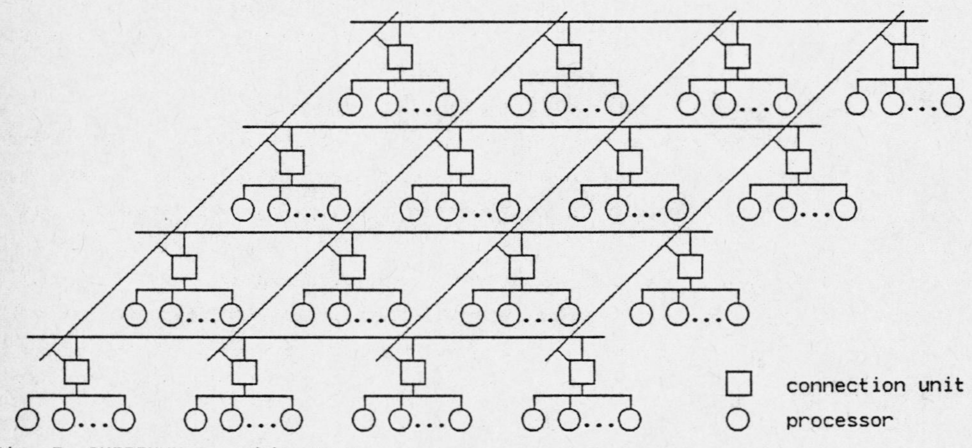

Fig. 3: SUPRENUM-1 architecture

Let ν_1 or ν_2 respectively be the amount of relaxations per grid in transition from the finer to the coarser grids or vice versa. Interpolations are only carried out for the fine points of a grid $(G_i - G_{i-1})$. In odd-even relaxation it is sufficient to perform the defect computing for the restriction only on the coarse points of G_i (i.e. G_{i-1}). The mean effort required per grid point and V-cycle is, in general, with suitable constants σ, τ

$$(\nu_1+\nu_2)*rel+\sigma*int+\tau*res \ . \tag{4.1}$$

In general, $\sigma=(a^d-1)/a^d$. In our asymptotic considerations we shall treat it as variable bounded by constants. In the simulated examples, this effort is reduced by the exclusively considered simple cases $\sigma+\tau=1$ and a_0=rel=int=res to

$$a_0*(\nu_1+\nu_2+1) \ . \tag{4.2}$$

With (4.2) and $(a^i-1)^d$ as number of inner grid points of G_i the computational work for V-cycle (c=1) and W-cycle (c=2) on a single processor system is [ST82]:

$$A(1) = \sum_{i=1}^{n} c^{n-i}(a^i-1)^d a_0(\nu_1+\nu_2+1) \tag{4.3}$$

and in general with (4.1) and $\alpha={}^a\log c$:

$$A(1) = \begin{cases} \Theta(a^{\alpha n}) & \text{if } a^d<c \text{ i.e. } d<\alpha \\ \Theta(na^{dn}) & \text{if } a^d=c \text{ i.e. } d=\alpha \\ \Theta(a^{dn}) & \text{if } a^d>c \text{ i.e. } d>\alpha \ . \end{cases} \tag{4.4}$$

The effort required for direct problem solving on the coarsest grid G_1 is not correctly treated in (4.3) which is of minor importance to larger problems. In the interesting standard cases, $a^d>c$ always holds because of a=2, c≤2 and d=2 or d=3 .

In general, the following holds for an arbitrary parallel algorithm

$$A(P) = \Omega(a^{\alpha n}) \ . \tag{4.5}$$

The computational work in the coarsest grid is namely hardly distributable. With N as number of points of the finest grid (i.e. $N=\Theta(a^{dn})$) we obtain from (4.4) and (4.5)

$$S(P) = O(n) = O(\log N) \qquad \text{if } d\le\alpha \ .$$

Parallelization is thus not useful in the case $d\le\alpha$. We therefore are always assuming for our further investigations $d>\alpha$.

5. Parallelization and an Asymptotically Optimal Mapping of the Problem

A given d-dimensional cubiform domain R is first decomposed by a-partition of the edges of already generated subdomains in the direction of all coordinate axes. The coordinate directions are subjected to an arbitrary but defined cycle of a-partition until the number of subdomains is in accordance with the number of connecting points of the 1st level. The subdomains are uniquely assigned to the connecting points in a way that will be discussed later. We now treat the subdomains in the 2nd level in the same way continuing the cycle of using the axis directions for partitioning. Partition and assignment are made successively for all levels in the same way until exactly one subdomain corresponds to each processor. A processor performs the computations for all grid points belonging to its subdomain. We assume that the subdomains are closed in direction of increasing coordinates and otherwise open (with obvious exceptions at the boundary). In this way we obtain a disjoint decomposition of the domain. All subdomains are of equal size. Therefore, the processors have the same number of points with respect to the order. Because of the cyclic use of axes in partition the shape of the obtained subdomains is nearly cubiform. For each grid it is true that the number of meshes is identical on two edges of a subdomain or is different by the factor a.

We now specify an assignment for subdomains of a level with b as dimension of the bus system and a^p connecting points to a bus. In the case b=1 assignment is arbitrary since on a bus all connecting points have identical neighboring relationships. For b=2 fig. 4 gives a schema for the assignment of a number of subdomains immediately following each other in an axis direction. For blocks of this type following each other in another axis direction, mapping is done in the same way onto the already generated blocks of connecting points by connecting corresponding elements of blocks such as in fig. 4. This is continued until all axis directions have been included. For higher dimensions b>2 the method can be generalized adequately. The transport work for all grids containing sufficient points is thus distributed to a minimum of $a^{(b-1)p}$ buses. Since there are exactly $ba^{(b-1)p}$ buses in this level, mapping is optimal with respect to the order, i.e. with the inaccuracy factor $\leq b$. This factor can be improved by choosing, if possible, for different axis directions of the problem different bus directions doing the main work. This is however of no importance to our asymptotical investigations.

We also know mappings where the neighborhoods of an axis direction of the problem are for fine grids already optimally distributed to the bus system. These mappings are however unsuitable for coarse grids and therefore not useful for the asymptotical consideration. A mapping that produces a completely balancing of transport work over all buses in any case and for all grids is not yet known. For practical use we shall later investigate more simple mapping principles that produce slightly worse results in some cases.

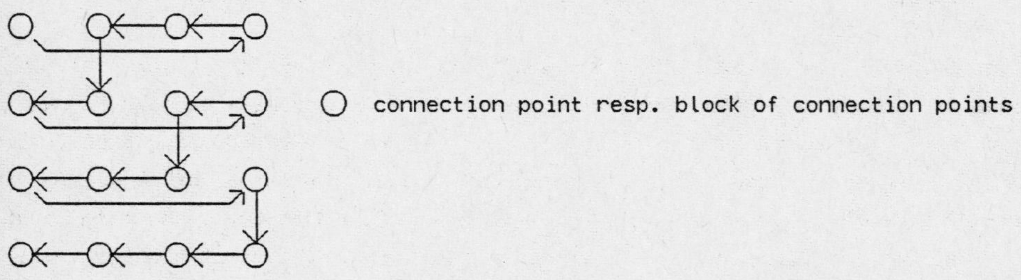

Fig. 4: Mapping of a sequence of neighbored subdomains to a 2-dimensional system of buses.

6. Complexity of Parallel Multigrid Methods on Bus Systems

First we consider the method described in section 5 and thus get lower bounds for the obtainable speedup. We assume such a situation where the order of the transport work in the first level is greatest. We now deduce the conditions under which this situation occurs.

Let t_j be the mean number of a-partitions of an axis direction of the problem for the j-th level of the system:

$$t_j = b_j p_j / d \quad .$$

Let $A_b(j)$ be the transport work during the processing of the i-th grid in the j-th level. This corresponds to a constant number of boundary faces of each subdomain. If for a grid every connecting point has at most one grid point, the number of all grid points, $\Theta(a^{id})$, must be distributed to all buses of the j-th level (3.2). Otherwise the number of connecting points of a bus has to be multiplied by the boundary size of a subdomain of the i-th grid on the j-th level. We obtain

$$A_b(j) = \begin{cases} \Theta(\lceil a^{id-\sum_{k=1}^{j} b_k p_k + p_j} \rceil) & \text{for } i \le \sum_{k=1}^{j} t_k + O(1) \\ \Theta(a^{p_j+(d-1)(i-\sum_{k=1}^{j} t_k)}) & \text{for } i > \sum_{k=1}^{j} t_k + O(1) \end{cases} \quad .$$

Our assumption is fulfilled if

$$A_b(j)/A_b(j+1) = \Omega(1) \quad .$$

For coarse grids (small i) the following is sufficient:

$$id-\sum_{k=1}^{j} b_k p_k + p_j \ge id - \sum_{k=1}^{j+1} b_k p_k + p_{j+1} \qquad \text{i.e.} \qquad p_j \ge p_{j+1}(1-b_{j+1}) \quad .$$

These conditions are always fulfilled.

For fine grids (great i) the following is sufficient:

$$p_j + (d-1)(i-\sum_{k=1}^{j} t_k) \ge p_{j+1} + (d-1)(i-\sum_{k=1}^{j+1} t_k)$$

i.e.

$$p_j \ge p_{j+1}(1-b_{j+1}(d-1)/d) \tag{6.1}$$

Let us assume (6.1) for our further considerations of lower bounds of the speedup. For $b_{j+1}>1$ and $d>1$ the conditions are always fulfilled. For $b_{j+1}=1$ or $d=1$ we obtain

$$dp_j \ge p_{j+1} \quad .$$

For the SUPRENUM-1 structure ($b_1=2, b_2=1$) this means

$$dp_1 \ge p_2 \quad . \tag{6.2}$$

For a real system the condition $A_b(j) \ge A_b(j+1)$ is especially interesting. In this case $dp_j \ge p_{j+1}+\text{const}$ is then sufficient. This constant mainly results from the definition of t_{j+1} and the situation at the boundary. For SUPRENUM-1 we found const$\le 1.5d$ if $p_1 \ge 3$ and $d=2,3$. For the mainly interesting problems ($d \ge 2$) these conditions will often be fulfilled in real systems.

With A_c and A_b as computational work or transport work for subdomain boundaries respectively we obtain as overall work of our parallel multigrid algorithm

$$A(P) = \Theta(A_c(P)+A_b(P)) \quad .$$

By even balancing over a maximum number of processors and with $\Theta(a^{id})$ as computational work for one V-cycle in the i-th grid, the following holds:

$$A_c(P) = O(\sum_{i=1}^{n} c^{n-i}\lceil a^{id}/P\rceil) \quad .$$

Hence by (3.3):

$$A_c(P) = O(\sum_{i=1}^{n} a^{\max(\alpha(n-i),\alpha n+i(d-\alpha)-(b+\beta)p)})$$
$$= O(n+a^{\alpha n}+a^{dn-(b+\beta)p}) \quad . \tag{6.3}$$

According to assumption we must consider only the first system level for the transport work. The load is distributed to $\Theta(a^{(b-1)p})$ buses as soon as sufficient grid points are available. Therefore, in the i-th grid the point set to be transported per bus shows at most a constant number of points for $id\leq(b-1)p$ and $O(a^{id-(b-1)p})$ for $(b-1)p\leq id\leq bp$. For greater i $t_1=bp/d+O(1)$ partitions have been done in one axis direction of the problem that lead to $\Theta(a^{bp/d})$ hyperfaces of the size $\Theta(a^{i(d-1)})$. By $j=\lceil(b-1)p/d\rceil$ and $k=\lceil bp/d\rceil$ we obtain:

$$A_b(P) = O(\sum_{i=1}^{j} c^{n-i} + \sum_{i=j+1}^{k} c^{n-i}a^{id-(b-1)p} + \sum_{i=k+1}^{n} c^{n-i}a^{i(d-1)+bp/d-(b-1)p}) \tag{6.4}$$

In the case j=n, the following holds:

$$A_b(P) = O(n+a^{\alpha n}) \quad . \tag{6.5}$$

In this case it is required that b>1. Furthermore, $p=dn/(b-1)+O(1)$. Hence

$$dn-(b+\beta)p \leq 0 \quad \text{if p is sufficiently large .}$$

From (6.3) and (6.5) it follows that

$$A(P) = O(n+a^{\alpha n}) \quad .$$

With (4.4) we finally obtain:

$$S(P) = \begin{cases} \Omega(a^{dn}/n) = \Omega(N/\log N) & \text{if } \alpha=0 \\ \\ \Omega(a^{dn}/a^{\alpha n}) = \Omega(N^{(d-\alpha)/d}) & \text{if } \alpha>0 \quad . \end{cases} \tag{6.6}$$

This is however the best speedup obtainable by parallelization (cf. [Br81] and (4.5)).

Because of (3.3) and (6.6) the efficiency of such an algorithm is so bad that this result is unimportant to practice. Therefore, we think it is useful to consider in the following only efficient algorithms for evaluating the power of our architecture.

An algorithm can only be efficient if the number of used processors is $\Theta(P)$. Without loss in generality we can therefore assume $k\leq n$. Furthermore, for the finest grid transport work should not exceed computational work with respect to the order:

$$a^{n(d-1)+bp/d-(b-1)p} \leq O(a^{dn-(b+\beta)p}) \quad .$$

Hence

$$p \leq nd/(b+d(1+\beta))+O(1) \tag{6.7}$$

and thus $n-k=\Theta(n)$. For the transport we then obtain:

$$A_b(P) = O(\sum_{i=1}^{j} a^{\alpha(n-i)} + \sum_{i=j+1}^{k} a^{\alpha n-(b-1)p+i(d-\alpha)} + \sum_{i=k+1}^{n} a^{\alpha n+bp/d-(b-1)p+i(d-1-\alpha)})$$

$$A_b(P) = O(n + a^{\alpha n} + a^{\alpha n+p(1-\alpha b/d)} + \begin{Bmatrix} a^{\alpha n+p(1-\alpha b/d)} & \text{if } d-1<\alpha \\ na^{\alpha n+p(1-\alpha b/d)} & \text{if } d-1=\alpha \\ a^{n(d-1)+p(b+d-bd)/d} & \text{if } d-1>\alpha \end{Bmatrix}) \tag{6.8}$$

The term n is for $\alpha>0$ smaller than the second term and for $\alpha=0$ smaller than the fourth term of (6.8). The third term of (6.8) is for $d-1\le\alpha$ obviously not greater than the fourth term. This is also true for $d-1>\alpha$, as far as $p\le nd/b$ holds. For an efficient algorithm the fourth term should however not exceed computational work (see (6.3)):

$$n(d-1)+p(b+d-bd)/d \le dn-(b+\beta)p+O(1) \quad .$$

Hence $p\le nd/(b+d(1+\beta))+O(1)$, so that the third term is of no importance to efficient algorithms. We disregard these terms and obtain with (6.3):

$$A(P) = O(a^{\alpha n} + \begin{cases} a^{\alpha n+p(1-\alpha b/d)} & \text{if } d-1<\alpha \\ n a^{\alpha n+p(1-\alpha b/d)} & \text{if } d-1=\alpha \\ a^{n(d-1)+p(b+d-bd)/d} & \text{if } d-1>\alpha \end{cases} + a^{dn-(b+\beta)p}) \quad . \tag{6.9}$$

In the following we are going to show that an arbitrary parallel algorithm requires at least the work (6.9), as far as the assumptions (I) and (II) of section 1 are valid.

The first and the last term of (6.9) are inevitable as computational work of the coarsest resp. finest grid. The second term specifies in the case $d-1>\alpha$ the minimum work for exchange of the boundary data for processing the finest grid. Therefore, the second term is necessary.

Under the assumptions (I) and (II) we now consider an arbitrary parallel algorithm. For the i-th grid let $a^{bp(i)}$ connecting points of first system level work. p(i) must no longer be an integer, and there are no assumptions about the location of these connecting points. Without loss in generality we can however assume $p(i+1)\ge p(i)$ since otherwise an at least equally suitable algorithm exists that fulfills this assumption. In the transition from the i-th grid to the (i+1)-th grid at least a transport is required which corresponds to the product of the point number of the i-th grid, $\Theta(a^{id})$, and the fraction of the newly added connecting points

$$(a^{bp(i+1)}-a^{bp(i)})/a^{bp(i+1)} = 1-a^{-b(p(i+1)-p(i))} \quad .$$

In case of balancing over all buses we have at least a transport A_t of the value:

$$A_t = \Omega(\sum_{i=j}^{n-1} a^{\alpha(n-i)-(b-1)p+id}(1-a^{-b(p(i+1)-p(i))}))$$
$$= \Omega(a^{\alpha n-(b-1)p+j(d-\alpha)} \sum_{i=j}^{n-1} (1-a^{-b(p(i+1)-p(i))})) \quad .$$

For each continuous, concave, monotonically increasing function $f(x)$ with $f(0)\ge 0$ the following holds:

$$\sum_{i=1}^{n} f(x_i) \ge f(\sum_{i=1}^{n} x_i) \qquad \text{for } x_i\ge 0 \quad .$$

$f(x)=1-a^{-x}$ shows these properties.

Hence by $p(n)=p$

$$A_t = \Omega(a^{\alpha n-(b-1)p+j(d-\alpha)}(1-a^{-b(p-p(j))})) \quad .$$

$a^{bp(j)}\le a^{dj}$ always holds (at most as many connecting points as grid points). For $j=\lfloor bp/d\rfloor-1$ we obtain $p-p(j)\ge\text{const}>0$ for sufficiently great p. Hence

$$A_t = \Omega(a^{\alpha n+p(1-\alpha b/d)}) \quad .$$

This proves our assertion in the case $d-1<\alpha$.

The case $d-1=\alpha$ is still to be investigated. As above we obtain for any j with

$p-p(j)\geq const>0$

$$A_t \geq \Omega(a^{\alpha n-(b-1)p+j}) \quad .$$

For making this algorithm not worse than the projection algorithm of section 5, the following must hold:

$$a^{\alpha n-(b-1)p+j} \leq O(na^{\alpha n+p(1-\alpha b/d)})$$

or with $\alpha=d-1$

$$j \leq pb/d + {}^a\log n + O(1) \quad .$$

Consequently, $p-p(i)\leq const$ must be true for all $i\geq pb/d+{}^a\log n + O(1)$. Therefore, the transfer of data for any of these grids especially for the finest grid has the same order of size. From (6.7) follows

$$n-(pb/d+{}^a\log n) \geq n(1-b/(b+d(1+\beta)))-{}^a\log n + O(1)$$
$$\geq \Theta(n) \quad .$$

Therefore, in this case an exchange of boundary data of an order like in (6.9) is required for an arbitrary algorithm that shall not exceed (6.9).

Let us denote the logarithms to the base a of the terms of (6.9) successively by T_1, T_2 and T_3. The terms T_1 and T_2 are caused by the transport. T_1 and T_3 represent the computational work. T_1 is constant and T_3 is decreasing in p. To find for p a function of n for which the best efficient algorithm with respect to the order is obtained, it is not useful to make T_3 smaller than T_1. If for such p T_2 is already considerably greater, the equation $T_2=T_3+O(1)$ will lead to the best p. Since the third term determines in any case the order of A(P), $A(P)=\Theta(N/P)$ is true and therefore $S(P)=\Theta(P)=\Theta(a^{(b+\beta)p})$. For determining S and p we distinguish four cases.

Case 1: $T_1=T_3+O(1)$, $T_1\geq T_2+O(1)$.

We obtain the equation $\alpha n = dn-(b+\beta)p + O(1)$. Hence

$$S(P) = \Theta(N^{(d-\alpha)/d}) \quad \text{for} \quad p = n(d-\alpha)/(b+\beta)+O(1) \quad . \tag{6.10}$$

This case occurs if one of the following conditions holds:

(1) $d-1<\alpha$ and $d/b\leq\alpha$

(2) $d-1=\alpha$ and $d/b<\alpha$

(3) $d-1>\alpha$ and $d(d(1+\beta)-\beta)/(d(1+\beta)+b) \leq \alpha$.

Remark 6.1: In case 1 neither assumption (I) nor (II) was used since computational work is always exceeding. Therefore, the result (6.10) cannot be improved by any parallel algorithm (nor by a more suitable computer architecture).

Case 2: $T_2=T_3+O(1)$, $T_2\geq T_1+O(1)$, $d-1<\alpha$.

We obtain the equation $\alpha n+p(1-\alpha b/d) = dn-(b+\beta)p+O(1)$. Hence

$$S(P) = \Theta(N^{(b+\beta)(d-\alpha)/(b(d-\alpha)+d(1+\beta))}) \text{ for } p=nd(d-\alpha)/(b(d-\alpha)+d(1+\beta))+O(1). \tag{6.11}$$

The case occurs under the condition $d-1<\alpha$ and $d/b\geq\alpha$.

Case 3: $T_2=T_3+O(1)$, $T_2\geq T_1+O(1)$, $d-1=\alpha$.

We obtain the equation $\alpha n+p(1-\alpha b/d)+{}^a\log n = dn-(b+\beta)p+O(1)$. Hence

$$S(P) = \Theta((N/\log^d N)^{(b+\beta)/(b+d(1+\beta))}) \quad \text{for } p = (n-{}^a\log n)d/(b+d(1+\beta))+O(1). \tag{6.12}$$

The case occurs under the condition $d-1=\alpha$ and $d/b\geq\alpha$.

Remark 6.2: In the cases 2 and 3 we used the assumptions (I) and (II).

Case 4: $T_2=T_3+O(1)$, $T_2 \geq T_1+O(1)$, $d-1>\alpha$.

We obtain the equation $n(d-1)+p(b+d-bd)/d = dn-(b+\beta)p+O(1)$. Hence

$$S(P) = \Theta(N^{(b+\beta)/(b+d(1+\beta))}) \quad \text{for } p = nd/(b+d(1+\beta))+O(1) \ . \qquad (6.13)$$

This case occurs under the condition $d-1>\alpha$ and $d(d(1+\beta)-\beta)/(d(1+\beta)+b) \geq \alpha$.

Remark 6.3: In case 4 we used only the assumption (I).

Remark 6.4: For V-cycle iterations the primarily interesting cases ($d \geq 2$) are all included in case 4.

We summarize the assignments of conditions and cases in the following tables:

condition	$d/b < \alpha$	$d/b = \alpha$	$d/b > \alpha$
$d-1 < \alpha$	case 1	cases 1 and 2	case 2
$d-1 = \alpha$	case 1	case 3	case 3

condition	$\frac{d(1+\beta)-\beta}{d(1+\beta)+b}\, d < \alpha$	$\frac{d(1+\beta)-\beta}{d(1+\beta)+b}\, d = \alpha$	$\frac{d(1+\beta)-\beta}{d(1+\beta)+b}\, d > \alpha$
$d-1 > \alpha$	case 1	cases 1 and 4	case 4

7. Simpler Mappings and their Influence to Complexity Results

In section 5 we made at each system level for all axis directions of the domain R nearly the same number of partitions to obtain the required number of subdomains. This led to a complicated mapping of subdomains onto system components. Now, let us assign all partitions of one direction to one bus direction. We assume $b_j \leq d$. It is still intended to obtain possibly cubiform subdomains. From all buses of one direction of the j-th system level all transports with respect to those boundaries of subdomains are performed which neighbor in those axis directions which belong to a specific subset of directions. These subsets are uniquely assigned to the b_j bus directions and they form a disjoint decomposition of the set of all d directions. The subsets should possibly be of equal size, i.e. they have $\lfloor d/b \rfloor$ or $\lceil d/b \rceil$ elements. We consider only bus directions that belong to the smaller subsets because they have a greater transport work. Let in the first system level be t the maximal number of partitions in an axis direction belonging to that class. In these directions $\lfloor d/b \rfloor a^t+O(1)$ inner boundary hyperfaces and $\Theta(a^{t\lfloor d/b \rfloor})=\Theta(a^p)$ subdomains are then generated. Hence

$$t = p/\lfloor d/b \rfloor+O(1) \quad .$$

After a completion of definition for mapping in the lower system levels, we can obtain similar conditions as earlier for the situation that the highest system level defines the transport in its order. This shall not be discussed in this context. For the SUPRENUM-1 architecture the further mapping is evident by $b_2=1$, and with respect to the generation of possibly cubiform subdomains in the processors the condition (6.2) has to be replaced by the approximate formula $(2d-2)p_1 \geq p_2 + d \log d$ if $d=2,3$

and $a=2$. This result has been confirmed by simulation for $P=256$ and $P=4096$ (fig. 8).

The number of buses to be used for the i-th grid is $\Theta(a^{\min(i(d-\lfloor d/b \rfloor),p(b-1))})$ and the number of hyperfaces is $\Theta(a^{\min(i,t)})$. Let

$$s = \lfloor p(b-1)/(d-\lfloor d/b \rfloor) \rfloor \text{ if } b>1 \quad \text{and} \quad s=0 \text{ if } b=1 .$$

We obtain $s \le t$ and

$$A_b(P) = O\left(\sum_{i=1}^{n} a^{\alpha(n-i)+i(d-1)+\min(i,t)-\min(i(d-\lfloor d/b \rfloor),p(b-1))} \right)$$

$$= O\left(\sum_{i=1}^{s} a^{\alpha n+i(\lfloor d/b \rfloor-\alpha)} + \sum_{i=s+1}^{t} a^{\alpha n+i(d-\alpha)-p(b-1)} + \sum_{i=t+1}^{n} a^{\alpha n+i(d-1-\alpha)+p/\lfloor d/b \rfloor-p(b-1)} \right) .$$

<u>Remark 7.1</u>: In the case $b=1$ we obtain no deviation from (6.8) and thus from our earlier results. In the case $b|d$ and $b>1$, $t=pb/d+O(1)$ and $s=t+O(1)$ are true. In this case we obtain only for $d-1<\alpha$ and $d/b=\alpha$ a deviation from (6.8) but these conditions are contradictory.

In general

$$A_b(P) = O\left(n + \begin{Bmatrix} a^{\alpha n} & \text{if } \lfloor d/b \rfloor<\alpha \\ na^{\alpha n} & \text{if } \lfloor d/b \rfloor=\alpha \\ a^{\alpha n+p(b-1)(\lfloor d/b \rfloor-\alpha)/(d-\lfloor d/b \rfloor)} & \text{if } \lfloor d/b \rfloor>\alpha \end{Bmatrix} + a^{\alpha n+p(1-\alpha b/d+r(d-\alpha))} + \begin{Bmatrix} a^{\alpha n+p(1-\alpha b/d+r(d-\alpha))} & \text{if } d-1<\alpha \\ na^{\alpha n+p(1-\alpha b/d+r(d-\alpha))} & \text{if } d-1=\alpha \\ a^{n(d-1)+p(b+d-bd+rd)/d} & \text{if } d-1>\alpha \end{Bmatrix}\right) \qquad (7.1)$$

with $r = r(b,d) = 1/\lfloor d/b \rfloor-b/d$. Differences from (6.8) occur in the second term and by the additional expression $r(b,d)$ in the third and fourth terms. As in section 6 the first and the third term can be disregarded.

We restrict the further discussion to the important case 4 of section 6. For the optimal p we then obtain the equation

$$n(d-1)+p(b+d-bd+rd)/d = dn-(b+\beta)p .$$

Hence

$$S(P) = \Omega(N^{(b+\beta)/(b+d(1+\beta+r))}) \quad \text{for } p=nd/(b+d(1+\beta+r)) . \qquad (7.2)$$

That case will occur for

(1) $\lfloor d/b \rfloor<\alpha$, $d-1>\alpha$ and $d(d(1+\beta+r)-\beta)/(d(1+\beta+r)+b) \ge \alpha$,

(2) $\lfloor d/b \rfloor=\alpha$, $d-1>\alpha$ and $d(d(1+\beta+r)-\beta)/(d(1+\beta+r)+b) > \alpha$,

(3) $\lfloor d/b \rfloor>\alpha$, $d-1>\alpha$.

<u>Remark 7.2</u>: For $b|d$ we obtain $r=0$ so that in accordance with remark 7.1 no important change occurs in case 4.

<u>8. Application of the Results to Some Examples of Algorithms and Structures</u>

From section 6 we know that in many cases the increase of $b=b_1$ leads to increased performance. The bounds (6.1) for system size are however such weak that in practice $b_j=1$ for $j>1$ is completely sufficient. Therefore, only two types of architecture are of special interest, namely trees ($b_j=1$ for $j=1,\ldots,m$ and $m\geq 1$) and SUPRENUM-1 ($b_1=2$, $b_2=1$, $m=2$). We restrict our consideration to SUPRENUM-1.

Let us now discuss our results for the most important problem classes ($d=2,3$; $a=2$ and $\alpha=0,1$). From (6.12) and (6.13) it follows for the algorithm of section 5:

$d=2,\ \alpha=0$ (V-cycle): $S(P) = \Theta(N^{1/2})$ if $p=n/(2+\beta)+O(1)$

$d=2,\ \alpha=1$ (W-cycle): $S(P) = \Theta(N^{1/2}/\log N)$ if $p=(n-{}^2\log n)/(2+\beta)+O(1)$

$d=3,\ \alpha=0,1$: $S(P) = \Theta(N^{(2+\beta)/(5+3\beta)})$ if $p=3n/(5+3\beta)+O(1)$.

The assignment of a constant $\beta=\beta_2$ to a real system first seems to present problems in the case $d=3$ because of $p_2=\beta_2 p_1+\tau_2$. In case of fixed values of p_1 and p_2 many pairs β_2,τ_2 fulfill this equation. While β has a functional effect on the speedup, τ_2 only influences the constant factor of the speedup.

In case of a fixed cluster size we have $\beta=0$ and therefore $S(P) = \Theta(N^{2/5})$ for large problems. Such a statement means that for a sufficiently large system with P processors the size of the problems to be handled with a given efficiency is bounded from below. In this case that lower bound is $N\geq\Omega(P^{5/2})$.

If we increase the system by other values for β, performance development is to be expected to be worse. The speedup is above $\Theta(N^{5/14})$ if the condition (6.2) is fulfilled ($\beta\leq d$).

Our simple mapping method of section 7 supplies for $d=2$ the same statements. For $d=3$ we obtain by $r(2,3)=1/3$ independently of β ($\beta\leq d$) only $S(P) = \Theta(N^{1/3})$.

9. Simulation Results

For simulation (i.e. numerical evaluation of our model without disregarding terms of lower order) we make some assumptions. As parameters of parallel algorithms we consider $a=2$, $\nu_1=\nu_2=1$ und $\alpha=0$ (V-cycle). We assume odd-even relaxation and simple operators for interpolation, restriction and discretization that only require the 2d next neighbors of a point. The parallelized form of the method is treated by means of the simpler mapping of section 7. As system we consider SUPRENUM-1. For the processors we assume as standard case $a_0=(d+1)/3$, $a_1=0$, $a_2=2^4$, $a_3=2^{-5}$ and $a_4=1$. Other parameter values are additionally specified in the figures. We assume processors of a power of 1 MFLOPS (million floating point operations per second) and a memory cycle of 100 nsec. $a_1=0$ requires an organization securing that the buses pick up the data where they have been generated during the algorithm. In simulations with values up to $a_1=2^{-5}$ we obtained no significant deviation of the results. $a_2=2^4$ corresponds to nearly 512 instructions per SEND and RECEIVE. We assume a mean time of about 200 nsec for an instruction. This value corresponds to a good implementation in a higher level language. $a_3=2^{-5}$ corresponds to a performance of a bus of nearly 200 Mbit/sec and 32 bit per data element. $a_4=1$ means a work of about 32 instructions per packet in the bus system.

Figures 5 and 6 show the strong influence of a_2, a_4 on the speedup. Together with fig. 7 they demonstrate that it is useless for the speedup to operate a system very inefficiently. The broken lines in fig. 7 run horizontally where the system behaves in accordance with asymptotical rules (e.g. for d=3: $E(P)=const*N^{1/3}/P$). Figures 9-14 investigate the influence of a_2, a_3 and a_4 in medium-size and large systems. Fig. 8 shows the limit between the areas where the upper or lower bus system does the main transportation work. The influence of a_2 and a_4 decreases if the system size increases and if a good efficiency is supposed. For our larger system the influence of a_2 is already unimportant. The greater the systems are the stronger their behavior is determined by a_3 (fig. 7, 11 and 12).

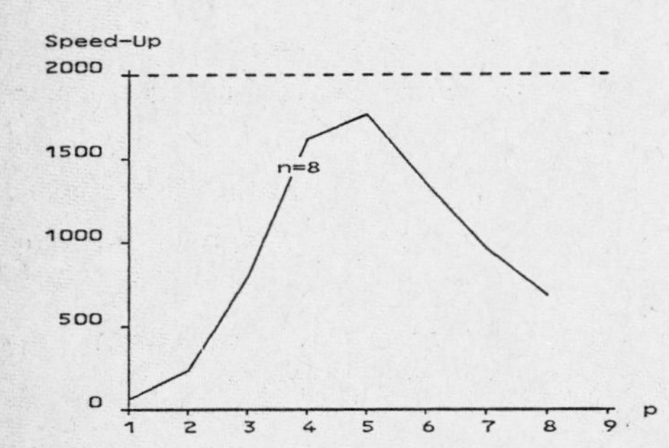

Fig. 5: Speedup in the standard case
$a_1=0$, $a_2=16$, $a_3=1/32$, $a_4=1$
d=3, $p_2=4$ ($P=2^p x 2^p x16$)

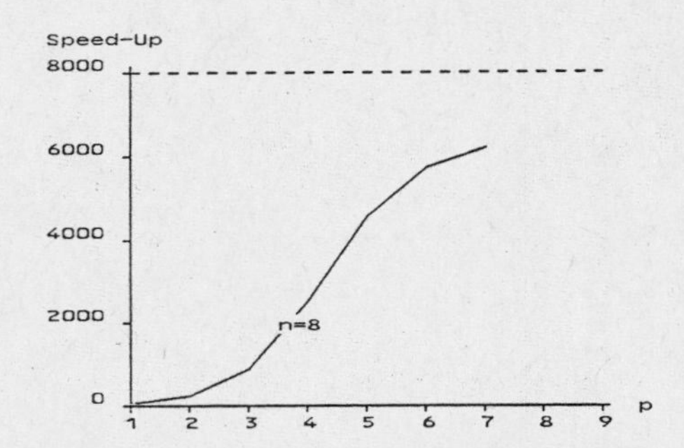

Fig. 6: Speedup without a_2, a_4
$a_1=0$, $a_2=0$, $a_3=1/32$, $a_4=0$
d=3, $p_2=4$ ($P=2^p x 2^p x16$)

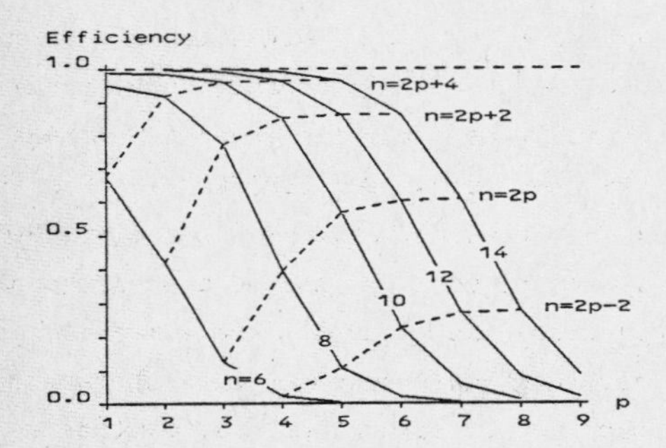

Fig. 7: Efficiency in the standard case
d=3, $p_2=4$ ($P=2^p x 2^p x16$)

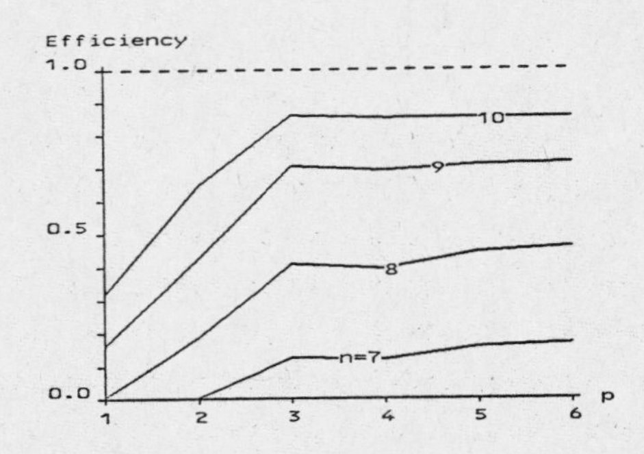

Fig. 8: Limit between areas of same transport behavior, d=3, P=4096,
$a_1=0$, $a_2=16$, $a_3=1/32$, $a_4=1$

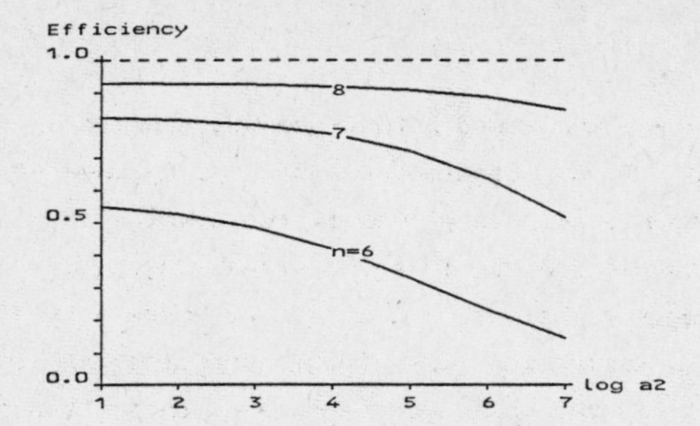

Fig. 9: Influence of a_2 in systems of medium size ($p=2, p_2=4, P=256$)

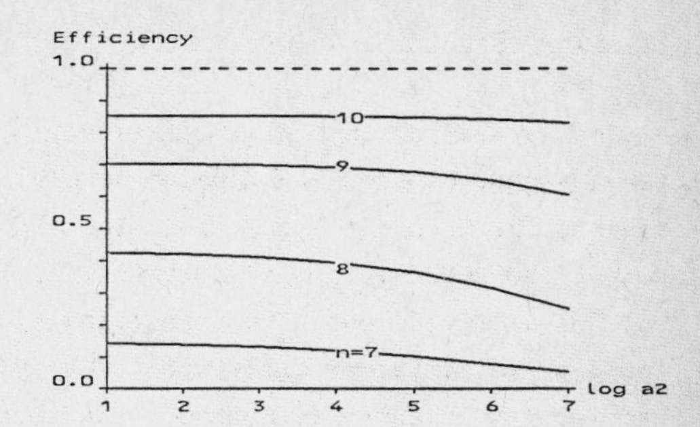

Fig. 10: Influence of a_2 in large systems ($p=4, p_2=4, P=4096$)

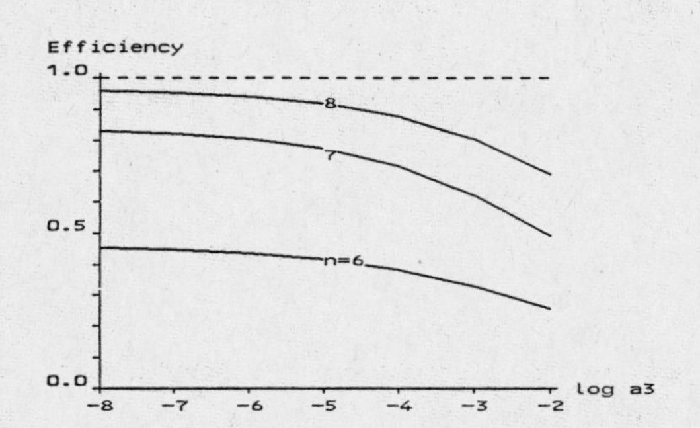

Fig. 11: Influence of a_3 in systems of medium size ($p=2, p_2=4, P=256$)

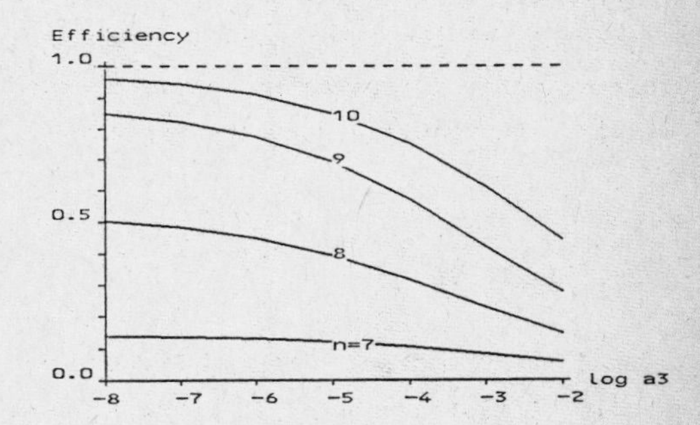

Fig. 12: Influence of a_3 in large systems ($p=4, p_2=4, P=4096$)

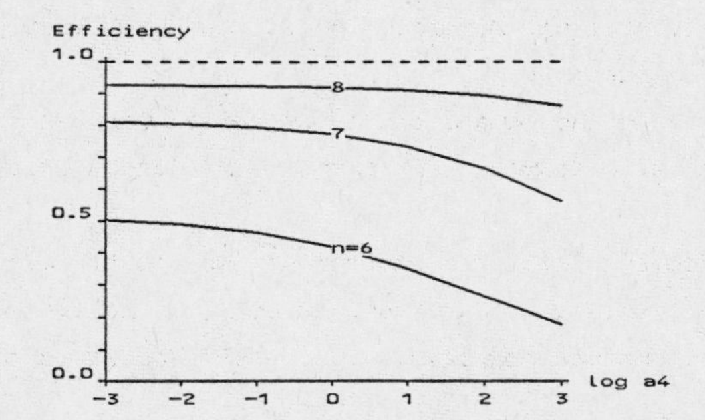

Fig. 13: Influence of a_4 in systems of medium size ($p=2, p_2=4, P=256$)

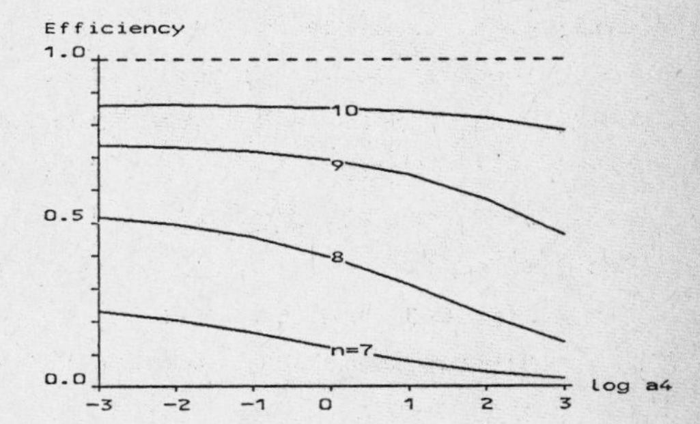

Fig. 14: Influence of a_4 in large systems ($p=4, p_2=4, P=4096$)

10. Final Remarks

The investigations of the present paper are concerned with large bus coupled systems of some hundred or thousand processors. It has been shown that, among the considered structure types, for static multigrid methods primarily two-level systems, such as SUPRENUM-1, are of interest. As a connection within a cluster a single powerful bus is sufficient. The connecting system for the set of all clusters should be at least a 2-dimensional bus system. For P processors the size of the problems to be handled efficiently is bounded from below. In the most important cases these bounds are $N \geq \Omega(P^2)$ or $N \geq \Omega(P^{5/2})$ for 2- or 3-dimensional problems respectively. For systems of a size that can be realized today, system performance is strongly influenced by the work required for processing the message headers.

Further investigations will be concerned a.o. with optimal mappings, values of constants in the asymptotical formulas and a model for system behavior in case of a predominant influence of the parameters a_2 and a_4. A more detailed documentation of simulation results is in preparation.

11. References

[Br81] A. Brandt: Multigrid Solvers on Parallel Computers; in M. H. Schultz (ed.): Elliptic Problem Solvers; Academic Press, 1981.

[CS85] T. F. Chan and Y. Saad: Multigrid Algorithms on the Hypercube Multiprocessor; Yale University, report no. YALEU/DCS/RR-368.

[CS83] T. F. Chan and R. Schreiber: Parallel Networks for Multigrid Algorithms; Yale University, report no. YALEU/DCS/RR-262.

[GM85] W. K. Giloi and H. Mühlenbein: Rationale and Concepts for the SUPRENUM Supercomputer Architecture; unpublished manuscript.

[GR84] D. B. Gannon and J. van Rosendale: On the Impact of Communication Complexity on the Design of Parallel Numerical Algorithms; IEEE Trans. on Comp., vol. C-33, no. 12, dec. 1984.

[KM85] O. Kolp and H. Mierendorff: Efficient Multigrid Algorithms for Locally Constrained Parallel Systems; Proc. of the 2nd Copper Mountain Conf. on Multigrid Methods; march 31 - april 3, 1985, Copper Mountain, Colorado; submitted to AMC.

[ST82] K. Stüben and U. Trottenberg: Multigrid Methods: Fundamental Algorithms, Model Problem Analysis and Applications; in Hackbusch and Trottenberg (eds.): Multigrid Methods, Proc. of the Conf. in Köln-Porz, nov. 23-27, 1981; Lecture Notes in Mathematics, Springer, Berlin 1982.

ON MULTIGRID AND ITERATIVE AGGREGATION METHODS FOR NONSYMMETRIC PROBLEMS

Jan Mandel

Computing Centre of the Charles University
Malostranské nám. 25
118 00 Praha 1, Czechoslovakia

Abstract. We prove a convergence theorem for two-level iterations with one smoothing step. It is applied to multigrid methods for elliptic boundary value problems and to iterative aggregation methods for large-scale linear algebraic systems arising from input-output models in economics and from a multi-group approximation of the neutron-diffusion equation in reactor physics.

1. Introduction

Usual multigrid convergence proofs require that the number of smoothing steps be sufficiently large to guarantee convergence, see, e.g., [1, 2,3,5,14]. But it was observed computationally that one smoothing step and h small enough are sufficient for fast convergence even in the nonsymmetric and indefinite case [1]. We proved this conjecture recently in [9] by a perturbation argument applied to the theory from [8,10] for the symmetric positive definite case and coarse grid problems derived variationally. In such case, the multigrid method converges as soon as all smoothing steps decrease the energy norm of error, which all reasonable smoothers do, cf. [5,10,14]. Convergence may be slow if suitable approximation properties do not hold, but it is guaranteed. One of the purposes of this paper is to provide such a guarantee for some nonsymmetric problems, although in terms of the spectral radius and for a two-grid method with one smoothing step only. The method of proof is completely different from [9] and it is based on a similarity between multigrid and so-called projection-iterative methods [4] as it was pointed out in [7].

We prove an abstract convergence theorem in Section 2. In Section 3, it is applied to multigrid methods. We show that the two-level method is guaranteed to converge for certain matrices of a positive type and the argument is stretched to give an h-independent result under suitable approximation assumptions. Next, we apply the theorem to the iterative aggregation method with block iterations for large-scale systems of linear algebraic equations with some positivity properties, which arise in economics /Section 4/ and in reactor physics /Section 5/.

For simplicity, general convergence results are formulated in a real finite dimensional space only. The space of linear operators which map U into V is denoted by $[U,V]$, and $[U]=[U,U]$. ρ is the spectral radius. Elements of R^n are identified with coordinate vectors and elements of $[R^n]$ with n by n matrices. We use an inner product $(.,.)$ in R^n different from the arithmetic inner product u^Tv. Adjoints relative to $(.,.)$ are denoted by an asterisk. Inequalities are to be understood by coordinates.

2. Abstract convergence results

Let H be a finite dimensional real Hilbert space. We are interested in iterative solution of the problem

$$Lu = Bu + f \tag{2.1}$$

with given $L,B \in [H]$ and $f \in H$. Let V be a linear space of lower dimension than H and $P \in [V,H]$, $R \in [H,V]$ two given mappings. P and R are called _prolongation_ and _restriction_, respectively. In the whole paper, we shall study the following _two-level method_:

$$R(L - B)(u^k + Pd) = Rf, \tag{2.2}$$

$$u^{k+1} = L^{-1}[B(u^k + Pd) + f]. \tag{2.3}$$

The equation (2.2) requires the solution of a lower dimensional problem in the space V for the unknown d. (2.3) is a linear iteration step corresponding to the splitting L - B.

Theorem 2.1. Suppose that L is invertible and

$$\text{Range } P \perp \text{Null } RL, \tag{2.4}$$

$$\text{Null } P = \{0\}, \tag{2.5}$$

and

$$q = \sup\{\|L^{-1}Bu\| : \|u\| = 1,\ R(L - B)u = 0\} < 1. \tag{2.6}$$

Then the problem (2.1) has a unique solution $\bar{u}$ and it holds

$$u^{k+1} - \bar{u} = T(u^k - \bar{u}),\quad \rho(T) \le \|(I - \Pi)T\| \le q < 1,$$

where $\Pi = P(RLP)^{-1}RL$ is the orthogonal projection onto the range of P, and $T(I - \Pi) = T$.

<u>Proof.</u> Assume that $RLPw = 0$. Then $Pw \in \text{Null } RL$, so $w = 0$ by (2.4) and (2.5). This proves that RLP is invertible. Denote $K = L^{-1}B$. We claim that $I - \Pi K$ is also invertible. Indeed, if $u - \Pi Ku = 0$, then $R(L - B)u = 0$. So from (2.6),

$$\|Ku\| \le q\|u\| = q\|\Pi Ku\| \le q\|Ku\|,$$

which gives $u = 0$. Now we prove invertibility of $R(L - B)P$, which is needed for the existence and uniqueness of the solution d of (2.2). Let $R(L - B)Pw = 0$. Then $u = Pw$ satisfies $\Pi(I - K)u = 0$, hence $u - \Pi Ku = 0$. It follows that $u = 0$ and, consequently, $w = 0$ by (2.5).

Denote $T = K(I - \Pi K)^{-1}(I - \Pi)$. We shall prove that

$$\|(I - \Pi)T\| \le q. \tag{2.7}$$

We start with the identity $(I - \Pi K)^{-1} = I + \Pi K(I - \Pi K)^{-1}$, which gives

$$Tu = K[(I - \Pi)u + \Pi Tu],$$

where the argument of K lies in $\text{Null } R(L - B) = \text{Null } \Pi(I - K)$. Now by orthogonality and by (2.6),

$$\|(I - \Pi)Tu\|^2 + \|\Pi Tu\|^2 \le q^2\left[\|(I - \Pi)u\|^2 + \|\Pi Tu\|^2\right]$$

with $q < 1$, which gives (2.7).

If $(L - B)u = 0$, then $u = Ku$, so also $u = (I - \Pi K)^{-1}(I - \Pi)Ku$. It follows that $u = 0$, as $\rho((I - \Pi K)^{-1}(I - \Pi)K) = \rho(T) < 1$. Thus the solution $\bar{u}$ exists and it is unique.

It remains to prove that the error is transformed by T. Denote $e^i = u^i - \bar{u}$. Then (2.2) becomes $R(L - B)(e^k + Pd) = 0$, which is in turn equivalent to $(I - \Pi K)Pd = - \Pi(I - K)e^k$. This equation has a unique solution d and it is satisfied by $Pd = -(I - \Pi K)^{-1}\Pi(I - K)e^k$. By a simple manipulation writing $\Pi(I - K)$ as $(I - \Pi K) - (I - \Pi)$, we obtain that $e^{k+1} = e^k + Pd = Te^k$.

Remark 2.1. The proof of (2.7) is adapted from [4]. The theorem can be extended to general Hilbert spaces and for nonlinear operators similarly as in [6].

Corollary 2.1. Assume that Null $R \perp$ Range P, Null $P = \{0\}$, and let $A \in [H]$ satisfy

$$\bar{q} = \inf \{ (u,Au): ||u|| = 1, RAu = 0 \} > 0.$$

If $0 < \omega < 2\bar{q}\,||A||^{-2}$, then the iterations defined by

$$RA(u^k + Pd) = Rf, \quad u^{k+1} = u^k + Pd - \omega [A(u^k + Pd) - f] \tag{2.8}$$

satisfy $u^{k+1} - \bar{u} = T(u^k - \bar{u})$, where $A\bar{u} = f$ and

$$\rho(T) \leq 1 - 2\omega\bar{q} + \omega^2 ||A||^2.$$

In particular, if $\omega = \bar{q}\,||A||^{-2}$, then $\rho(T) \leq 1 - \bar{q}^2 ||A||^{-2}$.

Proof. Put $L = I/\omega$ and $B = L - A$. Then

$$||L^{-1}Bu||^2 = ||(I - \omega A)u||^2 = ||u||^2 - 2\omega(u,Au) + \omega^2 ||Au||^2$$
$$\leq [1 - 2\omega\bar{q} + \omega^2 ||A||^2]\,||u||^2$$

for any u such that $RAu = 0$.

3. Application to multigrid methods

Consider a simple two-grid method with one smoothing step:

$$A_{2h}u_{2h} = I_h^{2h}(f_h - A_h u_h^k),\ u_h^{k+1/2} = u_h^k + I_{2h}^h u_{2h},\ u_h^{k+1} = u_h^{k+1/2} - \omega(A_h u_h^{k+1/2} - f_h), \tag{3.1}$$

where $A_h \in [V_h]$, $A_{2h} \in [V_{2h}]$, $I_h^{2h} \in [V_h, V_{2h}]$, $I_{2h}^h \in [V_{2h}, V_h]$, V_h and V_{2h} are spaces of grid functions equipped with suitably scaled arithmetical inner products $(.,.)_h$, $(.,.)_{2h}$, respectively. I_{2h}^h is a prolongation and I_h^{2h} is a restriction. We shall assume the following variational conditions, cf., e.g., [14]:

$$I_h^{2h} = (I_{2h}^h)^*, \text{ Null } I_{2h}^h = \{0\}, \; A_{2h} = I_h^{2h} A_h I_{2h}^h. \tag{3.2}$$

If we put $A = A_h$, $H = V_h$, $V = V_{2h}$, $P = I_{2h}^h$, and $R = I_h^{2h}$, then (3.1) becomes the iterative method (2.8) above.

It is known that u_h^k converge to the solution $\bar{u}_h$ of $A_h u_h = f_h$ if A_h is symmetric positive definite and $0 < \omega < 2/\rho(A_h)$, because then each step reduces the value of the discrete energy $J_h(u_h) = \frac{1}{2}(u_h, A_h u_h)_h - (u_h, f_h)_h$ unless $u_h^k = \bar{u}_h$. In the general case, we have no monitoring functional like J_h and the iterations may indeed diverge. We shall prove convergence in an important particular case of nonsymmetric problems. The space V_h is identified with R^n and A_h with its matrix.

Theorem 3.1. Assume (3.2) and let A_h be of a positive type, i.e., A_h has positive diagonal elements and non-positive off-diagonal elements. If A_h is irreducible and there exists $v_h > 0$ such that

$$A_h v_h \geq 0, \; A_h^T v_h \geq 0, \; A_h v_h + A_h^T v_h \neq 0,$$

then the iterations (3.1) converge for any $\omega > 0$ such that $I - \omega A_h \geq 0$.

Proof. As in the proof of Corollary 2.1, it suffices to prove $\|I - \omega A_h\|_h < 1$ in the norm induced by the inner product $(.,.)_h$. It holds

$$\|I - \omega A_h\|_h^2 = \|I - \omega A_h\|_{l^2}^2 = \rho(M),$$

where $M = (I - \omega A_h)^T (I - \omega A_h)$. Because $M \geq 0$, M is irreducible, and $Mv_h \leq v_h$, $Mv_h \neq v_h$, it follows from the Perron-Frobenius theory [16] that $\rho(M) < 1$.

If we know only that $u_h^T A_h u_h > 0$ for all $u_h \neq 0$, then Corollary 2.1 yields convergence of the method (3.1) for all sufficiently /but unrealistically/ small $\omega > 0$. We shall use it to obtain an estimate independent on h.

If A_h is a discretization of an elliptic operator, then it is natural to put

$$A_h = S_h + N_h, \tag{3.3}$$

where S_h is a discretization of the principal terms and N_h is considered a perturbation. For example, for the Dirichlet problem,

$$-\Delta u + b.\nabla u + c = 0 \text{ in } \Omega\ , \ u = 0 \text{ on the boundary of } \Omega\ , \tag{3.4}$$

S_h corresponds to $-\Delta$ or perhaps to $-\Delta + c$ if $c \geq 0$. We require that S_h be symmetric positive definite and define the "energy norm" by

$$|||u_h|||_h = [(u_h, S_h u_h)_h]^{1/2}.$$

We restrict ourselves to two dimensional problems like (3.4) and suppose that $||.||_h$ is an approximation of the $L^2(\Omega)$ norm and $|||.|||_h$ approximates a norm equivalent to the Sobolev $H^1(\Omega)$ norm.

Denote by C a generic constant which does not depend on h. We make the following assumptions:

$$\rho(S_h) \leq Ch^{-2}, \tag{3.5}$$

$$(u_h, N_h v_h)_h \leq C||u_h||_h |||v_h|||_h, \text{ for all } u_h,\ v_h, \tag{3.6}$$

$$||u_h||_h \leq Ch|||u_h|||_h, \text{ for all } u_h \text{ such that } I_h^{2h} A_h u_h = 0. \tag{3.7}$$

These conditions can be verified for usual finite element discretizations and under usual assumptions, see [9]. (3.5) follows from the assumption that all elements are approximately of the same size h. (3.6) is a consequence of the fact that N_h approximates lower order terms. The proof of (3.7) is more difficult and it uses the H^2 regularity of the adjoint problem, cf. also [1], Eq. (3.14).

Theorem 3.2. Assume (3.2), (3.3), and (3.5) - (3.7). Then the iterations defined by (3.1) satisfy $u_h^{k+1} - \bar{u}_h = T_h(u_h^k - \bar{u}_h)$ with $\rho(T_h) \leq C < 1$ for all sufficiently small h and suitable $\omega = \omega_h$.

Proof. By Corollary 2.1, it suffices to prove that $(u_h, A_h u_h)_h \geq C||A_h||_h$, $C > 0$, for all u_h such that $||u_h||_h = 1$ and $I_{2h}^h A_h u_h = 0$. For such u_h and h small enough,

$$(u_h, A_h u_h)_h = |||u_h|||_h^2 + (u_h, N_h u_h)_h \geq |||u_h|||_h^2 - C||u_h||_h |||u_h|||_h \geq Ch^{-2}$$

from (3.3), (3.6), and (3.7). Now from (3.6) and from (3.5) we obtain

$||N_h||_h \leq Ch^{-1}$, hence $||A_h||_h \leq Ch^{-2}$.

4. Application to iterative aggregation methods

Let $V = R^g$ and $H = R^n = R^{k_1} \times R^{k_2} \times \ldots \times R^{k_g}$. We shall use the notation $w = (w_i)$, $w_i \in R^1$, for $w \in V$, the block notation

$$u = (u_i),\ u_i \in R^{k_i},$$

for $u \in H$, and the corresponding block notation for linear operators, resp. matrices. Let $L, B \in [R^n]$ and consider the system $Lu = Bu + f$ with L block diagonal,

$$L = (L_{ij}),\ L_{ij} = 0 \text{ for } i \neq j, \tag{4.1}$$

and

$$L^{-1}B \geq 0. \tag{4.2}$$

For given weight vectors $s, y \in R^n$ such that

$$s^TL > 0,\ y > 0, \tag{4.3}$$

define $R \in [R^n, R^g]$ and $P \in [R^g, R^n]$ by

$$R\colon u \mapsto (s_i^T u_i),\quad P\colon w \mapsto (w_i y_i). \tag{4.4}$$

Such R and P are usually called <u>aggregation</u> and <u>disaggregation</u> operators. The method (2.2),(2.3) is now a version of the iterative aggregation method with <u>additive correction</u>, cf. [15] for the case $L = I$.

<u>Theorem 4.1.</u> Assume (4.1) - (4.4). If there exist $c_1, c_2 \in R^1$ such that

$$s^TB \leq c_1 s^TL, \tag{4.5}$$

$$L^{-1}By \leq c_2 y, \tag{4.6}$$

and $c_1c_2 < 1$, then $u^{k+1} - \bar{u} = T(u^k - \bar{u})$ with $\rho(T) \leq (c_1c_2)^{1/2} < 1$.

<u>Proof.</u> Denote $z^T = s^TL > 0$. For any vector x, denote by $\hat{x}$ the diagonal

matrix with x as the diagonal. We shall use Theorem 2.1 with the inner product in H defined by $(u,v) = u^T \hat{y}^{-1} \hat{z} v$.

If $u = Pw$ and $RLv = 0$, then from (4.4),

$$(u,v) = \sum w_i y_i^T \hat{y}_i^{-1} \hat{z}_i v_i = \sum w_i z_i^T v_i,$$

and we obtain using (4.1) that $z_i^T v_i = s_i^T L_{ii} v_i = 0$. Hence (2.4) holds. (2.5) is immediate from the definition of P and from $y > 0$. Now from the definition of the inner product, $(L^{-1}B)^* = \hat{y} \hat{z}^{-1} (L^{-1}B)^T \hat{y}^{-1} \hat{z}$. Thus the assumption (4.5) gives $(L^{-1}B)^* y \leq c_1 y$, which together with (4.6) implies

$$My \leq c_1 c_2 y, \quad M = (L^{-1}B)^* L^{-1}B \ .$$

Because $M \geq 0$ and $y > 0$, we have $\rho(M) \leq c_1 c_2$ by the Perron-Frobenius theory [16], so $||L^{-1}B||^2 = \rho(M) \leq c_1 c_2 < 1$.

<u>Remark 4.1.</u> If $L^{-1}B \geq 0$ and $\rho(L^{-1}B) \leq 1$, then there always exist y,s satisfying (4.3) and (4,5),(4.6) with $c_1, c_2 < 1$. If $L^{-1} \geq 0$, then the assumption (4.6) can be replaced by the stronger condition

$$By \leq c_2 Ly. \tag{4.7}$$

<u>Remark 4.2.</u> Systems arising from input-output models in economics are of the form $u = Au + f$ with A satisfying

$$A \geq 0, \ s^T A \leq cs^T, \ c < 1, \ s > 0, \tag{4.8}$$

cf. [11,15] and references therein. Let $y > 0$ and $Ay \leq y$. In [7], convergence was proved for the choice $L = I$ and $B = A$. Theorem 4.1 allows to extend this result to the case of block iterations with

$$I - A = L - B, \ I - A_{ii} \leq L_{ii} \leq I, \ L_{ij} = 0 \text{ for } i \neq j.$$

Indeed, for such L, (4.8) gives $s^T L > 0$, $L^{-1} \geq 0$, and

$$s^T B = s^T A + s^T (L - I) \leq cs^T + cs^T (L - I) = cs^T L,$$

which is (4.5) with $c_1 = c < 1$. Similarly, (4.7) holds with $c_2 = 1$.

The original version of the iterative aggregation method used the so-called <u>multiplicative correction</u>, cf. [7,11,15] and references therein.

It can be written in our context as follows:

$$R(L - B)P_k w = Rf, \quad u^{k+1} = L^{-1}(BP_k w + f), \tag{4.9}$$

where P_k is defined as in (4.4) but with $y = u^k$ changing in each iteration. This iterative process is nonlinear and we have the following local convergence theorem, which extends the result of [11] to block diagonal L.

Theorem 4.2. Assume (4.1), $s > 0$, $s^T B \leq c s^T L$, $c < 1$, $s^T L > 0$, $L^{-1} \geq 0$, $L^{-1}B \geq 0$, and $f > 0$. Then the iterations u^k defined by (4.9) converge to the unique solution $\bar{u}$ of $Lu = Bu + f$ if u^1 is sufficiently close to $\bar{u}$.

Proof. As in the preceding proof, put $z^T = s^T L$. Then $z^T L^{-1} B < z^T$ and $L^{-1}B \geq 0$ imply that $\rho(L^{-1}B) < 1$, hence $I - L^{-1}B$ is nonnegatively invertible and $\bar{u} \geq f > 0$. For $u^k > 0$, we have from the definition of P_k that $u^k = P_k(RLP_k)^{-1}RLu^k$. Hence (4.9) can be written as (2.2), (2.3) with $P = P_k$ after the substitution $w = (RLP_k)^{-1}RLu^k + d$.

Since $L^{-1} \geq 0$ and (4.7) holds with $y = \bar{u} > 0$, $c_2 = 1$, and (4.6) with $c_1 = c < 1$, we may apply Theorem 4.1 and Remark 4.1. Hence

$$u^{k+1} - \bar{u} = T_{u^k}(u^k - \bar{u}),$$

where $T = T_y \in [R^n]$ depends continuously on y. Because $\rho(T_{\bar{u}}) \leq c^{1/2} < 1$, it holds $||T_{\bar{u}}||_* < 1$ in some norm and $||T_{u^k}||_* < 1$ for u^k close enough to $\bar{u}$.

Remark 4.3. The assumption $f > 0$ can be weakened as in [11]. We conjecture that the method (4.9) converges under the assumptions of Theorem 4.2 for any $u^1 > 0$. This conjecture is supported by computational experiments and the method was found to be quite efficient in practice.

5. Application to the multi-group neutron diffusion equation

Let $H = R^n = R^m \times R^m \times \ldots \times R^m = (R^m)^g$. We shall use the block notation corresponding to this decomposition. Consider the problem $Lu = Bu + f$, where

$$L = (L_{ij}), \quad L_{ij} = 0 \text{ for } i \neq j, \tag{5.1}$$

$$L_{ii} \text{ are symmetric positive definite and } L_{ii}^{-1} \geq 0, \tag{5.2}$$

and

$$B \geq 0. \tag{5.3}$$

Such problems arise as discretizations of the multi-group approximation of the neutron transport in a nuclear reactor, see,e.g., [13,17]. Each component u_i is a discretization of a spatial distribution of neutrons in the energy group i. The diagonal blocks L_{ii} approximate diffusion operators and B_{ij} express the transfer of neutrons between the energy groups j and i. f is a source term. Such systems arise also in the solution of the related eigenvalue problem $Lu = \lambda Bu$.

We could proceed as in the preceding sections and aggregate each group into a single variable. We adopt a different approach here and aggregate the energy groups together. A similar method is used in practice [12]. Let the set of groups be decomposed into r disjoint subsets,

$$\{1,2,\dots,g\} = (1) \cup \dots \cup (r).$$

Put $V = (R^m)^r$ and define the restriction /aggregation/ operator by

$$R: u \longmapsto v,\ v_{(p)} = \sum_{i \in (p)} \mu_i u_i, \tag{5.4}$$

where

$$\mu_i > 0,\ i = 1,2,\dots,g, \tag{5.5}$$

are scalar weights and $v = (v_{(p)}) \in V$ in the block notation. The prolongation /disaggregation/ operator is defined by creating copies,

$$P: v \longmapsto u,\ u_i = v_{(p)} \text{ for } i \in (p). \tag{5.6}$$

<u>Theorem 5.1.</u> Suppose (5.1) - (5.6). If there exists $z \in (R^m)^g$, $z > 0$, such that

$$\sum_j L_{ii}^{-1} B_{ij} z_j \leq c_1 z_i \quad \text{for all } i = 1,2,\dots,g, \tag{5.7}$$

$$\sum_j \mu_j L_{ii}^{-1} B_{ji}^T z_j \leq c_2 \mu_i z_i \text{ for all } i = 1,2,\dots,g, \tag{5.8}$$

with $c_1 c_2 < 1$, then the iterations u^k defined by (2.2),(2.3) satisfy $u^{k+1} - \bar{u} = T(u^k - \bar{u})$, $\rho(T) \leq (c_1 c_2)^{1/2} < 1$.

<u>Proof.</u> We shall use Theorem 2.1 with the inner product in H defined by

$$(w,u) = \sum_i \mu_i w_i^T L_{ii} u_i .$$

From (5.4) and (5.6), we have for any $u \in H$, $v \in V$ using (5.1),

$$(Pv,u) = \sum_{(p)} \sum_{i \in (p)} \mu_i v_{(p)}^T L_{ii} u_i = v^T RLu.$$

This proves the orthogonality condition (2.4). (2.5) is obvious. It remains to estimate q in (2.6). Define

$$D \in [H],\ D: u \longmapsto v,\ v_i = \mu_i u_i \text{ for all } i.$$

Then D is diagonal, it commutes with L, and for any w,u, $(w,u) = w^T DLu$. Hence, we obtain $||L^{-1}B||^2 = \rho(M)$, where

$$M = (L^{-1}B)^* L^{-1}B = D^{-1}L^{-1}B^T DL^{-1}B.$$

It holds that $M \geq 0$ and (5.7),(5.8) give

$$L^{-1}Bz \leq c_1 z,\ D^{-1}L^{-1}B^T Dz \leq c_2 z.$$

Consequently, $\rho(M) \leq c_1 c_2 < 1$.

<u>Remark 5.1.</u> The conditions (5.7), (5.8) were assumed so that $||L^{-1}B|| < 1$ in the inner product defined above. They are stronger than the natural condition $\rho(L^{-1}B) < 1$. But the condition $||L^{-1}B|| < 1$ may be seen as natural in practice as well [12]. Also, if the aggregation is "well chosen", we may expect that q will be small.

6. Conclusion

In Theorem 2.1, we proved that the spectral radius of the iteration operator of the two-level method is bounded by q if the orthogonality condition (2.4) is satisfied. q was defined as the norm of the smoothing operator $L^{-1}B$ restricted to the subspace of "oscilatory" u characterized by $R(L - B)u = 0$. In the particular case when $L = I/\omega$, q can be bounded using the ellipticity constant $\bar{q}$ of $A = L - B$ on this subspace /Corollary 2.1/. Under approximation assumptions usual in multigrid methods, $\bar{q}$ and thus q can be estimated independently on h /Theorem 3.2/. In other cases, - a two-grid method for a system with the matrix of a positive type /Theorem 3.1/ and iterative aggregation methods with block diagonal L /Theorems 4.1 and 5.1/ - we found a suitable inner product and proved that $||L^{-1}B|| < 1$ using arguments based essentially on the positivity of $L^{-1}B$. Then $q < 1$, but this estimate is conservative, because q is often much smaller that $||L^{-1}B||$. So, we have guaranteed convergence of the methods and it would be desirable to find some better estimates of q. At present, we do not know of any reasonably simple and useful ones.

In [7], we proved that /in the notation on the proof of Theorem 2.1/

$$\rho(T) \leq \frac{||(I - \Pi)K||}{(1 - ||K||^2 + ||(I - \Pi)K||^2)^{1/2}} \tag{6.1}$$

if $||K|| < 1$. The quantity $||(I - \Pi)K||$ can be estimated more easily than q, but (6.1) does not give an h-indepentent result for the two-grid method. It should be noted here that the estimate (6.1) can be improved by taking the norms only on the subspace of "oscilatory" u such that $\Pi(I - K)u = 0$. The proof remains same as in [7]. Then (6.1) makes it possible to improve the result of Theorem 2.1. For simplicity, we did not include such modification of the theorem here.

Acknowledgements

I am indebted to Professors Ivo Marek and Steve McCormick for helpful and stimulating discussions. A part of this work was done while I was visiting the Department of Mathematics, University of Colorado at Denver, and the Mathematics Research Center, University of Wisconsin-Madison. The support of both institutions is acknowledged.

References

1. R.E. Bank: A comparison of two multilevel iterative methods for nonsymmetric and indefinite finite element equations. SIAM J. Numer. Anal. 18,724 - 743,1981.

2. R.E. Bank and T. Dupont: An optimal order process for solving finite elements equations. Math. Comp. 36,35-51,1981.

3. W. Hackbusch: On the convergence of multigrid iterations. Beitr. Numer. Math. 9,213-239,1981.

4. A.Ju. Lučka: Projection-Iterative Methods of Solving Differential and Integral Equations /in Russian/. Naukova Dumka, Kiev, 1980.

5. J.F. Maitre and F. Musy: Multigrid methods: convergence theory in a variational framework. SIAM J. Numer. Anal., to appear.

6. J. Mandel: A convergent nonlinear splitting via orthogonal projection. Apl. Mat. 29,250-257,1984.

7. J. Mandel: On some two-level iterative methods. In: K. Böhmer and H.J. Stetter /editors/, Defect Correction Methods, Computing Supplementum 5, Springer-Verlag, Wien, 1984.

8. J. Mandel: Algebraic study of multigrid methods for symmetric, definite problems. Appl. Math. Comput., to appear.

9. J. Mandel: Multigrid convergence for nonsymmetric, indefinite problems and one smoothing step. In: Preliminary Proceedings of the 2nd Copper Mountain Conference on Multigrid Methods, Copper Mountain, Colorado, April 1985 /mimeo/. Appl. Math. Comput., submitted.

10. J. Mandel, S.F. McCormick, and J. Ruge: An algebraic theory for multigrid methods for variational problems. SIAM J. Numer. Anal, submitted.

11. J. Mandel and B. Sekerka: A local convergence proof for the iterative aggregation method. Linear Algebra Appl. 51,163-172,1983.

12. I. Marek, personal communication, 1985.

13. I. Marek: Some mathematical problems of the theory of nuclear reactors on fast neutrons. Apl. Mat. 8,442-470,1963.

14. S.F. McCormick: Multigrid methods for variational problems: further results. SIAM J. Numer. Anal. 21,255-263,1984.

15. W.L. Miranker and V.Ya. Pan: Methods of aggregation. Linear Algebra Appl. 29,231-257,1980.

16. R.S. Varga: Matrix Iterative Analysis. Prentice-Hall, Englewood Cliffs, N.J., 1962.

17. E.L. Wachspress: Iterative Solution of Elliptic Systems and Applications to the Neutron Diffusion Equation of Reactor Physics. Prentice--Hall, Englewood Cliffs, N.J., 1966.

The Multigrid Method on Parallel Processors [1]

Oliver A. McBryan [2,3]

Eric F. Van de Velde [2]

Courant Institute of Mathematical Sciences,
New York University
New York, NY 10012

THE MULTIGRID METHOD AND ELLIPTIC EQUATIONS.

The basic elliptic equation we have studied is of the form:

$$\nabla \cdot (-\vec{K} \cdot \nabla P) (x,y) = F (x,y) .$$

Here K has discontinuities, possibly of order a thousand or more, across a given set of curves and in typical applications may represent a fluid density, permeability or dielectric constant. The right hand side F may contain arbitrary point and line sources. Boundary conditions may be Dirichlet, Neumann or mixed.

Discontinuities of coefficients imply discontinuities in the solution gradient. Discretization of the equation on a rectangular grid leads to bad pressure and velocity solutions at the front due to such discontinuities. For this reason it is essential to adapt the grids locally. In the resulting grids, discontinuities lie only along edges of triangles. The cost of grid generation is negligible compared to equation solution. In general, our grids consist of unions of rectangles and triangles, with triangles primarily used to fit boundaries and interior interfaces. For details of the grid construction methods used, we refer to our papers.[1,2,3]

We have used finite element methods to discretize these equations, though similar studies could be applied to finite difference methods. To provide sufficient accuracy, we allow high order elements up to cubics (on triangles) or bicubics (on rectangles). We have discussed the solution of singular elliptic equations by these techniques in [1,3] and the use of parallelism in the context of a tree of refinement grids elsewhere.[4] Solution using mutligrid methods was discussed in some of these papers, and also in.[5] In related papers,[4,6,7]

---

1. Presented to the 2nd European Multigrid Conference, Koln, Oct. 1985.
2. Supported in part by DOE contract DE-ACO2-76ER03077.
3. Supported in part by NSF grant DMS-83-12229.

we describe the parallel implementation of the *Conjugate Gradient Method* and of an *FFT-based Fast Poisson Solver* which we have used as a preconditioner for the Conjugate Gradient Method. The implementation of a parallel *Full Multi-grid Method* based on a five-point operator discretization of the equations is described here. Using either of these methods, the solution cost in total operations performed is essentially proportional to the number of unknowns, while at the same time allowing near optimal use of parallelism.

The total time for solution is usually dominated by the time spent in solving the resulting algebraic systems. Thus we will focus on parallel solution of the equations and refer to our papers for details of the numerical analysis and of the discretization approach.[1,2,3,8,9]

The basic multigrid idea[10,11,12] involves two aspects - the use of relaxation methods to dampen high-frequency errors and the use of multiple grids to allow low-frequencies to be relaxed inexpensively. A simple *Two-grid Iteration* involves a number of relaxations on a fine grid to reduce high-frequency errors in an initial guess of the solution, a projection of remaining errors to a coarser grid where they are solved for exactly, and then an interpolation back to the fine grid and addition to the solution there. This solution would now be exact but for errors introduced by the projection and interpolation processes. The solution is then improved by repeating the procedure.

The Two-grid Iteration is converted to the *Multigrid Iteration (MGI)* by recursively applying the 2-grid iteration in place of the exact solution on the coarse grid. The number of times that the coarse grid iteration is repeated before returning to the fine grid is important for convergence rates - typical values used are once, known as *V-cycles,* or twice, known as *W-cycles*.

Improved convergence can be obtained by choice of a good initial guess for the solution. A simple strategy would be to solve the equations on a coarse grid using the Multigrid Iteration, and interpolate the solution to the fine grid as the initial guess for the Multigrid Iteration there. Recursively applying this idea leads to the *Full Multigrid Iteration (FMG)* which performs a sequence of Full Multigrid solutions on increasingly finer grids, using the solution of each as an initial guess for the next.

Our studies are concerned with the case where there are many fine grid points per processor and we will assume this to be the case throughout the exposition.

HARDWARE AND SYSTEM SOFTWARE

In this section we introduce the 3 parallel computers used in our studies. We also discuss how parallelism is implemented in each of these systems. The Denelcor HEP was the first commercially available shared memory parallel computer for scientific computing. We discuss process creation and synchronization primitives on this system and monitors to use these primitives in a portable way. It is expected that, using different definitions for the monitors, the same program can run on other shared memory systems. The hypercube architecture is a leading design for parallel computers of message passing type. The Intel hypercube differs from the Caltech design mainly in its system software. The differing message passing systems and their associated communication overhead lead to differing behavior of programs on the two systems. While the hardware differences are important for debugging (the iPSC being easier to debug), they are less important from a conceptual point of view.

1. The Caltech Mark II Hypercube

The Cosmic Cube is a parallel processor developed by Geoffrey Fox and Charles Seitz[13,14] at Caltech. The Caltech Mark II Hypercube consists of 2^D ($D = 5$ *or* 6) independent processors, each with its own local memory. There is no shared memory available - the processors cooperate by message passing. Messages are passed over an interconnection network which is a *hypercube* in a space of dimension D. Processors are located at the vertices of the D-dimensional hypercube and adjacent vertices of the cube are connected by a communication channel along the corresponding edge. All data exchange between processors occurs in 8-byte packets along these cube edges which are asynchronous full duplex channels. In addition to the 2^D node processors, there is a host processor which acts as a control processor for the entire cube and also provides the interface between the cube and a user. All I/O to and from the cube must pass through the host, which is connected to one corner of the cube by an extra communication channel. The original Caltech design consists of a 64-node 6 dimensional hypercube utilizing Intel 8086/8087 processors with 128KB of memory at each node. This architecture has the advantage of being easily fabricated from standard components, and may be scaled up to much larger sizes (in powers of 2) with almost no change in design. Because of these features, machines of this type are likely to become widely available in the immediate future, whereas development of highly parallel global memory machines will take substantially longer.

A more advanced Caltech cube called the Mark III is now under development. This will have much faster processors at the nodes (Motorola 68020) and local memory per node will reach several megabytes. Other enhancements will be incorporated based on the experience with the prototype.

1.1. Caltech Hypercube Programming

There are two fundamentally different communication modes available on the hypercube. In the *Interrupt Driven Mode,* processors are interrupted by messages arriving from the communication channels. These messages are preceded by sufficient identification and destination information so that the processor can either forward them to another channel (if the current processor is not the destination) or process the incoming message (if the message has arrived at its destination). In the *Crystalline Operating System* messages are not preceded by address information. As a result, each processor has to know in advance exactly what communication pattern to expect. The latter system is unquestionably more efficient, although it is clearly also more restrictive. For the computations described in this paper the Crystalline Operating system was quite adequate. The parallelization of other algorithms (e.g. the local grid refinement algorithms discussed in our related papers[4,6]) will likely require some interrupt driven communication protocols. For the remainder of the discussion we will refer only to the Crystalline Operating System when discussing Caltech Hypercube software.

The software for the cube consists of an operating system kernel, a copy of which resides in each processor, as well as a run-time library providing user access to the communication facilities. Typically, identical copies of a user program are downloaded to all processors where they execute concurrently. All scheduling is accomplished through communication calls, so that some care is required to prevent locking situations from occurring.

As discussed previously, the D-cube has 2^D vertices with D edges radiating from each. Thus each processor sees D channels connecting it to its neighbors. The cube nodes are numbered in the range $[0,2^D-1]$, such that the D-digit binary representations of physically adjacent nodes differ only in 1 bit. The channels emanating from a node may then be numbered 0, 1, .., $D-1$ according to which bit differs in the binary node representations at either end of the channel. There is also an extra channel from node 0 to the intermediate host (referred to as the IH below) through which all communications to and from the cube pass. Data to be communicated between processors is sent in 8-byte packets, which are sufficient to encode all scalar data types. A set of system calls are available to node-resident programs which implement the required communication primitives for these packets. Similar system calls are available on the host to provide communication with the cube.

One particular routine is very useful in the simulation of many physically interesting problems - such as those derived from discretizations of partial differential equations on regular grids. An important feature in such discretizations is that there is typically only nearest neighbor connectivity among the variables of interest. For efficient use of the hypercube, it is then very desirable to map the grid onto the cube in such a way that neighboring grid points (in two or three dimensional space) are mapped onto adjacent nodes of the cube. Communication overhead will be minimized by such a mapping.

Accomplishing such a mapping is difficult and in general impossible - for example there is no such mapping of a 3 dimensional grid onto a 5-cube since the grid requires a local connectivity of 6 at each node. A general purpose routine called *whoami*() has been developed by John Salmon at Caltech[15] based on *binary gray codes,* which generates a suitable mapping of the above type in most cases where one is possible. The *whoami*() call is usually executed at the start of any grid-oriented program, and in addition to creating a suitable mapping of the grid to the cube nodes it returns communication channel information for each of the *grid neighbors* of each processor. This allows the programmer to think entirely in *grid space* rather than in the less intuitive *edge space* of the cube.

A hypercube program consists of two separate programs: an *Independent Host Program* and an *Element Program*. The Independent Host Program never interferes with the core of the computations. These are described by the Element Program, identical copies of which are executed in all processors of the hypercube simultaneously. The only function of the Independent Host lies in its role as interface between the hypercube and the outside world, hence for I/O.

2. The Intel iPSC Hypercube

The Intel Corporation has recently marketed the first commercial realization of the hypercube design, based largely on the Caltech Cosmic Cube. The machine, known as the iPSC, comes in three models - the d5, d6 and d7. These have respectively 32, 64 and 128 processors. The individual processors are the Intel 80286/80287 with up to 512Kb of memory, and the interconnections are provided by high-speed Ethernets, using an Intel Ethernet chip. The intermediate host machine, which is both the control processor and the user interface, is an Intel 310 microcomputer running a UNIX system (Xenix). In addition to the Ethernets along cube edges, a global communication channel is provided from the intermediate host machine to the individual processors. This feature is useful for debugging and to a limited extent for control purposes. Besides the UNIX system on the host, software for the system consists of a node-resident kernel providing for process creation and debugging along with appropriate communications software for inter-processor exchanges, and for host to processor direct communication. Combined computing power of a 128-node system can be over 5 MFLOPS, which along with the 64 Mbytes of memory available, provides a relatively powerful computer.

2.1. iPSC Programming

The software environment for the Intel iPSC is distinctly different from the Crystalline Operating System described above. To begin with, the operating system supports multiple processes at each cube node, identified by their process identity number *pid*. All communication primitives can address an arbitrary process on an arbitrary node. The underlying message passing system includes automatic routing of messages between any two processes. This frees the user from developing complex routing schemes in his software, but at the expense of some extra communication overhead.

A further flexibility is the availability of both synchronous and asynchronous communication modes. The system supports a concept of *virtual channel,* unrelated to the physical channels connecting nearest neighbor nodes. A process can communicate with several other processes simultaneously by opening several virtual channels and then exchanging messages using asynchronous communication calls. All messages have a user-defined integer attribute, called *type,* which is assigned by the sender. A receiver may request messages by type, but not by source process or source node. Fortunately the range of the type attribute is large enough ([0,32767]) to allow the source of a message to be encoded in its type. Messages of any size up to 16384 bytes may be sent, although the overhead for message transmission severely discourages sending small messages, a point which we return to in the next section. To send a message the message pointer and length are supplied along with the destination node and process, and the *type* attribute. To receive a message, a type and a message buffer and desired length are supplied, and on receipt of the message the actual length, source node and source process identity *(pid)* are returned. To support asynchronous transmissions, it is possible to determine if a previous message has completed on a specific virtual channel and to determine if there is a message of a specific type pending at a node.

2.2. Computation and Communication Costs

Two characteristics of the current iPSC design are the slow communication rate and the high overhead for short messages. In fact messages of length 0 and 1024 bytes take essentially the same time. As a measure of the slowness we note that a message of length 16384 bytes takes 12 seconds to traverse a nearest-neighbor ring of 128 processors, or over 17 seconds using a ring in random (sequential) order. The cost of sending a message of length 1 byte to a neighboring processor is approx 5.3 ms while longer messages require about 5.5 ms per 1024 byte segment. These numbers are approximate and were obtained by sending 30 consecutive messages from node 0 to its 6 neighbors on a 6d cube. This slow communication speed is way below the hardware limits of the Ethernet connections and suggests that much time is wasted in operating system overhead. Despite this fact we have found that the iPSC can be used with high efficiency on a wide range of problems because of the substantial memory available per node. To indicate the processor speed, we note that a C *for* loop with no body requires about 11 micro-secs per point, while a loop with a typical floating point computation such as $a = a+b*c$ requires about 67 micro-secs per point. Thus we rate the processor at about .03 Mflops though this estimate might vary by a factor of about 2 in different situations. We summarize processor speed characteristics in Table 1.

Table 1: iPSC Performance

C **for** loop: empty body	10.9 μs *per point*
C loop to copy real numbers	15.7 μs *per point*
C **for** loop $a = a+b*c$	67.4 μs *per point*
send 0 bytes	5.3 *ms*
send 1024 bytes	5.9 *ms*
send 16384 bytes	90. *ms*

2.3. Comparison of Communication Costs

Generally we assume that the cost to transfer a segment of k real numbers between two neighboring processors is of the form:

$$ST(k) = \alpha + \beta k \ .$$

This is accurate for the Caltech Hypercube, but is a simplification for the Intel iPSC since the formula does not model the communication cost correctly over the whole range of permissible message lengths. From the table we notice that messages shorter than 1024 bytes (256 reals) all take essentially the same time. This is an important case which we have included in our analyses by using different values α_{long} and β_{long} for long messages, and α_{short} and β_{short} for short messages.

We have derived estimates for the coefficients α and β from detailed experiments by measuring the time necessary to send a message around a 128 node nearest neighbor ring. From these we have deduced that, with times measured in microseconds,

$$\alpha_{short} = 6625 \ , \quad \beta_{short} = 8.28 \ ,$$

$$\alpha_{long} = 3477 \ , \quad \beta_{long} = 22.5 \ .$$

These numbers are in sharp contrast with the cost γ to perform a typical arithmetic operation, which from Table 1 is seen to be of order 30 microseconds. In particular the ratio $\alpha_{short}/\gamma = 220$ indicates that communication of single data items is hundreds of times slower than a corresponding computation. Another parameter that appears in the analysis of some algorithms is the length λ words of a buffer used to accumulate short messages for communication in a single packet. Ideally λ should be chosen such that $\alpha/\lambda < \beta$. On the iPSC we have used $\lambda = 4096$.

For comparison we present here the corresponding data for the Caltech Hypercube. In that case a single value of α and β suffice to cover the whole range, and we find:

$$\alpha = 92 , \quad \beta = 40 ,$$

Thus message startup overhead becomes small as soon as even a few words are communicated between nodes, although communication rates for very long messages are about twice as slow as on the iPSC. Computation rates γ for the Caltech Hypercube are comparable to those for the iPSC. The buffer of length λ is not required at all.

We also use γ as a measure of integer computation speed, for example it is used to measure the cost of copying arrays. In addition we use N to denote the size of vectors or the dimension of arrays and we use $P = 2^D$ to denote the number of processors.

3. The Denelcor HEP

We will describe the HEP-1 computer (1-PEM) very briefly, referring the interested reader to the paper of Jordan[16] for details.

The HEP computer consists of a number of processors, called PEMs, and a number of memory modules, connected by a complex data switching network. The individual processors are themselves pipelined processors, allowing several different instruction streams to execute in parallel. For the purpose of this paper, each HEP PEM may be thought of as an (approximately) 8-processor MIMD machine with a global shared memory. There is software/hardware support for creating virtual parallel processes. Up to 50 processes may execute concurrently, although the physical parallelism involved is, as we have said, only about 8. Process creation overhead is low, making it efficient in many cases, for example, to assign separate processes to subroutine calls. There is also support for controlling simultaneous access to memory by different processes, implemented using a special access bit which is appended to every memory word. Processes may read, set and clear these access bits in operations that are guaranteed to be indivisible by the system. The largest HEP computer built to date has been a four PEM configuration.

Parallel programming differs most from serial programming in its need for synchronization between processes. Synchronization requires sharing of data in some form or other between processes, either through a global memory or through some form of message passing. Processes involved in such data sharing must operate within *critical sections*, i.e. in an environment that guarantees to lock out all other processes. Critical sections provide serious bottlenecks in parallel code because of their inherently serial nature, and they can dominate execution time unless great care is taken to minimize their impact. To make this data sharing and process synchronization reasonably portable among a range of machines *monitors* are used. In the following paragraphs we present some of the HEP primitives for parallel programming and high level portable monitors, implemented using the primitives. For a complete discussion of these issues we refer to the paper of Jordan[16] for the primitives and to the paper of Lusk and Overbeek[17] for the monitors.

The parallelism in HEP programs is obtained through the use of parallel subroutine calls. In HEP FORTRAN this is done through a statement with syntax:

CALL CREATE (name,parameter list) .

When such a call occurs, a process is created to execute the subroutine "name" and control is handed over immediately to the calling process. The access bits which are attached to every word in the HEP memory, are manipulated asynchronously with reads and writes on variables. For example,

CALL AWRITE(var,val)

waits until the access bit of variable *var* is empty and then writes the value *val* into variable *var*. Analogously,

$$val = AREAD(var)$$

waits until the access bit is full to read the value of variable *var* and then assigns the value to *val*.

We now give some simple examples of the style of programming we have been using on the HEP, utilizing the concept of monitors to implement critical sections in a hardware-independent fashion. A *monitor* is a combination of data structures, initializers and code used to prevent interference between processes. Only one process at a time may be in a monitor. By implementing monitors as macros, they can be designed to be as portable as possible and, yet capable of generating efficient machine-specific FORTRAN after pre-processing (the above HEP primitives only occur in the macro definitions).

Some typical monitors we have used are:

create(func,arg1,arg2, ...,)
getsub(name,sub,max_sub,nprocs)
barrier(name,nprocs)
lock(j)
unlock(j)

The *create*() monitor is used to execute a concurrent function or subroutine call. It behaves similarly to an ordinary subroutine call except that a separate process is generated to run the subroutine. On completion of the subroutine, the process dies.

The other monitors are used to control concurrently executing processes. For example, the *getsub*() monitor is used by *nprocs* cooperating processes to acquire a unique subscript *sub* in the range $1 \leq sub \leq max\_sub$. When all subscripts in this range have been returned, further calls to *getsub*() will return a *sub* of 0. Because more than one subscript may be required, instantiations of *getsub*() are allowed with different values for *name*. The *getsub*() monitor ensures that two processes are never given the same subscript, and also arranges that processes that have been returned a 0 subscript are held until all the processes have reached that state. This approximates as closely as possible the semantics of a DO loop on a serial machine. Using *getsub*(), a self-scheduling parallel DO loop may be written, with arbitrary code inside the loop, and with no static connection between the number of processes to be used and the size of the loop upper bound.

The *barrier*() monitor holds processes that encounter it until *nprocs* processes are waiting, at which point all the processes are released. Again barriers are *named* since multiple instances may be required simultaneously. As mentioned above, the *getsub*() monitor includes an implicit barrier simulating the exit from a DO loop, and consequently explicit *barrier*() calls are not required between loops (e.g. as in the FFT Fast Solver) when programming with *getsub*().

On the HEP these monitors are written in terms of shared global variables. The extra read/write access bit available with all HEP variables is easily used to ensure the indivisibility of the basic operations underlying these monitors - for example the incrementing of the *sub* variable. On other machines, these monitors may be implemented quite differently. The monitors can even be sensibly defined on serial machines so that parallel codes may first be debugged serially, in the same form that they will be used on the parallel processor.

THE PARALLEL MULTIGRID METHOD

1. The Multigrid Method on the Hypercubes

We have developed a parallel multigrid algorithm and tested it on both the Caltech and Intel Hypercubes. Multigrid Methods are near-optimal for a wide range of computations on serial or vector processors. Our test problem was a Poisson equation with Dirichlet boundary conditions on a square, discretized with a 5-point finite difference operator, but the implementation is considerably more general. A multigrid algorithm for the Poisson equation in 3 dimensions on the Caltech Hypercube has been developed independently by Clemens Thole.[18] The algorithms presented in this section are valid for any architecture in which the processors can be arranged as a two dimensional periodic or non periodic array, depending on the nature of the boundary conditions of the elliptic problem.

1.1. Distributed Grids for Multigrid

Having discretized the PDE using finite differences, the resulting equations on any grid level involve the solution or error on a rectangular grid of points. Parallelizing multigrid amounts to distributing these grids over the processors and implementing the communications needed as a result between processors. We accomplish the distribution by representing the solution (or error) on each grid level as an *areally distributed matrix*. This distribution divides the matrix into rectangular regions, each of which is stored locally to one processor. Assuming the matrix has been divided in $p_x p_y = P$ regions, every processor contains one such region.

It is important to map these in such a way that North, East, South and West neighbors of any region are mapped to physical neighbors on the hypercube. Grids are mapped onto the Hypercube using the *whoami* facility developed by J. Salmon.[15] The routine uses Binary Gray Code (BGC) sequences[19,20] to number processors. These are sequences of numbers such that consecutive terms in the sequence differ by only one bit in their binary expansions.

As mentioned earlier, nodes on the hypercube with processor numbers that differ in only one bit in their binary expansion are physical neighbors. If we interpret the terms in a BGC sequence as processor numbers, then nearest neighbors in the sequence will also be physical neighbors. This allows one dimensional grids to be mapped onto the hypercube using only nearest neighbor connections. The generalization to higher dimensional grids is done by mapping each dimension to a suitable subcube and using the 1-dimensional mappings for those subcubes.

Certain applications require periodic grids to be mapped to the cube. We thus require a BGC sequence which is also periodic. A subclass of Gray codes, the Binary Reflected Gray Codes (BRGC), have the desired property.[15] In the sequel we will always use a canonical BRGC defined recursively as follows. With one bit the BRGC sequence is 0 1 . With d-bits we take the $d-1$ bit sequence, followed by the same sequence reversed in order and with each dth bit set. Thus with two bits we have the sequence 00 01 11 10 and with four bits the sequence is:

0000 0001 0011 0010 0110 0111 0101 0100
1100 1101 1111 1110 1010 1011 1001 1000

This is the standard processor ordering used by the Caltech Hypercube research group.[21] With a BRGC ordering the subcubes used for a grid whose dimensions are each a power of 2 will all be periodic, allowing periodic grids of such dimensions to be mapped to the cube. The BRGC sequence also has the property that two elements that are a power of 2 apart differ in at most two bits.[22]

We surround each of the regions by an artificial boundary, which we use to store duplicate values of the grid function at neighboring points from other regions. We will use the term *extended subgrid* to denote such a subgrid with boundary, and will refer to the non-boundary points of the subgrid as its *interior*. We assume that the grids to be used are of dimensions $2^{l_x} p_x \times 2^{l_y} p_y$ for some l_x, l_y. With this choice, the interior of a level l subgrid assigned to a processor may be identified with a subset of the extended level $l+1$ subgrid assigned to that processor. Similarly the interior of the level $l+1$ subgrid in a processor is contained in the extended subgrid of level l.

1.2. The Shuffle Operation

Along the boundary of the subgrids at any level some communication will be required by the multigrid relaxation operation. For example, a relaxation at boundary points of a subregion will require the boundary data of the adjoining subregion. Here the artificial boundary introduced above, plays its role: it is clear that if the duplicate values of the neighboring grid points are available the relaxation can proceed normally. For a 5-point operator, a boundary width of 1 suffices. For more general operators however, the boundary width might have to be increased to extend over a distance of more than one grid point.

To keep the duplicate values updated, communication is necessary between the processors. We call the operation that performs this update the *shuffle operation*. By executing this operation, one can assure that each boundary edge at the start of any operation (relaxation or intergrid transfer) holds a copy of the closest interior edge of the adjacent subgrid. The shuffle of the whole duplicate boundary requires the update of the east, south, west and north edges and of each of the corner nodes. In the edge updates there are two different kinds of exchanges. Assuming the matrix representing the grid function is stored by rows, we can exchange north and south boundaries with their respective neighbors at once. The east and west exchanges however are more complicated since the values are not stored contiguously in the local memories. To avoid a possibly large startup cost with the transfer of each value (as on the iPSC), a buffer needs to be provided to collect all the values before the actual exchange takes place. The corner nodes are not actually needed in a five-point problem, but in more general situations are accommodated by adding an extra corner node shuffle. The only possible problem here is the updating of corner nodes which *a priori* have to be copied from a non-neighboring processor. From Figure 1 it is clear however, that after the edge updates have been performed, the corner node values are available in neighboring processors and that a simple nearest neighbor communication can also update these values. The shuffle operation is thus seen to be the basic communication routine necessary for multigrid to work on distributed grids.

It is easily seen that asymptotically, for large number of grid points per processor, communication is proportional to the perimeter of each of the regions. The computation however is proportional to the area of the same regions. Thus in order to minimize communication overhead in the algorithm the ratio of area over perimeter needs to be maximized - this is referred to as the *area-perimeter law*. When comparing communication costs for the areal distributions displayed in Figure 2 it can be expected that performance will decrease from top to bottom.

Intergrid transfers between multigrid levels (injection and interpolation) are entirely local to each processor because of the containment relationships between the grids presented in the previous section. Thus no communication is involved in these steps. However before any updated grid values are used it is essential to perform a shuffle operation to preserve the integrity of the duplicate boundary data. Thus we use the shuffle operation not only before each relaxation but also before every intergrid transfer.

1.3. The Error Calculations

To control termination of the iterations in adaptive multigrid algorithms, norms of error values need to be calculated. Since these are global values, the computation of these needs interprocess communication. As we will show, it is natural for this computation to use a mapping of the hypercube processors onto a binary tree or, more precisely, a D-ary tree, where D is the cube dimension. In the following discussion we choose the root of the tree to be at node 0, but any other point can be used as easily.

The observation that allows a balanced binary tree to be mapped to the hypercube, is that a D-cube may be regarded as two $(D-1)$-cubes with corresponding processors from each connected by an extra channel. Numbering the processors in the two $(D-1)$-cubes as $p_0, \ldots, p_0+2^{D-1}-1$, and $p_0+2^{D-1}, \ldots, p_0+2^D-1$, respectively, we connect processor p_0+p in the first subcube with processor p_0+p+2^{D-1} in the second subcube. The binary tree on the D-cube is then defined recursively as having the lowest numbered processor as its root and the two $(D-1)$-cubes as its left and right sub-trees. Note that this is a logical binary tree only - some processors occur several times at tree nodes. For example, processor 0 occurs D times, being the root of D sub-trees. Each processor is located exactly once at a leaf of the tree.

An alternative tree mapping is to represent the network as an unbalanced D-ary tree. This mapping is based on the obvious fact that D connections emanate from every node. This tree is a physical tree - each node in the tree corresponds to a unique processor. The tree may be defined either by giving the children of each node, or by specifying the parent of each node. The children of a node are the processors whose node numbers are obtained by setting in turn each of the low-order unset bits in the D-digit binary representation of the node number. Alternatively, the parent of a node is located by unsetting the lowest-order set bit in the binary representation of the node number.

These two logical trees are in reality two different ways of representing the same set of connections. Depending on the application it can be helpful to consider trees from either viewpoint.

We use this tree mapping to implement the error calculation. The semantics of this operation needs discussion. Each processor calls an *error*() routine simultaneously, providing as arguments the grid patch local to each processor. The *error*() routine returns in *every* processor with the value of the global error of the complete distributed grid function used to represent the solution. No processor is distinguished in these semantics.

The implementation of the routine does of course distinguish processors. The error operation is complicated because it involves communication between the processors to sum local errors, and further communication to broadcast the final sum back to all of the nodes. On cube architectures, this should be done by summing over a *tree* of processors, since if one processor is used to accumulate all partial errors, it will result in a critical section. A D-ary tree is mapped onto the processor network with node 0 as the root. The parent of a node is defined by zeroing the lowest non-zero bit of the node number while the children of a node are obtained by setting in turn exactly one of the low-order zero bits of the node number. The error routine then takes the form:

```
real error(gridfun):

mysum = local_error(gridfun)

for each child
        read child_sum from child
        mysum = mysum + child_sum
end for

if (node≠0) send mysum to parent
if (node≠0) read sum from parent

for each child
        send sum to child
end for

return sum
```

With this implementation the cost of an error calulation will be $O(N/P)+O(\log P)$, with the $O(N/P)$ reduced further if vector hardware is available on the nodes.

1.4. The Relaxation Operation

The relaxation operation is straightforward to implement in each of the subregions. Standard Gauss-Seidel relaxation is discarded on numerical grounds of slower convergence. It introduces artificial discontinuities along the boundaries of the subregions as the parallel version uses the original values of the boundary duplicates. Note that in the limit of subregions consisting of only one grid point, this would reduce to a Jacobi relaxation. Instead we utilize a Red-Black ordering of the grid points. This can be implemented such that the parallel algorithm will, after each iteration step, give the same result as the corresponding serial algorithm. The procedure consists of shuffling, relaxing the red points, shuffling again and then relaxing the black points.

1.5. Intergrid Transfers

The error (the residual equation to be precise) is projected to the coarse grid using half-injection - the errors at the black grid points are identically zero after the previous relaxation step. For this step as discussed previously, it is necessary that the extended fine grid subregions contain the non-extended coarse grid regions.

Having solved the error equation on the coarse grid the solution on the fine grid has to be updated by addition of a suitable interpolation of the computed coarse grid error. We have used linear interpolation at this point. To do this in parallel without communication, the extended coarse grid subregion must contain the non-extended fine grid subregion. We update the boundary duplicates in the fine grid subregions following the interpolation using a shuffle operation.

2. The Multigrid Method on the Denelcor HEP

2.1. The Relaxation Operation

In our HEP studies we have also taken the standard five-point discretization operator for the Poisson equation on a rectangular grid. We apply a red-black coloring to the grid points, alternating between the colors as one proceeds through the grid by rows. We smooth by applying Gauss-Seidel relaxation first to all of the red points, in some order, then to all of the black points. With this procedure all of the red points are uncoupled from each other for the 5-point operator - i.e. only black points contribute to the new value of x_i at a red point i. Similarly the black points are uncoupled from each other. This makes the Gauss-Seidel sweep over the red points completely parallelizable, and each red point (or a set of such) may be given to a separate process, and similarly for the black points. However the red points must be completed before starting on the black points. As was the case in our hypercube algorithm, the results after each iteration are identical to those that would be obtained by a serial algorithm. The underlying reason is of course the decoupling obtained by the colouring of the grid points.

We have also implemented the multigrid scheme using several other smoothing operations, including Weighted Point Jacobi with weight parameter ω of .8 and also a Line Gauss Seidel smoother. For the Point Jacobi operator, the solution at iteration step $n+1$ is defined in terms of that at iteration step n by:

$$x_i^{(n+1)} = \omega \, (f_i - \sum_{j \neq i} A_{i,j} \, x_j^{(n)}) \, / \, A_{i,i} \; + \; (1 - \omega) \, x_i^{(n)}.$$

Since only the values $x_i^{(n)}$ from the previous iteration are used, this operation is fully parallelizable. In our implementation we distribute all the points of each row to the same process, giving distinct rows to different processes whenever possible. In the Red-Black Line Gauss Seidel all points of each row are simultaneously updated. We alternately label rows as red or black. Distinct rows are given to distinct processes - all the red rows being treated first, then the black in order to achieve parallelism. Within each row we use a sequential tridiagonal solver to implement the simultaneous solution. For each of the smoothing methods we have obtained more or less similar results in terms of successful use of available parallelism.

2.2. Intergrid Transfers

To project the residual equation to a coarser grid, we again use *half-injection*. Note that the residual at all fine grid black points vanishes identically because of the immediately preceding smoothing operation - this was also true of the red points after smoothing, but the subsequent black smoothing will disturb some of these red point residuals. After computing the coarse grid error, we return the error corrections to the fine grid using *linear interpolation*. The control over when to transfer to a coarser or finer grid is adaptive, but we have also implemented fixed multigrid schemes on the HEP with similar speedups.

2.3. Synchronization Issues

Each of the red and black smoothers, the half-injection, the projection and the linear interpolation is coded as a parallel algorithm. In our implementation, each process asks for a row of grid points to work on, does the requested operation on these and then goes on to ask for a new row. The bookkeeping for this is done in a monitor. Between each of the operations (e.g. between the red and the black relaxation) a barrier is set requiring all processes to complete their work before proceeding to the next step, though in fact as soon as adjacent rows of points are completed one could proceed to the next phase in that region if desired.

On very coarse grids, there is typically not enough work for all processes. For example, if the number of rows of grid points drops below the number of processes, then some processes will remain idle. At this point one can subdivide the rows into smaller pieces and finally assign individual points to processes. However sooner or later one encounters difficulty in using all of the processes effectively, which seems to be an inherent characteristic of multigrid methods. One solution is to terminate the grid coarsening at a higher level, computing the solution on that level by allowing the Gauss-Seidel iterations to proceed to convergence. This will provide a slower algorithm on a serial machine but may be faster on parallel machines.

We point out that on shared memory systems like the HEP, the calculation of the error is almost no issue. The global error can easily be summed in a globally shared variable, provided access to that variable is done asynchronously. The same semantics as for the non-shared memory designs can be used here.

3. Results and Timings for Multigrid

In Figure 3 we compare timings for 3-level V-cycle and W-cycle multigrid iterations with finest grid of dimension 64×64, and also a 4-level V-cycle iteration on a 128×128 grid. These computations were performed on the Caltech hypercube with from 2 up to 32 nodes. The curves represent speedup attained with increasing number of processors (the 128 grid problem would not fit on 2 processors). We notice that the speedup for the W-cycle is worse than that for the V-cycle. This is due to the fact that in the W-cycle algorithm, one spends a larger segment of the computation time on the coarser grids. In spite of this, it can be expected that for some equations it may be profitable to use the W-cycle instead of V-cycle because of its faster convergence rate. This issue however is too dependent on the particular type of equations solved to be discussed in this context. The V-cycle curve for the finer grid shows substantially improved performance compared to the coarser grid. All of these curves correspond to relatively very coarse grids since the number of grid-points per processor is at most 512 on the finest grid.

In Figure 4 we illustrate the effect of Area/Perimeter considerations on the Caltech Hypercube with 32 processors. The hypercube was organized as logical rectangular meshes of dimensions 32×1, 16×2 and 8×4, and we display the resulting computation time for a multigrid iteration on a 128×128 grid with 2 grid levels. The improvement in timing as the subregions approach squares is obvious. The success of this demonstration is related to the fact that even for short messages communication costs on the Caltech Hypercube are almost proportional to message length.

To illustrate that communication overhead decreases with an areal distribution as problem size increases, we compare in Figure 5 the scaled time of a multigrid iteration as a function of grid size. The scaled time is obtained by dividing the iteration time by a quantity proportional to the number of grid points and is therefore proportional to effective work per grid-point for multigrid *including* communication overhead. Specifically we compare the times for 16 iterations on a 64 by 64 grid, 4 iterations on a 128 by 128 grid and 1 iteration on a 256 by 256 grid on the 32 node Caltech Hypercube. Scaled time is over twice as small for the finest grid as for the coarsest.

Our multigrid results for the Intel iPSC, are cast in a somewhat different form. We solved problems with varying grid sizes on a 128 node iPSC. To compare the iteration times, we calculated the number of *useful* floating point operations (additions or multiplications, but not memory accesses) that were executed in each of the runs. We divided these by the time it took for an iteration to complete. This defines an attained *floprate* for each run. In Figure 6 we plot the attained floprate for V-cycles as a function of grid size, going from 4096 unknowns to over 1.6 million unknowns. We see that for the largest problems we attained about 3.75 megaflops.

We present our results for the Weighted Jacobi smoother on the Denelcor HEP, in the form of speedup versus number of processes used in Figure 7. It displays results for a 128*128 grid computation on the HEP, where the iteration proceeds down to 2*2 grids. Speedups for 1 to 19 processes on a HEP-1 are presented, and as can be seen, are optimal. In Figure 8, the actual times for the same problem when the iterations are cut off at the 2*2, 4*4, 8*8, 16*16, 32*32 and 64*64 grid levels. Each curve is a plot of CPU time versus number of processes. As less levels are used the speedup over one process increases (the curve is steeper), but the actual timing becomes worse because relaxation is not an effective way to solve the equations on the coarsest grids. Using a direct solver on the coarsest grid would likely improve these results, provided it took advantage of parallelism.

Figure 9 and Figure 10 repeat the same results, but this time using the Red-Black Gauss Seidel relaxation operator. The Red-Black smoother can be seen to give a much faster solution algorithm than Jacobi.

References

1. O. McBryan, "Computational Methods for Discontinuities in Fluids," *Lectures in Applied Mathematics*, vol. 22, AMS, Providence, 1985.
2. O. McBryan, "Elliptic and Hyperbolic Interface Refinement in Two Phase flow," in *Boundary and Interior Layers*, ed. J. J. H. Miller, Boole Press, Dublin , 1980.
3. O. McBryan , "Shock Tracking for 2d Flows," in *Computational and Assymptotic Methods for Boundary Layer Problems*, ed. J.J. Miller, Boole Press, Dublin , 1982.
4. O. McBryan and E. Van de Velde, "Parallel Algorithms for Elliptic Equations," *Commun. Pure and Appl. Math.*, Oct 1985.
5. O. McBryan, "A Multi-grid Finite Element Solution of Discontinuous Elliptic Equations," *Proceedings of International Multigrid Conference*, Elsevier, Copper Mountain, Colorado, 1983 . to appear
6. O. McBryan and E. Van de Velde, "Parallel Algorithms for Elliptic Equation Solution on the HEP Computer," *Proceedings of the First HEP Conference*, University of Oklahoma, March 1985.
7. O. McBryan and E. Van de Velde, "Algorithms and Archictectures: Pde Solution on Parallel Computers," *Proceedings of the 2nd SIAM Conference on Parallel Processing for Scientific Computation, Norfolk, Nov. 1985*, SIAM, to appear.
8. J. Glimm and O. McBryan, "A Computational Model for Interfaces," Courant Institute Preprint, May 1985.

9. O. McBryan, "Fluids, Discontinuities and Renormalization Group methods," in *Mathematical Physics VII*, ed. Brittin, Gustafson and Wyss, pp. 481-494, North-Holland Publishing Company, Amsterdam, 1984.

10. A. Brandt, "Multi-level adaptive solutions to boundary-value problems," *Math. Comp.*, vol. 31, pp. 333-390, 1977.

11. W. Hackbusch, "Convergence of multi-grid iterations applied to difference equations," *Math. Comp.*, vol. 34, pp. 425-440, 1980.

12. K. Stuben and U. Trottenberg, "On the construction of fast solvers for elliptic equations," *Computational Fluid Dynamics*, Rhode-Saint-Genese, 1982.

13. Charles L. Seitz, "The Cosmic Cube," *Communications of the ACM*, vol. 28, No. 1, pp. 22-33, Jan. 1985.

14. G. C. Fox and S. W. Otto, "Algorithms for concurrent processors," *Phys. Today*, vol. 37,5, pp. 50-59, May 1984.

15. John Salmon, "Binary Gray Codes and the Mapping of a Physical Lattice into a Hypercube," Caltech Concurrent Processor Report (CCP)Hm-51, 1983.

16. H. F. Jordan, "HEP Architecture, Programming and Performance," University Of Colorado Preprint, 1984.

17. E. L. Lusk and R. A. Overbeek, "Implementation of Monitors with Macros," Technical Memorandum ANL-83-97, Argonne National Lab. Mathematics and Computer Science Div. , December 1983.

18. C. Thole, "Experiments with Multigrid on the Caltech Hypercube," GMD Internal Report, October 1985.

19. E. N. Gilbert, "Gray Codes and Paths on the n-Cube," *Bell System Technical Journal*, vol. 37, p. 915, May 1958.

20. Martin Gardner, "Mathematical Games," *Scientific American*, p. 106, August 1972.

21. "Caltech/JPL Concurrent Computation Project Annual Report 1983-1984," Caltech Report, December 1984.

22. T. Chan, Copper Mountain, Colorado, April 1985. Remark at 2nd Multigrid Conference

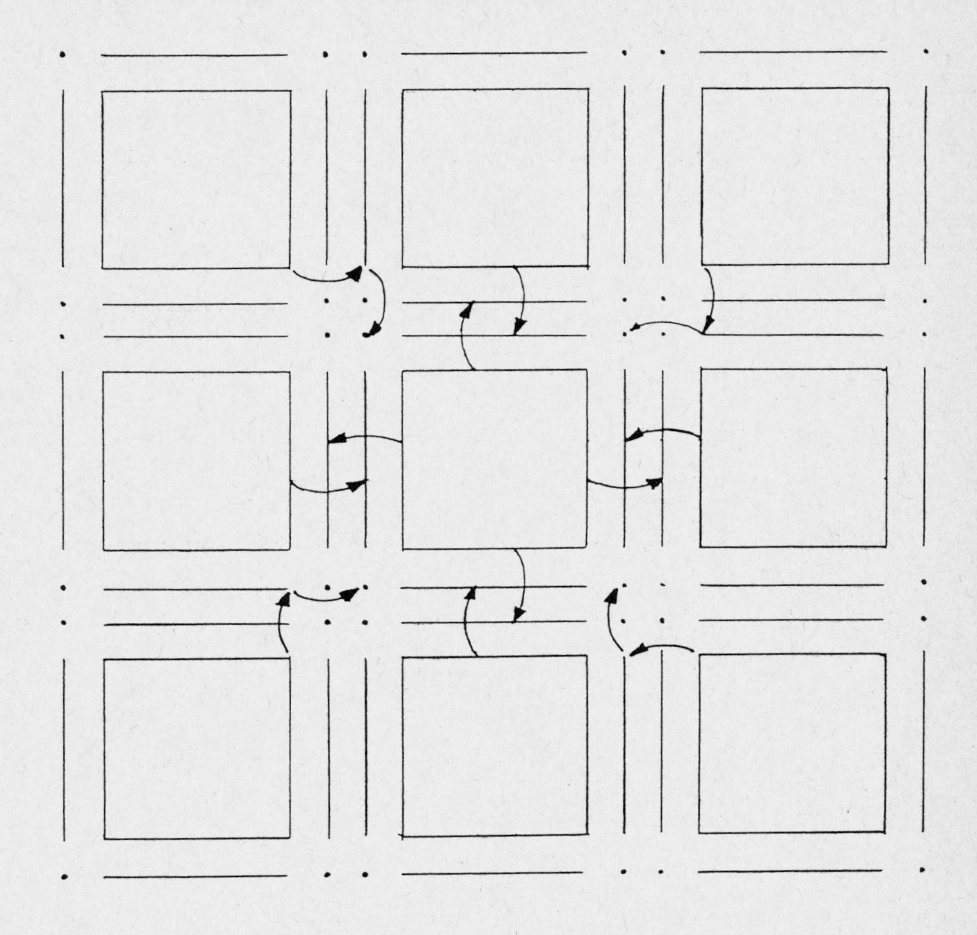

Fig. 1: The Shuffle Operation on an Areally Distributed Grid.

Fig. 2: Assignment of equal area grid blocks to 32 processors but with different area/perimeter ratios.

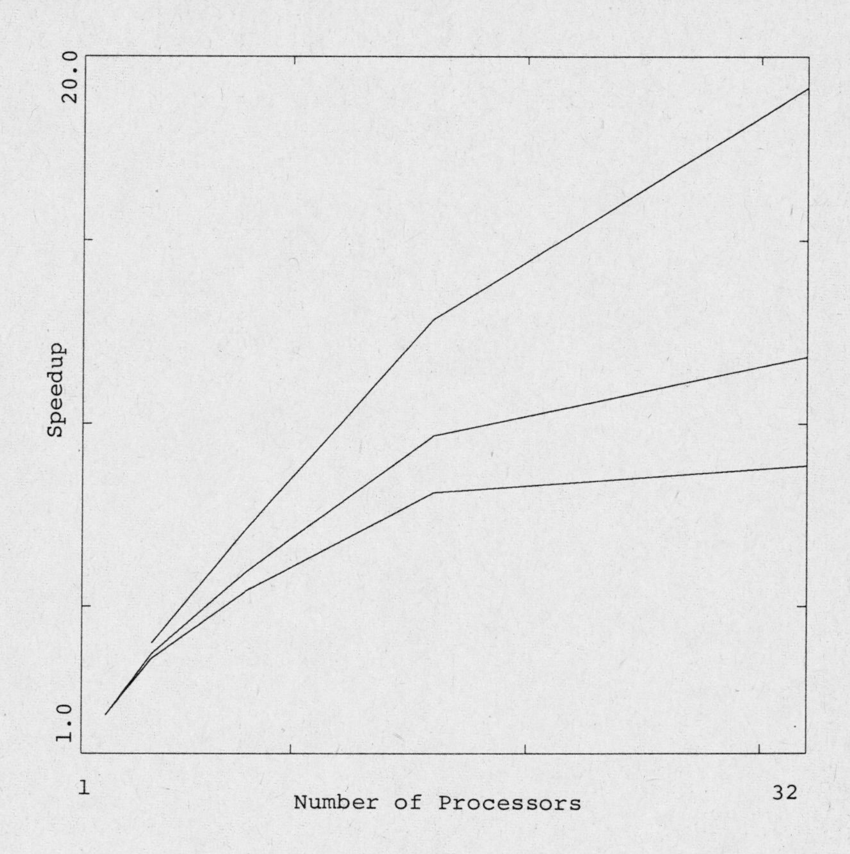

Fig. 3: The top curve is speedup versus Number of Processors for V cycle Multigrid on a 128 by 128 grid with 4 levels on the Caltech Hypercube. The two lower curves are for V and W cycle Multigrid on a 64 by 64 grid with 3 levels.

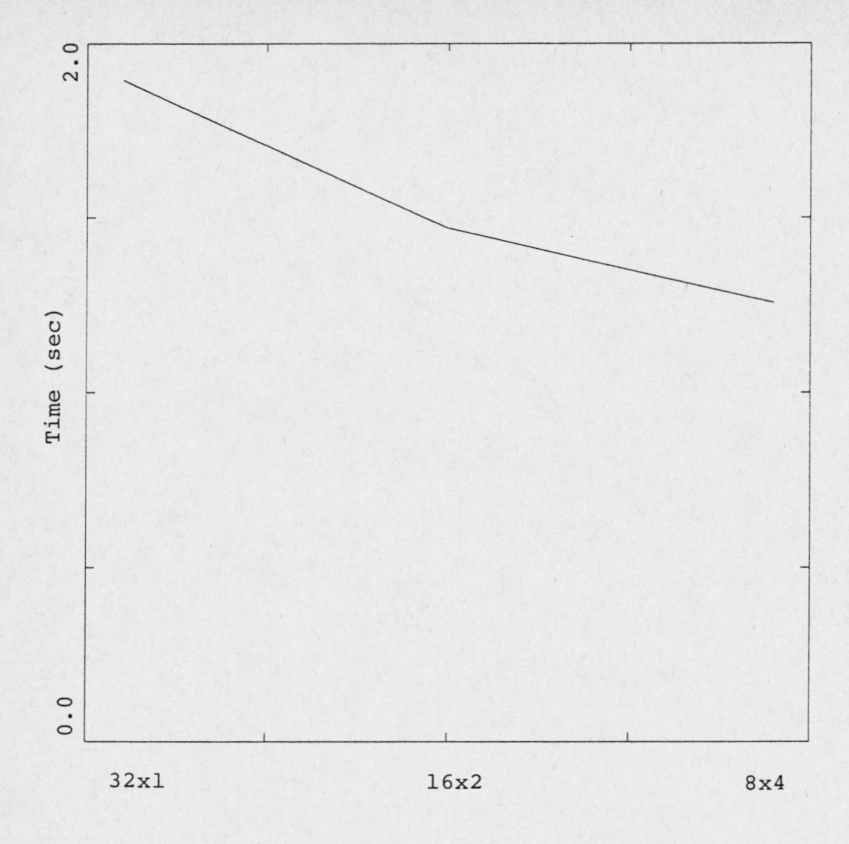

Fig. 4: Area-Perimeter considerations for Multigrid on the 32 node Caltech Hypercube. We compare times for a V cycle multigrid iteration on a 128 by 128 grid with 2 levels, when the domain is split in 32 by 1, 16 by 2, 8 by 4 rectangles.

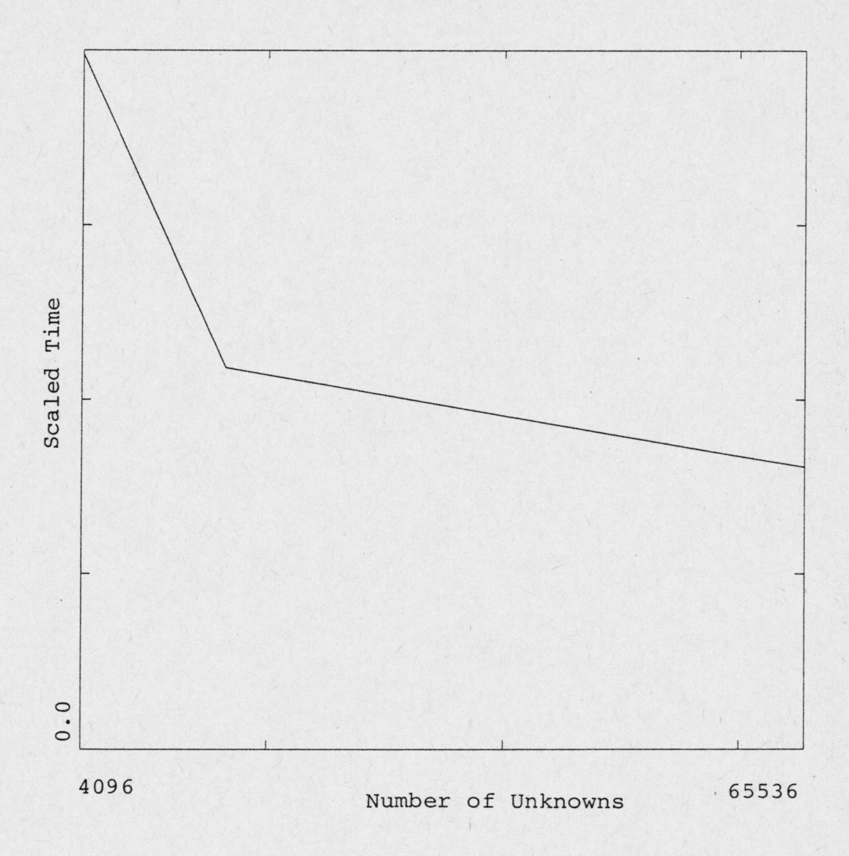

Fig. 5: Multigrid, 3 levels, V cycle on the Caltech Hypercube. We compare execution times of different sized problems by scaling the times by the number of floating point operations executed.

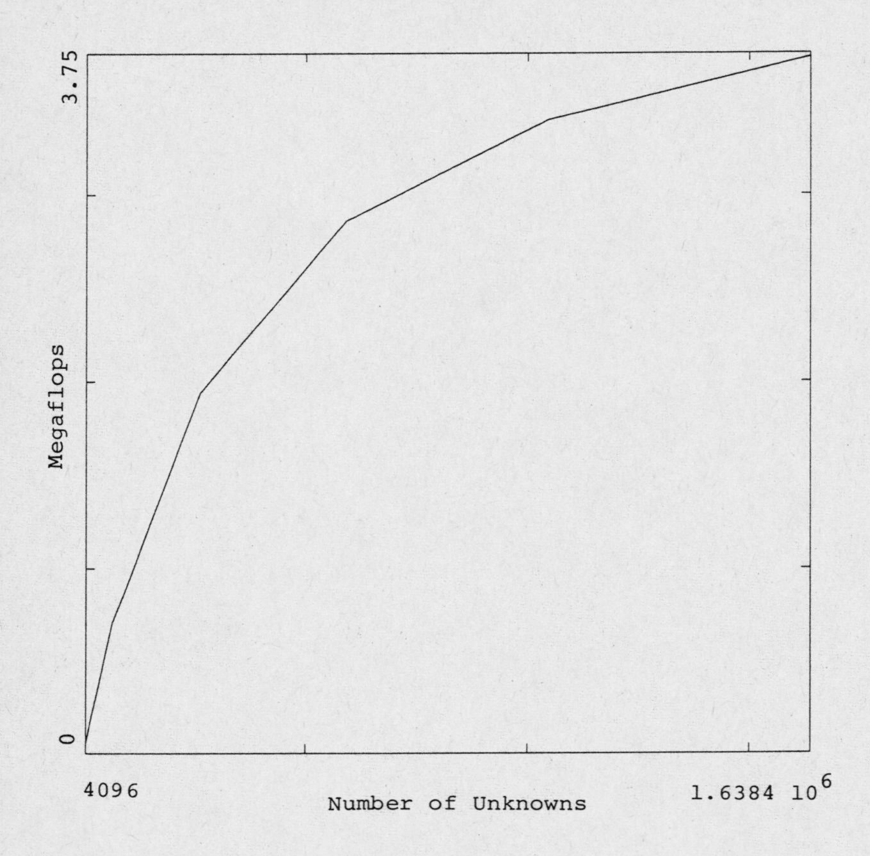

Fig. 6: Performance of 2 level V Cycle Multigrid Method on the 128 node iPSC. Number of Floating Point Operations executed per second as function of Problem Size.

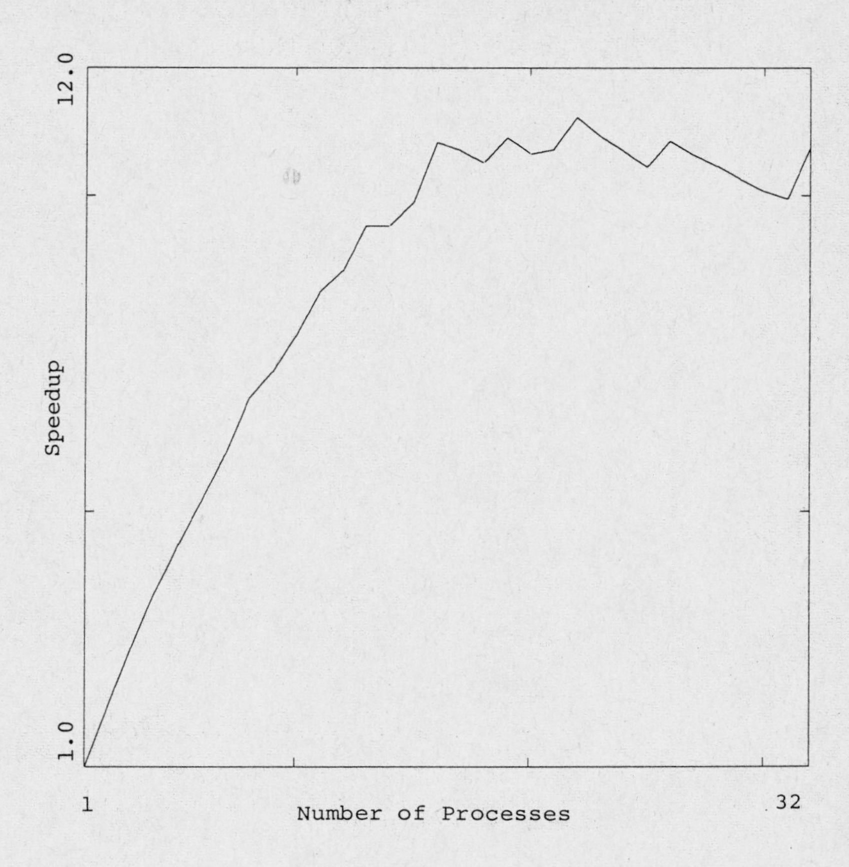

Fig. 7: Speedup versus Number of Processes for the Multigrid algorithm using a Jacobi relaxation on a 128 by 128 grid with 7 levels on the Denelcor HEP.

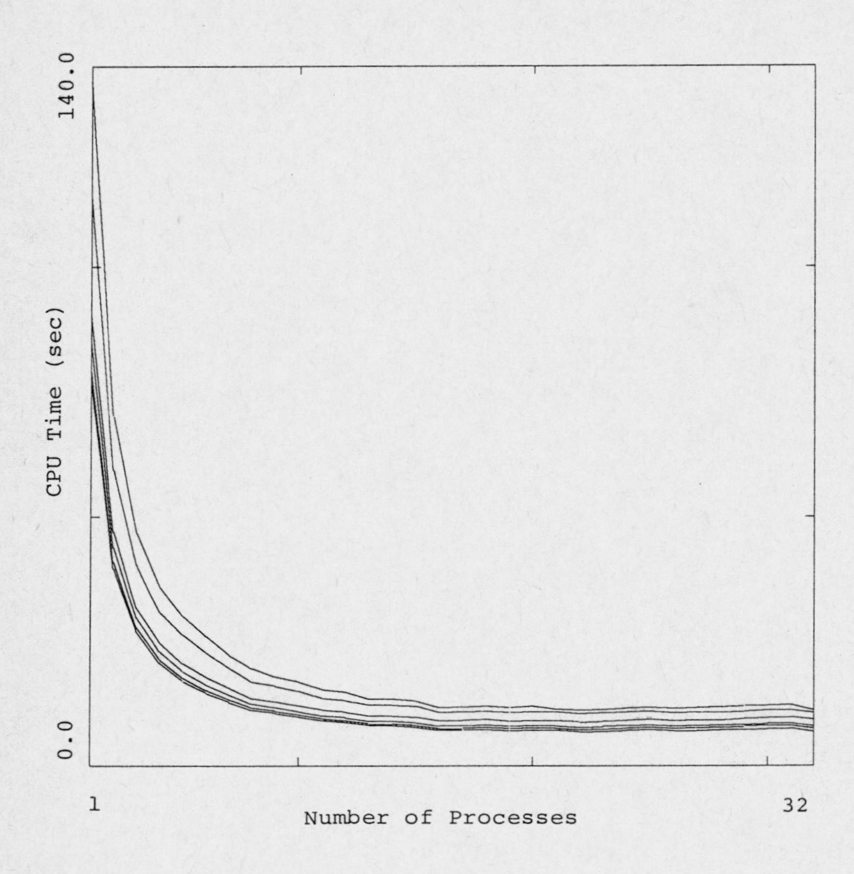

Fig. 8: CPU Time versus Number of Processes for the Multigrid algorithm using a Jacobi relaxation on a 128 by 128 grid on the Denelcor HEP. We compare times for the algorithm using 2 up to 7 levels.

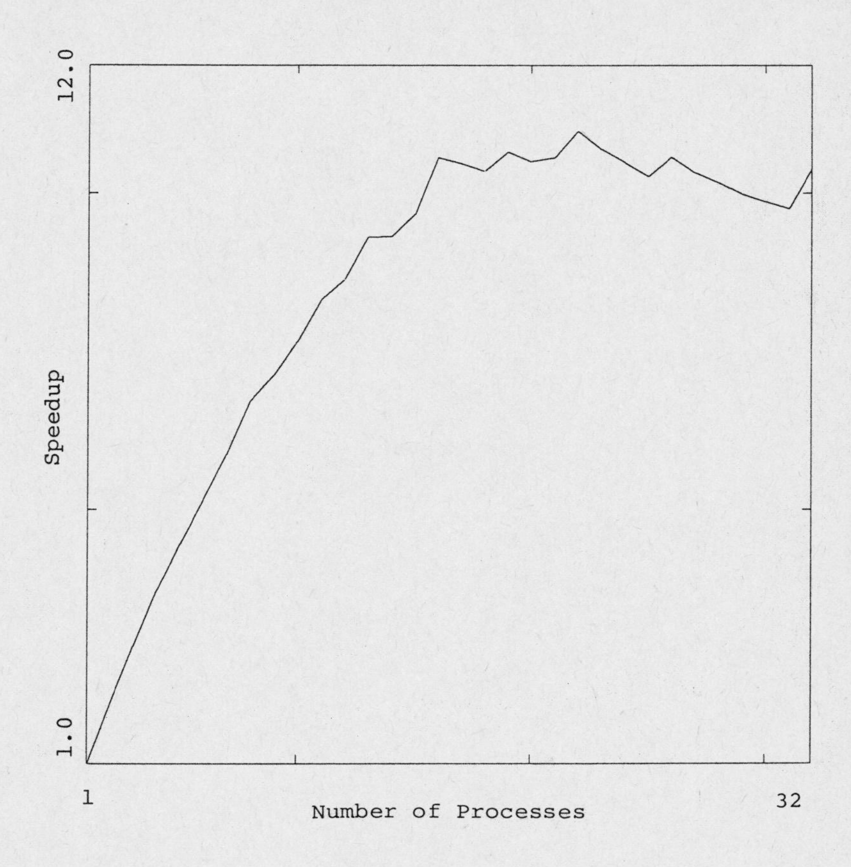

Fig. 9: Speedup versus Number of Processes for the Multigrid algorithm using a Red Black Gauss Seidel relaxation on a 128 by 128 grid with 7 levels on the Denelcor HEP.

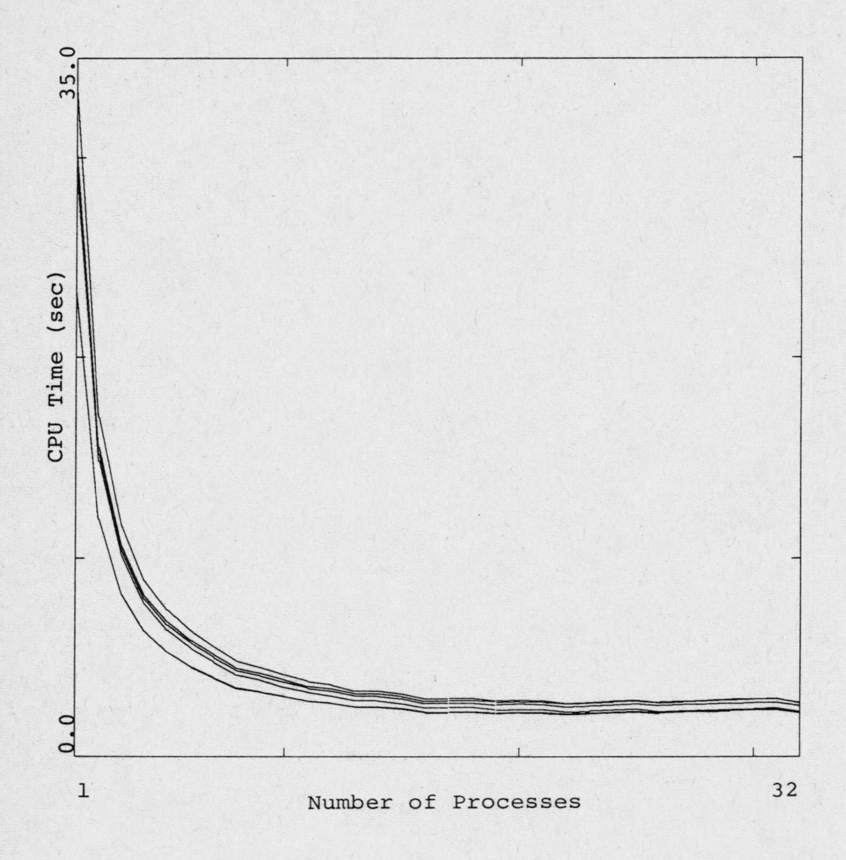

Fig. 10: CPU Time versus Number of Processes for the Multigrid algorithm using a Red Black Gauss Seidel relaxation on a 128 by 128 grid on the Denelcor HEP. We compare times for the algorithm using 2 up to 7 levels.

On the Treatment of Singularities in the Multigrid Method

U. Rüde

Chr. Zenger

Technische Universität München

1. Introduction

Structural singularities cause severe problems in the numerical solution of elliptic boundary value problems. On the one hand the occurence of singularities in the solution results in a bad convergence of the discrete solution to the solution of the differential equation and - to some extent as a consequence of this fact - in the context of multigrid methods, singularities deteriorate the rate of convergence of the iteration scheme.

As a model problem we refer to the interface problem of the diffusion equation, where the diffusion coefficient is piecewise constant on rectangular subdomains. This problem was investigated in detail in [1], where appropriate multigrid schemes were constructed.

The differential equation is given by

$$-\nabla(D(x,y)\ \nabla u(x,y)) = f(x,y) \quad \text{on } \Omega \qquad \text{+ boundary conditions} \tag{1.1}$$

$D(x,y)$ is piecewise constant on rectangular subdomains of Ω.

We restrict our attention to the so called four corner juncture, the region where four rectangles with possibly different diffusion coefficients meet (fig 1).

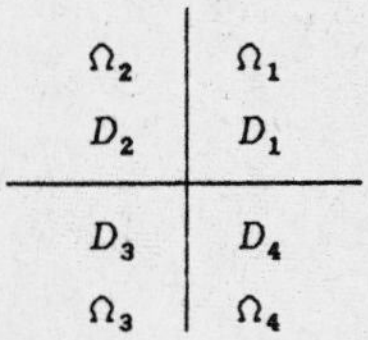

Ω_2	Ω_1
D_2	D_1
D_3	D_4
Ω_3	Ω_4

Figure 1.

The solution at the juncture point (the origin of our coordinate system) is usually singular. The analytic solution in the neighbourhood of the origin was investigated in detail by Kellogg [2]. He showed that the solution in the neighbourhood of the singularity is given in polar coordinates by

$$u(r,\varphi) = w_1(\varphi)r^{\alpha} + w_2(\varphi)r^{2-\alpha} + w_3(r,\varphi), \tag{1.2}$$

where $\alpha = \alpha(D_1, D_2, D_3, D_4)$ is a function of the diffusion constants and $0 \leqslant \alpha \leqslant 1$. The function $w_3(r,\varphi)$ is of $O(r^2)$ in the neighbourhood of the origin.

Note that the Dirichlet problem in an L-shaped region is a special case of our problem. We only have to choose $D_1 = D_2 = D_3 = 1$ and $D_4 \to \infty$. The boundary values in Ω_4 have to be choosen $\equiv 0$.

But the L-shaped region with $\alpha = 2/3$ produces a comparably weak singularity. In fact if we choose $D_1 = D_3 = 1$ and $D_2 = D_4 = \varepsilon$ for $\varepsilon \to 0$, one can show that $\alpha \to 0$, which, in the limit case, gives a solution that is discontinuous at the origin.

This situation was studied extensively in [1]. It was observed that if the multigrid scheme is applied in a naive way, the convergence rate deteriorates and the process may even become singular.

Moreover, it can be shown that the convergence to the exact solution also deteriorates. The order of convergence at a fixed point in the interior of Ω is only $O(h^{2\alpha})$ and so the order of convergence becomes arbitrarily small if $\alpha \to 0$. This order is the same for all usual finite difference or finite element schemes, if no special attention is paid to the singularity.

2. Discretization of the Problem.

The usual and most simple discretization is given by the five point difference scheme, where the differential operator in (1.1) is approximated by

$$\begin{aligned} \frac{1}{h^2} [\bar{D}_1 (u(x+h,y) - u(x,y)) + \bar{D}_2(w(x-h,y) - u(x,y) \\ + \bar{D}_3 (u(x,y+h) - u(x,y)) + \bar{D}_4 (u(x,y-h) - u(x,y)] \end{aligned} \tag{2.1}$$

For simplicity we assume that the origin is a grid point in an equidistant mesh. Then the constants $\bar{D}$ in (2.1), if derived in a standard way, are given by

$$\bar{D} = \begin{cases} D_i \text{ if both corresponding points are in the closure of } \Omega_i \quad \text{(see fig. 1)} \\ (D_i + D_j)/2 \text{ if both points are in the intersection of the closures of } \Omega_i \text{ and } \Omega_j. \end{cases}$$

The values of $\bar{D}$ in the neighbourhood of the four corner juncture are presented in fig 2.

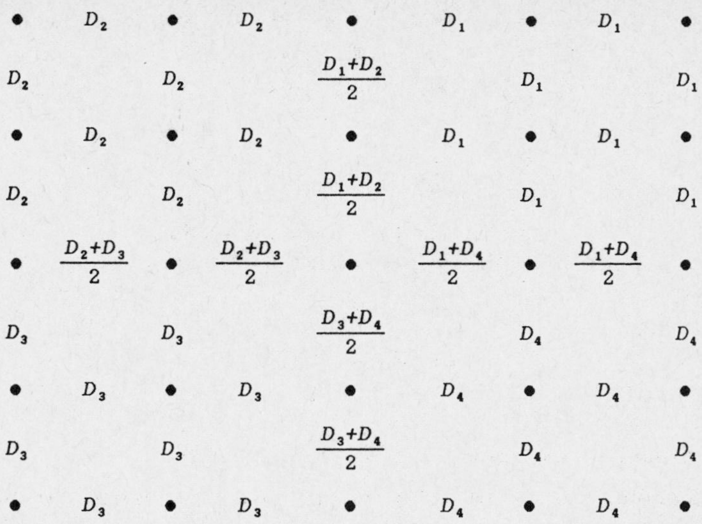

•	D_2	•	D_2	•	D_1	•	D_1	•
D_2		D_2		$\frac{D_1+D_2}{2}$		D_1		D_1
•	D_2	•	D_2	•	D_1	•	D_1	•
D_2		D_2		$\frac{D_1+D_2}{2}$		D_1		D_1
•	$\frac{D_2+D_3}{2}$	•	$\frac{D_2+D_3}{2}$	•	$\frac{D_1+D_4}{2}$	•	$\frac{D_1+D_4}{2}$	•
D_3		D_3		$\frac{D_3+D_4}{2}$		D_4		D_4
•	D_3	•	D_3	•	D_4	•	D_4	•
D_3		D_3		$\frac{D_3+D_4}{2}$		D_4		D_4
•	D_3	•	D_3	•	D_4	•	D_4	•

Figure 2.

Consider now $D_1 = D_3 = 1$ and $D_2 = D_4 = 0$. This case is illustrated in fig 3, where constants $\bar{D} = 0$ are neglected. In [1] it was already observed that in this case the numerical solutions in Ω_1 and Ω_3 are strongly coupled whereas the analytic solution has two independant branches in Ω_1 and Ω_3. From this it follows that also in the case $D_2 = D_4 \neq 0$, but small, the numerical solution is a bad approximation of the analytic solution, as was already stated in the preceding section. In [1] it was proposed to suppress the strong coupling by multiplying the four constants $\frac{1}{2}$ in fig 3 around the origin by some small number δ. But it is clear that the bad order of convergence to the analytic solution cannot be changed in general by such a heuristic modification. In the next section we propose the same type of modification but we give a precise value for δ depending only on the constant α in (1.2). We shall see that in this case the order of convergence for fixed points in the interior of Ω is $O(h^2)$, as would be expected in the case of smooth solutions.

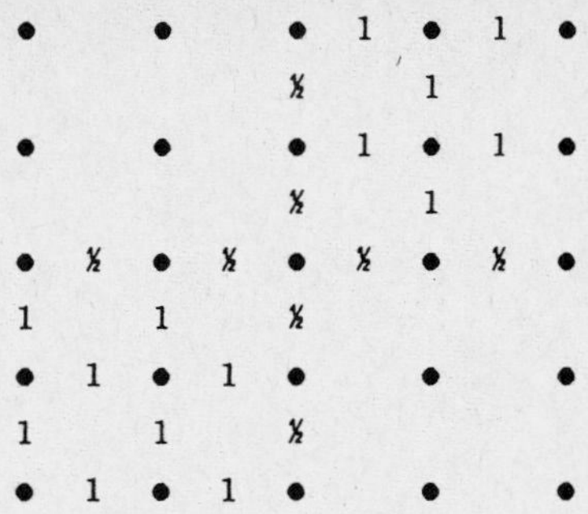

```
●       ●       ●   1   ●   1   ●
                ½       1
●       ●       ●   1   ●   1   ●
                ½       1
●   ½   ●   ½   ●   ½   ●   ½   ●
1       1       ½
●   1   ●   1   ●       ●       ●
1       1       ½
●   1   ●   1   ●       ●       ●
```

Figure 3.

3. A Modified Discretization Scheme.

Using a technique introduced in [4] for the L-shaped membrane problem it can be shown that the solution of the standard difference equation in the neighbourhood of the four corner juncture is given by

$$u_h\ (w,\varphi) = w_1(\varphi)\, r^{\alpha} + h^{2\alpha}\, c\ w_1(\varphi) r^{-\alpha} + \overline{w}(h,r,\varphi)\, h^2, \tag{3.1}$$

where α and $w_1(\varphi)$ were introduced in (1.2), and where the function $\overline{w}(h,r,\varphi)$ is bounded for $h \to 0$ for all fixed values of r, φ with $r \neq 0$.

If we now use the modification of the preceding section, multiplying the four constants $\overline{D}$ next to the four corner juncture with a constant δ one can show that we get the same error expansion (3.1), where now $c = c(\delta)$ depends on δ.

We see that in general the modification does not result in a better order of convergence with the exception of the special case $c(\delta) = 0$. If $c(\delta) = 0$ then the order of convergence to the analytic solution is $O(h^2)$ as would have been expected if $u(r,\varphi)$ were smooth. The only exception is the point $r = 0$, where (3.1) does not give any information because $\overline{w}(h,r,\varphi)$ may be unbounded for $h \to 0$.

At this time we are not able to give an analytic expression for the value of δ, but it can be shown that δ depends only on α which can be computed from D_i, i=1,2,3,4 by solving Kelloggs equations in [2].

The values of δ, plotted in fig 4 (and given in table 2), were computed by a numerical calculation: For a given α, δ was chosen in such a way that the predicted $O(h^2)$-error was observed in the computed solution of the modified finite difference scheme.

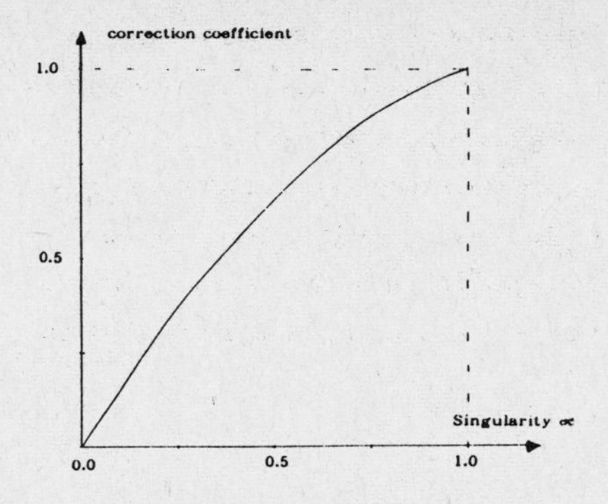

Figure 4.

The computations in the next section show that the modification results in a rather dramatic gain of accuracy. Moreover, if we use the modification also for the coarser grids in the multigrid scheme we observe an improved rate of convergence of the multigrid iteration. Note, that this presupposes that the four-corner-juncture is a grid point of all grids. This may be too hard a restriction in practical cases, and then we have to introduce additional modifications investigated in [1]. But usually, even in the case of many interfaces, it is possible to place the intersection points on grid points of the finest grid. And this is sufficient to get the improved rate of convergence of the finite difference solution to the analytic solution.

4. A Multigrid Algorithm for the Diffusion Problem.

We have implemented the improved discretizations in a multigrid algorithm. In this section we show that not only the accuracy of the final solution on the finest grid is improved, but that also, due to the improved representation of the continuous problem on all levels, the convergence rate of the multigrid method becomes almost independent of the strength of the singularity.

4.1. The Coarse Grid Problem.

If the unmodified discretizations are used for a problem with strong singularities, the discrete solution is not only a bad approximation of the analytic solution, but also coarser grids will yield poor approximations for finer grid solutions. If the singulatities are bad, this is one of the reasons, why the multigrid method shows poor convergence: From the coarse grids only bad corrections to the fine grid solutions can be obtained. In [1] quite an involved discussion shows how satisfactory coarse grid discretizations can be constructed from the fine grid equations. In their case this effort is necessary in order to obtain coarse grid equations that approximate the fine grid problem well.

We now have improved discretizations that yield order h^2 accurate solutions for all meshsizes. Thus coarser grids with these discretizations give good approximations of the fine grid solutions, and simply applying these modified discretizations on all grid levels will be shown to yield satisfactory multigrid convergence rates, no matter how bad the singularity is. The discrete equations on all grid levels are thus constructed directly from the continuous problem, and not recursively from the discretizations on the next finer level. as mentioned above this requires that the discontinuities must be resolved by gridlines on all multigid levels.

4.2. The Interpolation.

Here we follow the argumentation given in [1]. The coarse grid correction process can only give reasonable results if the interpolation supports the fine grid solution behavior. Along discontinuities this means that the interpolation must model the discontinuous first solution derivative. Thus we will use weighted averages of neighboring points, where the weights are proportional to the diffusion coefficients. An interpolated gridfunction will then have the correct jumps in the first differences.

In our implementation the interpolation along horizontal and vertical lines is performed first. The value at points in the middle of four coarse grid points is calculated in a second step using two of the new fine grid values. At the four corner juncture the *modified* diffusion coefficients are used as weights in the interpolation.

4.3. The Restriction.

The magnitude of the residuals is proportional to the magnitude of the diffusion coefficients, thus where the diffusion coefficients jump, residuals of different scales are adjacent. A straightforward residual weighting (full weighting) will then combine quantities having different scales: This is not likely to give good results.

In our code we compensate the different residual scales. We bring all residuals to the same scale dividing by the diffusion coefficients, then use full weighting, and finally, on the coarse grid, bring them again to the right scale by multiplying with the appropriate diffusion coefficient. Where the diffusion coefficient is discontinuous, the avarage of the left and right limit values is used for these calculations. Note that this procedure needs to be performed explicitly only along lines of discontinuity.

4.4. The Smoother.

With this coarse grid correction all non-local solution features are well approximated on the coarser grids. The relaxation needs only take care of local effects, and all standard smoothers can be expected to work well. Thus we simply use a Gauss-Seidel smoother with lexical ordering of the gridpoints.

4.5. The Multigrid Cycle.

Here we again follow a standard suggestion and use *V-cycles* with two smoothing steps before, and one smoothing step after the coarse grid correction. The above components have also been used for a full multigrid method. In this case two *V-cycles* on each level (before proceeding to the next finer one) have been sufficient for a solution accuracy below truncation errors.

5. Numerical Results

In this section we present numerical experiments showing the efficiency of the multigrid algorithm described in the above section. Our results refer to the test problem of the first section. This problem is very critical in its solution behavior. We have tested examples with singularities as bad as $r^{0.01}$. This corresponds to jumps in the diffusion coefficients of magnitude 10^4. Kershaw's test problem for example, as treated in [3] is by far not as critical. Although there the jumps in the diffusion coefficients are 10^5, locally the solution cannot behave worse as in an L-shaped region, where the singularities are of the type $r^{2/3}$.

In the presented examples the finest grid has 33 by 33 points (we have also solved this problems for grids with up to 1025 by 1025 points), and the coarsest 3 by 3 points, so that there are 5 different levels.

In the following two tables we show the reductions of the residuals by *V-cycles* for different types of singularities. For the first table we have used standard discretizations, for the second the proposed modifications have been used. Clearly the rates of convergence are improved. In the tables we list the Euclidean norms of the residuls after the first six *V-cycles*. The iteration starts with the gridfunction being 0 throughout the region.

Table 1.
Calculation without corrections.

Iteration	0	1	2	3	4	5	6
α							
1	9653.6103	564.3647	42.1793	3.4804	0.3094	0.0290	0.0027
2/3	8952.5570	532.7327	42.7498	5.0925	0.9158	0.1861	0.0384
1/2	8762.5078	527.9929	50.0224	10.4054	2.8726	0.8005	0.2230
1/10	8366.3711	519.3783	80.6519	32.1966	13.2571	5.3608	2.1648
1/100	8267.1735	514.5115	82.2797	33.5934	14.0478	5.7692	2.3662

Table 2.
Calculation with corrections.

Iteration α	0	1	2	3	4	5	6	coefficient
1	9653.6	564.36	42.179	3.4804	0.3094	0.0290	0.0027	1
2/3	8952.5	531.49	40.720	3.7718	0.4626	0.0716	0.0123	0.803
1/2	8762.5	525.04	42.166	5.2274	1.0417	0.2365	0.0541	0.6527
1/10	8366.3	508.50	45.188	9.8458	3.3819	1.1867	0.4097	0.155
1/100	8267.1	500.40	41.463	7.4101	2.4281	0.8606	0.3049	0.0157

Still more impressive is the improvement of the solution accuracy. For our 'worst case' example with solution behavior $r^{0.01}$ we show the development of the solution and the error (with respect to the analytical solution) at various stages of the multigrid algorithm. Fig 5 shows the analytic solution, fig 6 the start solution before the first *V-cycle.*

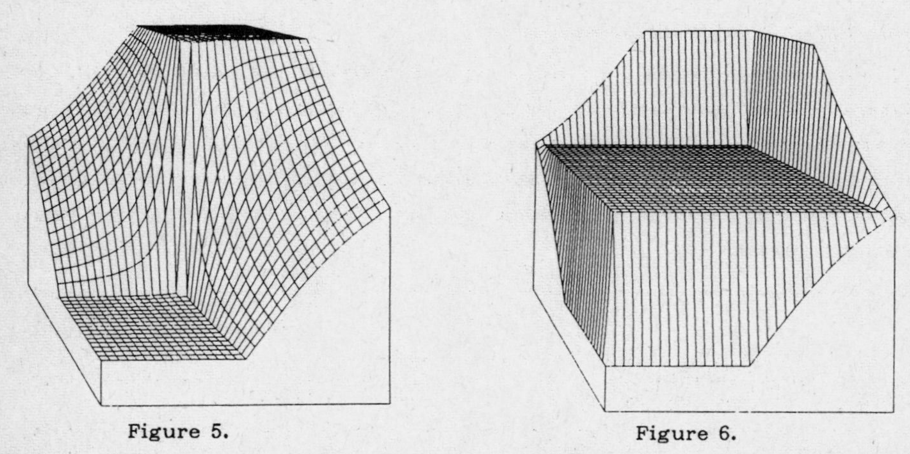

Figure 5. Figure 6.

In the figures 7 to 11 the left column (a) always refers to the case when the straightforward, unmodified discretizations are used, while the right column (b) shows the results for the improved schemes.

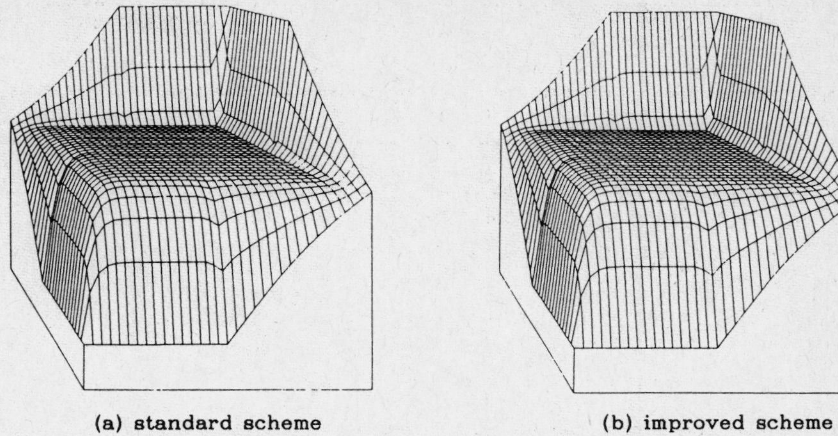

(a) standard scheme (b) improved scheme

Figure 7: Solution after 2 relaxations.

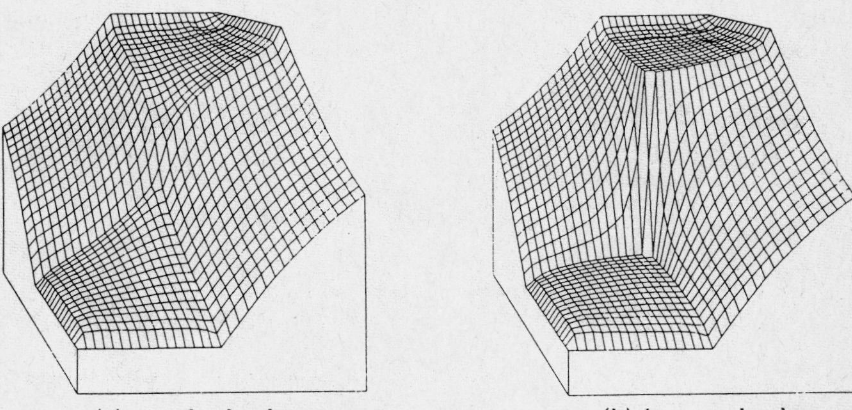

(a) standard scheme (b) improved scheme

Figure 8: Solution after first V-cycle.

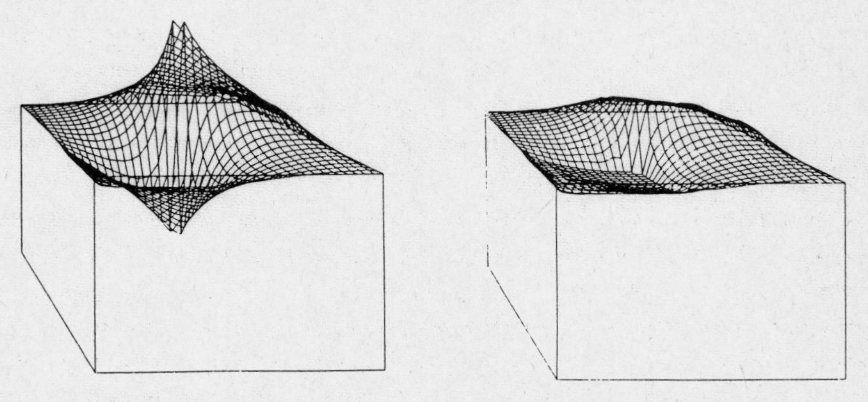

(a) standard scheme (b) improved scheme

Figure 9: Error wrt. analytical solution after first cycle.

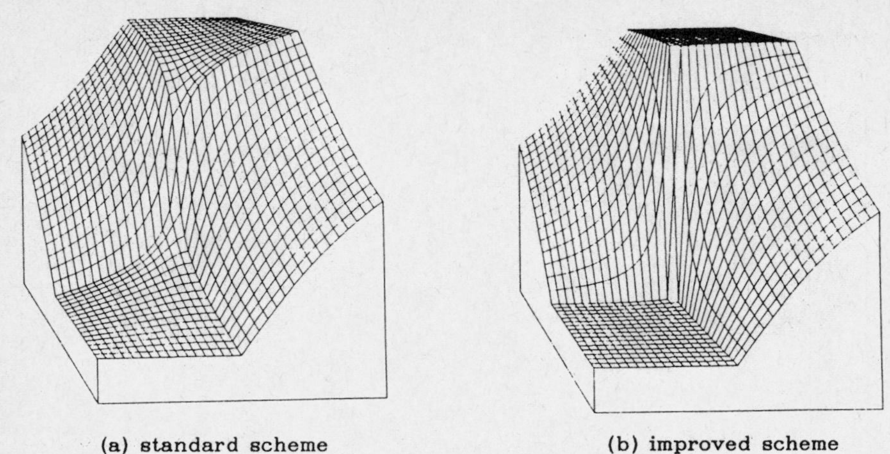

(a) standard scheme (b) improved scheme

Figure 10: Solution after 6 cycles.

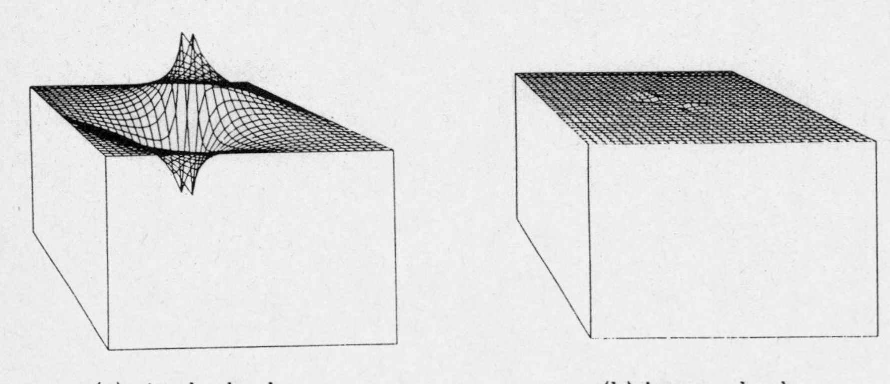

(a) standard scheme (b) improved scheme

Figure 11: Error wrt. analytical solution after six cycles.

From the figures one can see how much the singularity is smeared when the standard discretizations are used. With the modified equations the singular solution behavior is modelled much better. As the modified equations are used on all grids, also the coarse grid corrections support this solution type, which is essential for fast convergence.

In [1] it is suggested that the diffusion coefficients at the singularity are changed to some ε, which may be any small number. We wish to remark that this is only qualitatively correct. In this case the $O(h^{2\alpha})$ error term is still harming the solution accuracy, though of course not as bad as with the naive discretization. We have tryed the $r^{0.01}$ example with a correction coefficient 5×10^{-5} instead of the correct value 0.01157. Here the error at the point $(i,j)=(20,20)$ in the grid is 1.07×10^{-2} instead of 2.38×10^{-5}. Thus with the correct coefficient the error is more than 10^3 times smaller. In order to obtain high accuracy the use of the correct coefficient is indeed essential.

6. Acknowledgement.

The authors are indepted to H. Gietl for valuable discussions.

References.

1. Alcouffe, R.E., Brandt, A., Dendy, J.E. (Jr.), and Painter, J.W., "The Multigrid Method for the Diffusion Equation with Strongly Discontinuous Coefficients," *SIAM J. Sci. Stat. Comput.* 2, pp. 430–454 (1981).

2. Kellogg, R.B., "On the Poisson Equation with Intersecting Interfaces," *Technical Note of the University of Maryland,* **BN-643** (1970).

3. Kettler, R., "Analysis and Comparison of Relaxation Schemes in Robust Multigrid and Preconditioned Conjugate Gradient Methods.," in *Lecture Notes in Mathematics 760: Multigrid Methods, Proceedings of the Conference Held at Köln Porz, November 23-27, 1981,* ed. Hackbusch, W., Trottenberg, U., Springer Verlag, Berlin (1982).

4. Zenger, C., and Gietl, H., "Improved Schemes for the Dirichlet Problem of Poisson's Equation in the Neighbourhood of Corners.," *Numerische Mathematik* **30**, pp. 315-332 (1978).

A COMPARISON OF SEVERAL MG-METHODS FOR THE SOLUTION OF THE TIME-DEPENDENT NAVIER-STOKES EQUATIONS

W. Schröder, D. Hänel
Aerodynamisches Institut, RWTH Aachen

Abstract

In this paper a comparative study of the two main multigrid concepts for the time-dependent compressible Navier-Stokes equations is presented. In the direct multigrid method the solution advances in time on all grid levels with the accuracy of the finest grid. This method can be used for explicit or implicit difference schemes. The indirect multigrid method is applied within an implicit difference scheme to accelerate the iteration process between two time levels. The explicit MacCormack scheme and the Switched Evolution/Relaxation scheme are used as relaxation schemes. All methods are second-order accurate in space. First, the methods are applied to the time-dependent compressible Couette flow problem. In this case the indirect method using flux-vector splitting for the convective terms and central approximations for the viscous terms of the Navier-Stokes equations shows the best rate of convergence. Compared to a single fine grid solution obtained with the explicit MacCormack scheme the multigrid solution reduces the computational work by more than a factor of ten. Secondly, the indirect multigrid method is used to determine complex subsonic and supersonic flows. In the subsonic case the flow past a flat plate for different Reynolds and Mach numbers $10^3 \leq Re \leq 10^5$, $0.5 \leq Ma \leq 0.9$ is computed; in the supersonic case the flow past a wedge for a Reynolds number $Re = 10^4$ and Mach numbers $1.5 \leq Ma \leq 3.0$ is determined. In order to increase the rate of convergence second-order accurate flux-vector splitting with a flux limiter which takes into account the very different space steps has to be employed. The investigation shows that the number of time steps required to obtain a converged solution is nearly independent of the number of grid points for the subsonic case as well as for the supersonic case.

1. Introduction

There exists a number of methods of solution for the Euler and Navier-Stokes equations which are based on the multigrid concepts as proposed by Brandt [1]. The aim of the multigrid strategy is to reduce the computational work on the finest grid. This is achieved by different procedures. In this study two multigrid approaches for time-dependent problems are compared with each other. In the first method, called the direct or parabolic multigrid method, the solution advances in time on the finest and on the coarser grids while the accuracy of the finest grid is preserved on all grid levels. The methods described in [2,3,4,5] belong to this kind of multigrid procedure. The second method called the indirect method, uses the multigrid strategy to solve effectively the large system of finite difference equations arising for an implicit discretization by an iterative process. An example of such a procedure is described in [6].

In [2] Ni developed a one-step distribution formula scheme which is based on the one-step Lax-Wendroff method as integration scheme. In order to propagate the local fine grid corrections to the solution rapidly throughout the entire computional domain coarse grids were incorporated in the algorithm. Thereby the rate of convergence during the transient period was increased while the low truncation errors associated with the fine grid discretization were maintained. Schmidt, and Jameson [3] applied the same multigrid method to compute transonic flows. Also, Chima, Johnson [4] used the multigrid technique to propagate the fine grid correction in the field as wide as possible with little computational work. As basic integration scheme they employed the explicit MacCormack method. In [5] Jameson applied a multigrid procedure with a four stage Runge-Kutta scheme to advance the solution of the Euler equations in time. In contrast to the methods mentioned above Jameson did not determine coarse grid corrections but computed the corrections for the actual variables. As mentioned before such multigrid concepts belong to the direct multigrid method for time-dependent problems either in the form of a correction scheme [2,3,4] or in the form of a Full Approximation Storage (FAS) scheme [5]. In [6] Mulder employed the indirect multigrid method. An implicit time discretization and upwind differences for the spatial discretization of the Euler equations were used. Van Leer's flux-vector splitting [7] was implemented as upwind differencing method. The relaxation scheme was the Switched Evolution/ Relaxation scheme of van Leer, and Mulder [8]. To accelerate the iterative process in every time step the multigrid method was used in form of the Correction scheme.

In all the papers mentioned the multigrid method converged much faster than the single grid method consisting of the basic integration scheme of the multigrid procedure. However, there has been no comparison between the different multigrid methods for time-dependent problems. In this investigation the rates of convergence of the direct and the indirect multigrid concept in the form of a FAS scheme will be compared with each other and with explicit and implicit single grid methods.The time-dependent compressible Couette flow problem is chosen as test case. Thereafter the subsonic flow past a flat plate and the supersonic flow past a wedge is determined with the indirect multigrid procedure.

2. Governing equations and spatial discretization

The governing equations are the time-dependent Navier-Stokes equations for a compressible ideal gas, written in conservative form:

$$\frac{\partial}{\partial t}\int_{\tau} U\, d\tau + \oint_{S} (H - R)\, \vec{n}\, dS = 0 \tag{1}$$

Here, U is the vector of the conservative variables, H includes the Euler fluxes and R the viscous terms. In this study only two-dimensional flows are considered, where

$$H = \begin{pmatrix} E \\ F \end{pmatrix} \quad , \quad R = \begin{pmatrix} S \\ T \end{pmatrix} \quad , \quad \bar{n}\, dS = \begin{pmatrix} dy \\ -dx \end{pmatrix} \tag{2}$$

with the components

$$U = \begin{vmatrix} \rho \\ \rho u \\ \rho v \\ e \end{vmatrix}, \quad E = \begin{vmatrix} \rho u \\ \rho u^2 + p \\ \rho u v \\ u(e+p) \end{vmatrix}, \quad F = \begin{vmatrix} \rho v \\ \rho v u \\ \rho v^2 + p \\ v(e+p) \end{vmatrix}, \quad S = \begin{vmatrix} 0 \\ \tau_{xx} \\ \tau_{xy} \\ u\tau_{xx} + v\tau_{xy} + \\ + q_x \end{vmatrix}, \quad T = \begin{vmatrix} 0 \\ \tau_{xy} \\ \tau_{yy} \\ u\tau_{xy} + v\tau_{yy} + \\ + q_y \end{vmatrix}$$

All variables have their usual meaning. The inflow and outflow boundary conditions are determinded by the characteristics of the Euler equations. The no slip condition is imposed on solid walls and undisturbed flow conditions for the far field. The discretization is carried out by the finite volume technique in a stretched Cartesian mesh. The variables U and the coordinates are defined at the center of a control volume τ , its corner points are defined by averaging the coordinates of the neighbouring points. This discretization is directly applicable on curvilinear coordinate systems. The difference equation of Eq. (1) is

$$L(U) = \delta_t U + Res(U) = 0 \tag{3}$$

where Res(U) corresponds to the discretized steady state operator of Eq. (1) and $\delta_t U$ to the discretized time derivative defined later on by the method of solution.

3. Multigrid method

For completeness a short description of the multigrid method will be given. The Full Approximation Storage (FAS) scheme is used [1]. The difference equation Eq. (3) on the finest grid $G_{k=m}$ may be expressed as

$$L_k U_k = f_k \tag{4}$$

with

$$L_k U_k = \frac{U_k(t^{n+1}) - U_k(t^n)}{\Delta t_k} + Res_k\, U_k(\tilde{t})$$

The index n counts the time and $\tilde{t}$ is either $\tilde{t} = t^n$ for an explicit or $\tilde{t} = t^{n+1}$ for an implicit scheme. For the injection from the fine grid G_k to the coarse grid G_{k-1}, designed by $I_k^{k-1} f_k$, simple point-to-point injection is used for the variables U_k and volume-weighted injection for the residual Res(U). The transfer from the coarse to the fine grid, $I_{k+1}^{k} U_k$, is computed by bilinear or cubic interpolation. On the coarser grid G_{k-1} the difference equation

$$L_{k-1} U_{k-1} = f_{k-1} + \tau_m^{k-1} \tag{5}$$

has to be solved with

$$f_{k-1} = I_k^{k-1} f_k \tag{6}$$

The local discretization error τ_m^{k-1}, conserving the accuracy of the fine grid solution U_k on the coarse grid G_{k-1}, is defined by

$$\tau_{k-1} = \tau_m^{k-1} = \tau_k^{k-1} + \tau_m^k \quad , \qquad \tau_k^{k-1} = L_{k-1}(I_k^{k-1} U_k) - I_k^{k-1}(L_k U_k) \tag{7}$$

For gradient boundary conditions a similar correction has to be derived. From the coarsest grid G_1 the solution is interpolated to the finer grids by a correction

$$U_{k+1} = U_{k+1} + I_k^{k+1}(U_k - I_{k+1}^k U_{k+1}) \tag{8}$$

In principle such a V-cycle is used for both methods described in the following sections.

3.1 Direct multigrid method

The direct method[1] is studied in connection with the explicit predictor-corrector scheme of MacCormack [9], used as method of solution and for smoothing the high frequency error components. On the grid G_k the scheme is written as

$$\begin{aligned} \tilde{U}_k^{n+1} &= U_k^n - \Delta t_k (\bar{\delta}_k H - \bar{\bar{\delta}}_k R) \\ U_k^{n+1} &= \frac{1}{2} [\tilde{U}_k^{n+1} + U_k^n - \Delta t_k (\vec{\delta}_k \tilde{H} - \vec{\vec{\delta}}_k \tilde{R})] + \Delta t_k \tau_k \end{aligned} \tag{9}$$

where the Euler terms are differenced backward, $\bar{\delta}_k$, and forward $\vec{\delta}_k$ resp.. The viscous terms R are approximated centrally. The local discretization error occurs only in the corrector step. The restriction of the time step Δt_k from a linear analysis is

$$\Delta t_k \leq \left(\frac{|u| + a}{\Delta x_k} + \frac{|v| + a}{\Delta y_k} + \frac{2\mu}{8} \left(\frac{1}{\Delta x_k^2} + \frac{1}{\Delta y_k^2} \right) \right)^{-1} \tag{10}$$

As the time step depends on the step sizes Δx, and Δy, larger time steps can be used on coarser grids. The direct method makes use of the observation and the solution proceeds in time on coarser grids G_{k-1} with the time step $\Delta t_{k-1} > \Delta t_k$ in the difference equation Eq. (4). In addition, the computational work on the coarser grid is reduced by the lower number of grid points. After a certain time the V-cycle is repeated to renew the discretization error, Eq. (7), since this correction term includes information of the high frequency error from the time when the finest grid was left. The discretization of the gradient boundary conditions on the coarser grids are corrected by the local dicretization error in the same way as the inner points.

3.2 Indirect multigrid method

The solution of an implicit difference scheme requires the inversion of large matrices on one time level which needs a large amount of computational work if elimination methods are used. The indirect multigrid method substitutes the elimination method by an iterative relaxation procedure which is accelerated by the multigrid method [1]. In comparison to the direct method the time step in the indirect multigrid method is frozen during the multigrid procedure. For time-accurate calculations the implicit difference equation (4) has to converge between two-time levels. If only the steady state solution is of interest a few iterations on every grid and at every time level are sufficient for convergence.

The efficiency of the multigrid method depends strongly on the smoothing properties of the relaxation scheme. For the conservation equations, discretized centrally with second order accuracy, the usual relaxation methods, like point Gauss-Seidel or line Gauss-Seidel, become unstable for higher Reynolds numbers and for inviscid flow. Therefore in our first computations we use the explicit MacCormack scheme with an artificial time derivative added as a relaxation scheme. Then Eq. (4) reads

$$\frac{U_k(t^{n+1,\nu+1}) - U_k(t^{n+1,\nu})}{\Delta\mu_k} + \frac{U_k(t^{n+1,\nu}) - U_k(t^n)}{\Delta t_k} + Res_k\, U_k(t^{n+1,\nu}) = f_k \tag{11}$$

where ν is the iteration index of the artificial time step $\Delta\mu_k$. Consequently, the calculation proceeds with an unrestricted time step Δt_m of an implicit method but with an explicit iteration procedure in every time step. To accelerate this iterative process we apply the direct multigrid method to Eq. (11), i.e. on coarser grids larger $\Delta\mu_k$ are used. The relaxation scheme indeed is sufficiently dissipative to work but for large time steps Δt_m the reduction of the residual in every V-cycle reduces to .95. For this reason the resulting work reduction factor is not satisfactory for a multigrid method.

Further improvement of the indirect method requires a more dissipative difference scheme to get smaller residual reduction factors. From the computation of inviscid flow it is well known that flux-vector splitting in connection with usual relaxation methods for elliptic equations - e. g. Gauss-Seidel or Jacobi relaxation - results in a sufficiently dissipative scheme to damp the high frequency error in a few sweeps [8]. Therefore the compressible Navier-Stokes equations are discretized using flux-vector splitting for the Euler terms and central differencing for the viscous terms. From the existing flux-vector formulations the one proposed by van Leer [7] is used for this study. The flux-splitting leads to expressions for the Euler fluxes like

$$E = E^+ + E^- \qquad F = F^+ + F^- \tag{12}$$

Van Leer's splitted fluxes are

$$P^{\pm} = \begin{bmatrix} \pm g c \left[\frac{1}{2} (M \pm 1) \right]^2 = f_1^{\pm} \\ f_1^{\pm} \left[u - t_x \left(\frac{u \mp 2c}{\gamma} \right) \right] \\ f_1^{\pm} \left[v - t_y \left(\frac{v \mp 2c}{\gamma} \right) \right] \\ f_1^{\pm} \left[((\gamma - 1)(t_x u + t_y v) \pm 2c)^2 / (2 (\gamma^2 - 1)) + \frac{1}{2} (t_y u^2 + t_x v^2) \right] \end{bmatrix} \tag{13}$$

where M is the local Mach number $\frac{t_x u + t_y v}{c}$ and for $E^{\pm} = P^{\pm}$ counts $t_x = 1$ and $t_y = 0$ and for $F^{\pm} = P^{\pm}$ counts $t_x = 0$ and $t_y = 1$. u and v are the velocity components and c is the speed of sound. If $M \geq 1$ then $P^+ = t_x E + t_y F$, $P^- = 0$ and for $M \leq -1$ holds $P^- = t_x E + t_y F$, $P^+ = 0$.

Further details of these formulations are given in the literature [7]. In the difference equation of Eq. (1) the derivatives of the Euler fluxes are chosen backward and forward, respectively. The changes in time in F^+ are e. g.

$$(\Delta F^+)^n = (F^+)^{n+1} - (F^+)^n = (A^+)^n (U^{n+1} - U^n) \tag{14}$$

where the Matrix A^+ is the Jacobian of E^+. Then the difference equation to be relaxed on the finest $\mathrm{Grid}_{k=m}$ is

$$L_k U_k = \left(\frac{I}{\Delta t} + \overleftarrow{\delta}_x A^+ + \overrightarrow{\delta}_x A^- + \overleftarrow{\delta}_y B^+ + \overrightarrow{\delta}_y B^- - \overleftrightarrow{\delta}_{xx} \frac{\partial S}{\partial U} - \overleftrightarrow{\delta}_{yy} \frac{\partial T}{\partial U} \right)_k^n \Delta U_k^{n,\nu} + \mathrm{Res}_k\, U_k^n = f_k \tag{15}$$

with

$$\Delta U_k^{n,\nu} = U_k^{n+1,\nu} - U_k^n$$

In the implicit part of equation (15) the Euler terms are approximated with first-order accurate upwind differences and the viscous terms with central differences. For the Euler terms in the steady state operator Res(U) in equation (15) the MUSCL approach (Monotonic Upstream Centered Schemes for Conservation Laws) [10] is employed. The values of the conservative variables of the grid points are extrapolated to the cell boundaries, e.g.

$$U^+_{i+1/2,j} = U_{i,j} + \zeta \frac{x - x_i}{x_i - x_{i-1}} D_{i,j} \tag{16a}$$

and then the fluxes are determined with the extrapolated variables.

$$E^{\pm}_{i\pm1/2,j} = E^{\pm}(U^{\pm}_{i\pm1/2,j}) \qquad F^{\pm}_{i,j\pm1/2} = F^{\pm}(U^{\pm}_{i,j\pm1/2}) \tag{16b}$$

If the variables are assumed to be piecewise constant distributed in each cell ($\zeta = 0$ in Eq. (16a)) the approximation of the Euler terms is first-order accurate, if they are assumed to be piecewise linear distributed in each cell ($\zeta = 1$ in Eq. (16a)) the discretization is second-order accurate. In the computations concerning the comparison of the methods the Res(U) terms contains the second-order upwind-biased differencing, resulting in a second-order accurate solution in the steady state. $D_{i,j}$ in equation (16a) represents the limiter derived by van Albada, van Leer, Roberts [11] which eliminates oscillations in regions where strong gradients occur. In this investigation $D_{i,j}$ was formulated for a stretched Cartesian grid

$$D_{i,j} = \left[\frac{2\,\overrightarrow{\Delta U}\,\overleftarrow{\Delta x}\,\overleftarrow{\Delta U}\,\overrightarrow{\Delta x} + \varepsilon}{(\overrightarrow{\Delta U}\,\overleftarrow{\Delta x})^2 + (\overleftarrow{\Delta U}\,\overrightarrow{\Delta x})^2 + \varepsilon} \right]_{i,j} \left[\frac{\overleftarrow{\Delta U}\,\overrightarrow{\Delta x} + \overrightarrow{\Delta U}\,\overleftarrow{\Delta x}}{\overrightarrow{\Delta x} + \overleftarrow{\Delta x}} \right]_{i,j} \tag{17}$$

with $\overrightarrow{\Delta U}$, $\overleftarrow{\Delta U}$ and $\overrightarrow{\Delta x}$, $\overleftarrow{\Delta x}$ representing forward and backward differences in the conservative variables U and in the Cartesian coordinates x and y, respectively. The quantity ε is a small number that prevents division by zero. In this study ε was chosen to be of order $O(\Delta x^4)$ and $O(\Delta y^4)$, respectively. This form of the limiter yields the largest rate of convergence and the best accuracy in the results. Other formulations of $D_{i,j}$, for example, the original expression given in [10] for a uniform grid, show a decrease of the rate of convergence and of the accuracy compared to the formulation given in equation (17). Eq. (15) is iterated with a collective Gauss-Seidel relaxation scheme (CSOR) in alternating directions. The time step Δt is increased with decreasing residual so that in the nearly converged state very large time steps can be employed. Using this form scheme (15) becomes a Switched Evolution/Relaxation (SER) scheme [8]. Since the iterative solution of equation (15) is very costly the multigrid FAS method is applied to accelerate the inversion of the large system.

4. Results

4.1. Comparison of the methods

To compare the different single grid and multigrid methods with each other the time-dependent compressible Couette flow (Re = 1000, Ma = 0.5) was computed. At the inflow boundary the profiles of the velocity components and of the density (ρ,u,v) are specified. Since we are only interested in computing the Couette flow with zero pressure gradient the static pressure on the inflow boundary is determined from the condition $p_x = 0$. At the outflow boundary the static pressure is given, the other variables are obtained by gradient boundary conditions. At the lower and upper wall all variables are specified except for the static pressure; it is computed by extrapolation. The multigrid procedures use 4 grid levels (17x17 grid points on the finest grid) in a V-Cycle with standard coarsening. The direct multigrid method consists of four time steps on the finest grid and of two time steps on the coarser grids. In the indirect multigrid procedure two relaxation sweeps are performed before the coarsest grid is reached, one relaxation on the coarsest grid itself and after the interpolation of the coarse grid error to the finer grid. The trace of the maximum residual of different methods of solution is plotted against the work units (~ number of operations) in Fig. 1. The comparison of the methods which use the explicit MacCormack scheme (curve 1,2,3) shows that the direct multigrid method has the best rate of convergence. In order to decrease the residual by 5 orders of magnitude the direct multigrid approach (curve 3) needs about 30 per cent of the computational work of the single grid method (curve 1). In [4] the same order of work reduction was found for a similar multigrid method. To compare the convergence history of an explicit multigrid procedure with an implicit single grid scheme the approximate factorization method of Beam and Warming [12] was coded to solve equation (4) for the Couette-flow problem. As shown in Fig. 1 the rate of convergence of this method (curve 4) is slightly better than the rate of the direct multigrid method with the MacCormack scheme (curve 3). The best rate of convergence, however, is achieved with the indirect multigrid method in conjunction with van Leer flux-splitting of the Euler terms (curve 5). In the calculation the residual is reduced in every V-Cycle by a factor of 0.7. The method gives a reduction of the computational work by a factor of more than ten for convergence down to machine zero compared to the single grid MacCormack method. For a finer grid with 33x33 grid points and 5 grid levels the results are qualitatively the same. The indirect multigrid approach is not only the method with the best rate of convergence but also the most robust multigrid procedure. In the direct multigrid approach the convergence process can be stopped by modifying the number of relaxation sweeps performed on every grid level. The high frequency modes are not sufficiently damped before transfering to a coarser grid. In the indirect multigrid procedure including the flux-splitting such a change in the input data has only an effect on the rate of convergence but not on the actual convergence behaviour. As a consequence of the results of the comparison of the different methods the indirect multigrid approach combined with flux-splitting was employed in further computations.

4.2 Applications of the indirect multigrid method

4.2.1 Flow past a flat plate

With the indirect multigrid method containing flux-splitting for the Euler terms further computations were performed. First the subsonic flow past a flat plate is considered as illustrated in Fig. 2. The boundary conditions are also indicated in Fig. 2. The Reynolds numbers range from 10^3 to 10^5 and the Mach numbers from 0.5 to 0.9. The V-Cycle consists of two, three and four grid levels with a coarsest grid of 9x9 grid points. Fig. 3 shows the three fine grids used for a computation with Re = 10^3. The indirect multigrid strategy employed is the same as described in section 4.1. A comparison of the convergence histories of a first-order accurate and a second-order accurate approximation to the Navier-Stokes equations shows that the second-order accurate scheme needs only about 50 per cent of the time steps of the first-order accurate scheme to converge. The first-order scheme is too dissipative to yield a high rate of convergence. To illustrate this, the computed distribution of the pressure coefficient c_p on the wall is plotted in Fig. 4 for a first- and a second-order accurate scheme (see section 3.2). For computations with second-order accurate differencing the maximum residual is plotted against the time steps for different fine grids in Figs. 5 and 6. The Reynolds number in Fig. 5 is $Re = 10^3$, the Mach number $Ma = 0.8$; an isothermal wall is assumed; in Fig. 6 $Re = 10^4$, $Ma = 0.9$, and the wall is adiabatic. The convergence histories of the indirect multigrid approach show that the number of time steps to converge to the steady state is almost independent of the number of grid points. This result, i. e. the independence of the rate of convergence of the number of space steps, would have been expected by solving an elliptic equation with the multigrid method but not for the solution of the two-dimensional Navier-Stokes equations.

4.2.2 Flow past a wedge

The supersonic flow past a wedge was investigated as depicted in Fig. 7. The boundary conditions are also given in Fig. 7. The flow was computed for a Reynolds number of $Re = 10^4$ and for Mach numbers $1.5 \leq Ma \leq 3.0$ with different wedge angles β. As for the subsonic flow in section 4.2.1 the V-Cycle of the indirect multigrid procedure has two, three and four grid levels ranging from 17x17 to 65x65 grid points on the finest grid. For supersonic flows with $M \geq 2$ the convergence history of the indirect multigrid method is not independent of the number of grid points if the limiter $D_{i,j}$ (Eq. (17)) is computed as in the subsonic flow with the conservative variables. For this reason $D_{i,j}$ (Eq.(17)) is determined by the differences of the characteristic variables.

$$U_{Ch,x} = \begin{bmatrix} \varrho - p/c^2 \\ -v \\ (u + p/\varrho c)/\sqrt{2} \\ (-u + p/\varrho c)/\sqrt{2} \end{bmatrix} \qquad U_{Ch,y} = \begin{bmatrix} \varrho - p/c^2 \\ u \\ (v + p/\varrho c)/\sqrt{2} \\ (-v + p/\varrho c)/\sqrt{2} \end{bmatrix} \tag{18}$$

The averaged gradients are transformed into gradients of the primitive variables ϱ, u, v, p to compute $U^{\pm}_{i\pm1/2,j}$ and $U^{\pm}_{i,j\pm1/2}$ [13]. This formulation of $D_{i,j}$ was successfully employed for subsonic flows, too. In Fig. 8 and in Fig. 9 residual traces of the computations for $Ma = 1.6$, wedge angle $\beta = 8°$ and $Ma = 2.0$, $\beta = 10°$ are plotted against the time steps for three fine grids. The result is that for this supersonic flow with the altered limiter the number of time steps to reach the steady state is also nearly independent of the number of grid points.

5. Conclusions

The two main multigrid concepts for solving the time-dependent conservation equations were formulated and tested for the solution of the full two-dimensional Navier-Stokes equations. For a test problem the direct multigrid concept in conjunction with the explicit MacCormack scheme showed a higher efficiency than the corresponding single grid solution. The comparison, however, to the single grid approximate factorization method of Beam and Warming revealed a slightly better rate of convergence of the implicit method. The best rates of convergence were achieved with the indirect multigrid approach including the flux-vector splitting by van Leer for the Euler terms combined with a collective Switched Evolution/Relaxation scheme. The application of this multigrid procedure to different subsonic and supersonic flow problems showed that the number of time steps to reach the steady state is almost independent of the number of grid points for this method.

6. References

[1] Brandt, A.: Guide to multigrid development. In "Lecture Notes in Mathematics" Vol. 960, pp. 220-312, Springer Verlag Berlin, 1981.

[2] Ni, R.H.: A Multiple-Grid Scheme for Solving the Euler Equations, AIAA Paper No. 81-1025, 1981.

[3] Schmidt, W., Jameson, A.: Applications of the Multi-grid Methods for Transonic Flow Calculations. In "Lecture Notes in Mathematics", Vol. 960, pp. 599-613, Springer Verlag Berlin, 1981.

[4] Chima, R.V., Johnson, G.M.: Efficient Solution of the Euler and Navier-Stokes Equations with a Vectorized Multiple Grid Algorithm

[5] Jameson, A.: Solution of the Euler Equations for Two Dimensional Transonic Flow by a Multigrid Method, presented at the International Multigrid Conference, Copper Mountain, 1983.

[6] Mulder, W.A.: Multigrid Relaxation for the Euler Equations. In "Lecture Notes in Physics", Vol. 218, pp. 417-421, Springer Verlag Berlin, 1985

[7] van Leer, B.: Flux-vector splitting for the Euler equations. In "Lecture Notes in Physics", Vol. 170, pp. 507-512, Springer Verlag Berlin, 1982.

[8] van Leer, B., Mulder, W.A.: Relaxation Methods for Hyperbolic Equations. In Proceedings of the INRIA Workshop on Numerical Methods for the Euler Equations for Compressible Fluids", Le Chesnay, France, Dec. 1983; to be published by SIAM.

[9] MacCormack, R.W.: The effect of viscosity on hypervelocity impact cratering. AIAA Paper No. 69-354, 1969.

[10] van Leer, B.: Towards the Ultimate Conservative Difference Scheme V. A Second-Order Sequel to Godunov's Method, J. Comp. Phys. 32 (1979), 101-136.

[11] van Albada, G.D., van Leer, B., Roberts, W.W.: A Comparative Study of Computational Methods in Cosmic Gas Dynamics, Astron. Astrophys. 108, 1982, pp. 76-84.

[12] Beam, R.M., Warming, R.F.: An implicit factored scheme for the compressible Navier-Stokes equations. Proceedings of the AIAA 3rd. Computational Fluid Dynamics Conference, Albuquerque, New Mexiko, 1977.

[13] Mulder, W.A., van Leer, B.: Implicit Upwind Methods for the Euler Equations. AIAA Paper No. 83-1930, 1983.

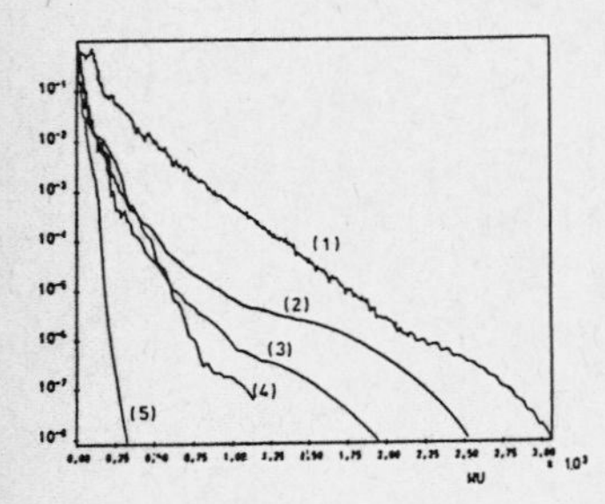

Figure 1: Maximum residual for the development of the compressible Couette flow (Re = 1000, Ma = 0.5, 17x17 grid points) as function of the work units.
Curve (1): single grid method, explicit MacCormack scheme
Curve (2): indirect multigrid method (4 grids) relaxation by the MacCormack scheme [9]
Curve (3): direct multigrid method (4 grids) using explicit MacCormack scheme [9]
Curve (4): single grid method, implicit factored scheme [12]
Curve (5): indirect multigrid method (4 grids) relaxation by collective Switched Evolution/ Relaxation (SER) including flux-splitting [7]

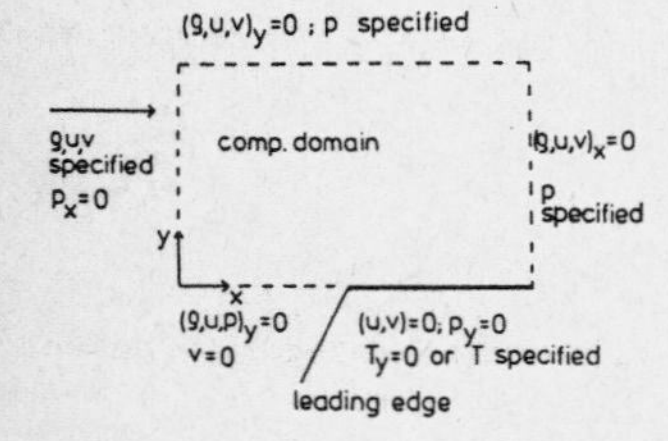

Figure 2: Viscous subsonic flow past a flat plate

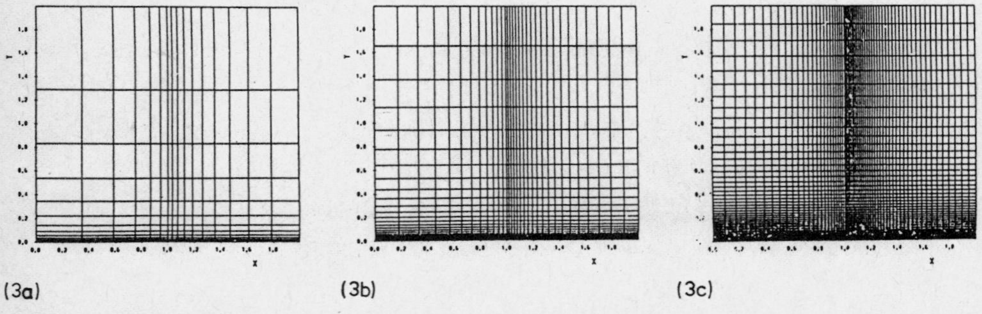

(3a) (3b) (3c)

Figure 3: Different fine grids
a) 17x17 grid points
b) 33x33 grid points
c) 65x65 grid points

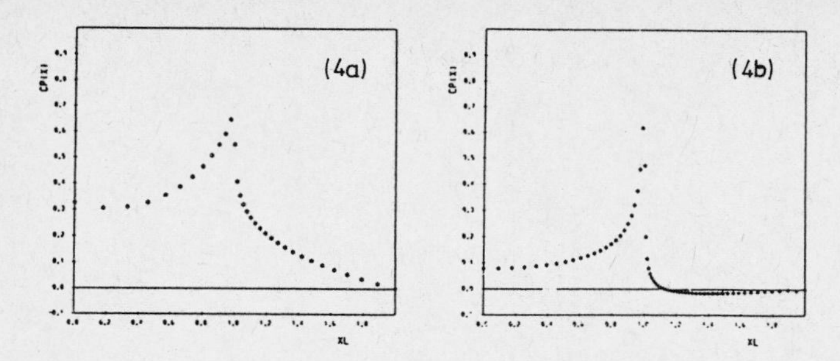

Figure 4: Pressure coefficient on the wall Re = 10^3, Ma = 0.5
a) first-order accurate scheme
b) second-order accurate scheme

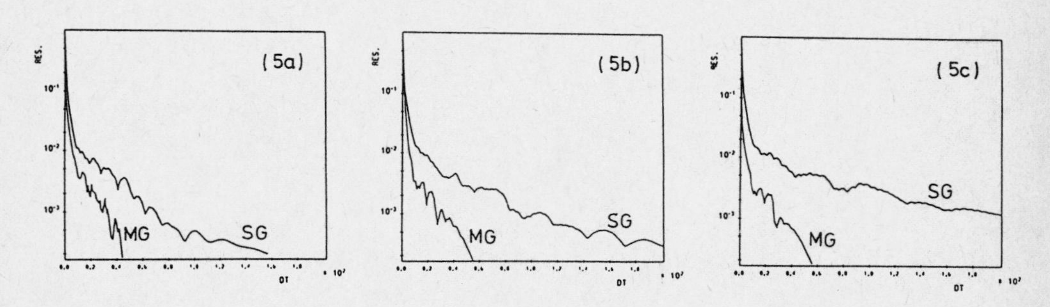

Figure 5: Maximum residual as function of the time steps for the viscous flow past a flat plate; Re = 10^5, Ma = 0.8; multigrid (MG) and single grid (SG) method
a) fine grid 17x17 grid points
b) fine grid 33x33 grid points
c) fine grid 65x65 grid points

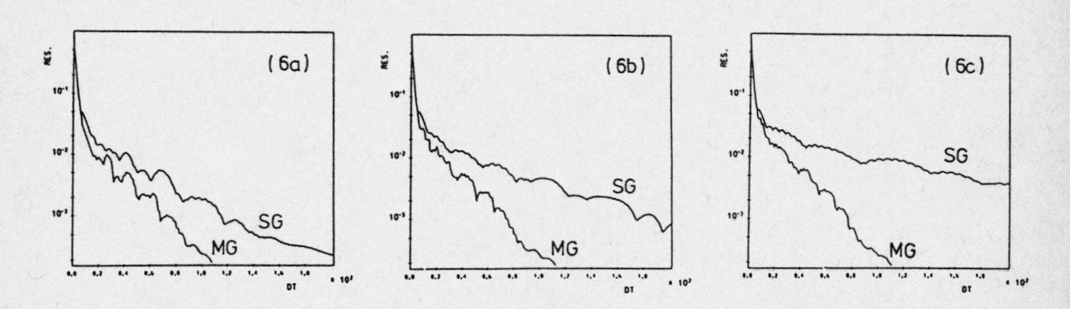

Figure 6: Maximum residual as function of the time steps for the viscous flow past a flat plate; Re = 10^4, Ma = 0.9; multigrid (MG) and single grid (SG) method
a) fine grid 17x17 grid points
b) fine grid 33x33 grid points
c) fine grid 65x65 grid points

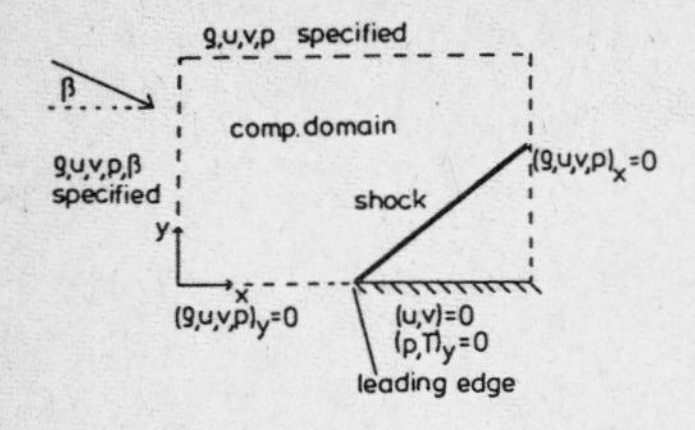

Figure 7: Viscous supersonic flow past a wedge

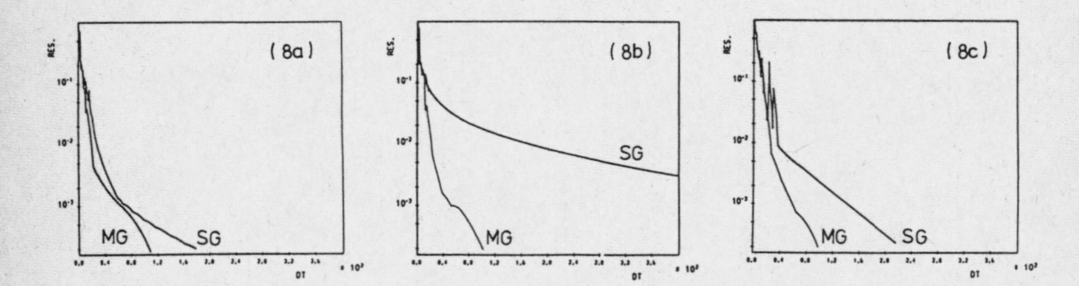

Figure 8: Maximum residual as function of the time steps for the viscous flow past a wedge; Re = 10^4, Ma = 1.6, β = 8° ; multigrid (MG) and single grid (SG) method
a) fine grid 17x17 grid points
b) fine grid 33x33 grid points
c) fine grid 65x65 grid points

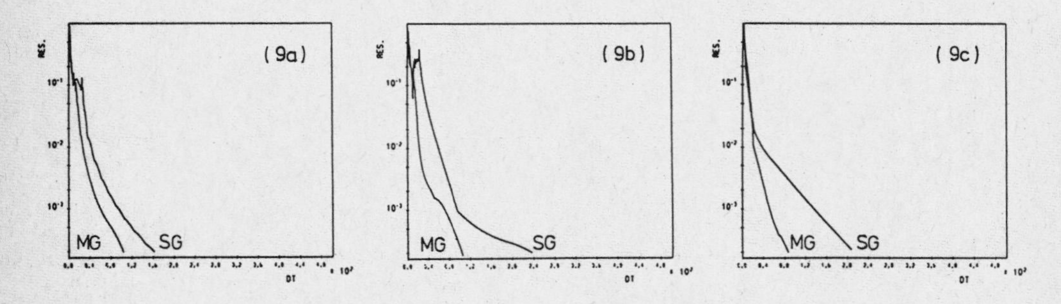

Figure 9: Maximum residual as function of the time steps for the viscous flow past a wedge; Re = 10^4, Ma = 2.0, β = 10° ; multigrid (MG) and single grid (SG) method
a) fine grid 17x17 grid points
b) fine grid 33x33 grid points
c) fine grid 65x65 grid points

SECOND ORDER ACCURATE UPWIND SOLUTIONS OF THE 2D STEADY EULER EQUATIONS BY THE USE OF A DEFECT CORRECTION METHOD

S.P.Spekreijse

CWI, Centre for Mathematics and Computer Science
P.O.Box 4079, 1009 AB Amsterdam, The Netherlands

ABSTRACT

In this paper a description is given of first and second order finite volume upwind schemes for the 2D steady Euler equations in generalized coordinates. These discretizations are obtained by projection-evolution stages, as suggested by Van Leer. The first order schemes can be solved efficiently by multigrid methods. Second order approximations are obtained by a defect correction method. In order to maintain monotone solutions, a limiter is introduced for the defect correction method.

1. INTRODUCTION

The steady state equations of inviscid flow, the steady Euler equations, are a nonlinear nonelliptic system of equations admitting solutions with discontinuities (shocks, contact discontinuities).

An important class of difference schemes for the (steady) Euler equations are the first order upwind schemes. These schemes are found by subdividing the domain of interest in disjunct control volumes (finite volume technique) and by assuming that the states in the volumes (or cells) are uniform (piecewise constant approximation). Then at each cell boundary two uniform states meet in a discontinuity. A unique flux, in the literature called the numerical flux [6], can be assigned after resolving the discontinuity by a set of elementary waves moving normal to the cell boundary or, in other words, after solving the one-dimensional Riemann-problem. This may be done exactly (Godunov [5]) or approximately (Osher [11], Roe [12], Van Leer [15], Steger & Warming [13]). The only difference between these schemes is the way of approximating the Riemann-problem. Therefore each first order upwind scheme is characterized by its numerical flux function.

Succesful application of the multigrid method for the solution of the nonlinear system obtained by first order upwind schemes has been reported by Mulder & Van Leer [10],[18] and Hemker & Spekreijse [8],[9]. They use respectively Van Leer's and Osher's approximate Riemann-solver.

The purpose of this paper is to improve solutions of first order upwind schemes. This is highly desirable because solutions of these schemes have some important shortcomings. Because of the hidden viscosity in these schemes, oblique (with respect to the mesh) shocks and contact discontinuities are smeared out disastrously. Furthermore, in the smooth part of the flow field the solution is only first order accurate, which is too low for practical purposes. Therefore we wish to improve the order of accuracy and to steepen oblique discontinuities without introducing over- or undershoot.

In the literature many second order upwind schemes with flux limiters are available [14],[17],[3],[1]. Without a flux limiter solutions of second or higher order schemes suffer from oscillations in the neighbourhood of discontinuities. Therefore flux limiters have been constructed and implemented in these schemes to prevent these oscillations. Such schemes, with their time dependent term, belong to the class of total variation diminishing (TVD) schemes [14]. The steady state solution of a second order TVD scheme possesses the improvements we wish to obtain.

The construction of time dependent first or second order (TVD) upwind schemes can be considered most conveniently (as suggested by Van Leer [16],[1]) in two stages: a projection stage and an evolution stage (MUSCL-approach). In the projection stage the states in the cells are interpolated to yield approximations of the states at the cell boundaries. If this approximation is made by extrapolation from both sides of a cell boundary then two uniform states meet in a discontinuity. In the evolution stage, at each cell boundary, an approximate Riemann solver (or equivalently a numerical flux function) is used to calculate the flux from these two uniform states. When a second order upwind scheme is constructed by this method, a flux limiter, developed to make the scheme TVD, only needs to be applied in the projection stage.

Because the solution of first order upwind schemes can be obtained efficiently (by the multigrid method), it seems that an iterative defect correction (DeC) method provides a simple way to obtain the steady state solution of a second order TVD scheme. Unfortunately however, such an iterative DeC process will converge very slowly or might not even converge at all. On the other hand, it is well known [2] that just one or more DeC iterations are enough to obtain a second order accurate approximation. Therefore we shall apply the DeC iteration steps only a few times. This makes the method cheap to apply but it also implies that the steady state solution of the second order TVD scheme will not be achieved. The flux limiter, developed to make the second order upwind scheme TVD, only ensures that the steady state solution will be monotone and not that a second order approximation obtained after a few DeC iteration steps is monotone as well. It is practical experience that wiggles may occur after a few defect corrections despite the use of a flux limiter. Therefore in the context of the DeC method, it is not appropriate to use a flux limiter.

On the other hand, after each DeC iteration step the new solution can be considered as the sum of the solution of the first order upwind scheme plus a correction. In order to prevent that the addition of the correction to the first order solution creates new local extrema (wiggles), this correction has to be modified (limited!). The modification of the correction is one of the main topics of this paper. Because the first order upwind scheme is a monotone scheme, this strategy ensures that after each DeC iteration step the solution will stay monotone, or in other words no over- or undershoot will occur. Shortly, we suggest to apply a limiter in the DeC method and not in the second order upwind scheme.

In section 2 we describe the first and second order upwind discretization of the 2D steady Euler equations in general geometries. The subdivision of the discretization by the projection and evolution stages is applied. A proof is given of the accuracy of these discretizations.

In section 3 the DeC method is described. A description of a simple limiter used in the DeC method is given.

In section 4 some numerical results concerning the resolution of a contact discontinuity and an oblique shock are presented.

2. SECOND ORDER FINITE VOLUME UPWIND DISCRETIZATION OF THE 2D STEADY EULER EQUATIONS

The 2D Euler equations can be written in conservative-vector form as

$$\frac{\partial}{\partial t} q + \frac{\partial}{\partial x} f(q) + \frac{\partial}{\partial y} g(q) = 0 , \tag{2.1}$$

on an open (irregular) domain $\Omega^* \subset \mathbb{R}^2$, where

$$q = \begin{bmatrix} \rho \\ \rho u \\ \rho v \\ e \end{bmatrix}, \quad f(q) = \begin{bmatrix} \rho u \\ \rho u^2 + p \\ \rho u v \\ u(e+p) \end{bmatrix}, \quad g(q) = \begin{bmatrix} \rho v \\ \rho u v \\ \rho v^2 + p \\ v(e+p) \end{bmatrix}. \tag{2.2}$$

Here, respectively ρ, u, v, p and e are density, velocity components in the x- and y-directions, pressure and total energy per unit volume. Furthermore, e may be expressed as

$$e = \rho(\epsilon + ½(u^2+v^2)) , \tag{2.3}$$

where the specific internal energy ϵ , is related to the pressure and density by the perfect gas law

$$p = (\gamma-1)\rho\epsilon , \tag{2.4}$$

with γ denoting the ratio of specific heats.

The physical domain Ω^* is subdivided into disjunct quadrilateral cells $\Omega^*_{i,j}$, $(i,j) \in \{ 1..M, 1..N \}$ in a regular fashion such that

i) $\Omega^* = \bigcup_{i,j} \Omega^*_{i,j}$,

ii) $\Omega^*_{i,j}$, $\Omega^*_{i\pm1,j}$, $\Omega^*_{i,j\pm1}$ are neighbouring cells ,

iii) $(x_{i+½,j+½} , y_{i+½,j+½}) = \bar{\Omega}^*_{i,j} \cap \bar{\Omega}^*_{i+1,j} \cap \bar{\Omega}^*_{i,j+1} \cap \bar{\Omega}^*_{i+1,j+1}$ is a common vertex of the cells $\Omega^*_{i,j}$, $\Omega^*_{i+1,j}$, $\Omega^*_{i,j+1}$ and $\Omega^*_{i+1,j+1}$.

It is clear that the vertices $\{ (x_{i+½,j+½} , y_{i+½,j+½}) \}$ define the subdivision of Ω^* completely.

Let (ξ,η) and (x,y) denote the cartesian coordinates in respectively the computational and physical space. In the computational space, we consider a rectangular domain Ω subdivided into equidistant square control volumes (or cells) $\Omega_{i,j}$, $(i,j) \in \{ 1..M, 1..N \}$ in such a way that $(h \cdot i , h \cdot j)$ is the mid-point of $\Omega_{i,j}$; h denotes the length of the edges. Assume the existence of a sufficiently smooth 1-1 relation between (ξ,η) and (x,y):

$$\begin{cases} \xi = \xi(x,y) \\ \eta = \eta(x,y) \end{cases} \Leftrightarrow \begin{cases} x = x(\xi,\eta) \\ y = y(\xi,\eta) \end{cases}, \tag{2.5}$$

such that each cell $\Omega_{i,j}$ corresponds with $\Omega^*_{i,j}$ by this mapping i.e. for all $(i,j) \in \{0..M, 0..N\}$

$$(x_{i+½,j+½} , y_{i+½,j+½}) = (x(\xi_{i+½} , \eta_{i+½}) , y(\xi_{i+½} , \eta_{i+½})) , \tag{2.6}$$

where $\xi_{i+½} = (i+½) \cdot h$ and $\eta_{j+½} = (j+½) \cdot h$. It can be easily seen that in the computational space (ξ,η) the Euler equations become

$$\frac{\partial}{\partial t}(Jq) + \frac{\partial}{\partial \xi}(y_\eta f(q) - x_\eta g(q)) + \frac{\partial}{\partial \eta}(x_\xi g(q) - y_\xi f(q)) = 0 , \tag{2.7}$$

with $J = x_\xi y_\eta - y_\xi x_\eta$.

The discretization of (2.1) on Ω^* is equivalent with the discretization of (2.7) on Ω. Because Ω is a rectangle subdivided into equidistant cells of lenght h, it is easier to obtain first and second order upwind discretizations of (2.7) on Ω than of (2.1) on Ω^*. In symbolic form we write (2.7) as

$$(Jq)_t + N(q) = 0\,, \tag{2.8}$$

and the steady Euler equations as

$$N(q) = 0\,, \tag{2.9}$$

Here $N: X \to Y$ is a nonlinear operator, $X \subset [L^2(\Omega)]^4$ is the space of possible fluid states and $Y = [L^2(\Omega)]^4$ is the Banach space of rates of change (of states).

Define the finite dimensional vector spaces X_h and Y_h by

$$X_h = Y_h = \{q_{i,j} \in \mathbb{R}^4 \mid i = 1..M, j = 1..N\}\,. \tag{2.10}$$

The relation between the spaces X and X_h , Y and Y_h is obtained by introducing $R_h : X \to X_h$ and $\bar{R}_h : Y \to Y_h$

$$(R_h\, q)_{i,j} = (\bar{R}_h\, q)_{i,j} = \frac{1}{h^2} \iint_{\Omega_{i,j}} q(\xi,\eta) d\xi d\eta\,, \tag{2.11}$$

for any $q \in [L^2(\Omega)]^4$. Thus $(R_h q)_{i,j}$ is the mean value of q in $\Omega_{i,j}$. A p-order accurate discretization of (2.9) is an associated problem

$$N_h^p(q) = 0\,, \tag{2.12}$$

where $N_h^p : X_h \to Y_h$ has the property that for all sufficiently smooth $q \in X$

$$(N_h^p\, R_h\, q)_{i,j} - (\bar{R}_h\, N\, q)_{i,j} = O(h^p)\,. \tag{2.13}$$

In this paper we will consider first and second order upwind schemes, so p=1 or p=2.

Notice that

$$(\bar{R}_h\, N\, q)_{i,j} = \frac{1}{h^2} \cdot \{ \int_{\Gamma_{i+½,j}} (y_\eta f - x_\eta g)(q((i+½)\cdot h,\eta)) d\eta -$$

$$\int_{\Gamma_{i-½,j}} (y_\eta f - x_\eta g)(q((i-½)\cdot h,\eta)) d\eta + \int_{\Gamma_{i,j+½}} (x_\xi g - y_\xi f)(q(\xi,(j+½)\cdot h)) d\xi -$$

$$\int_{\Gamma_{i,j-½}} (x_\xi g - y_\xi f)(q(\xi,(j-½)\cdot h)) d\xi \}\,, \tag{2.14}$$

where $\Gamma_{i+½,j}$, $\Gamma_{i-½,j}$, $\Gamma_{i,j+½}$, $\Gamma_{i,j-½}$ denote the boundaries of cell $\Omega_{i,j}$, defined by $\Gamma_{i+½,j} = \bar{\Omega}_{i,j} \cap \bar{\Omega}_{i+1,j}$, $\Gamma_{i,j+½} = \bar{\Omega}_{i,j} \cap \bar{\Omega}_{i,j+1}$ etc.

Now we will construct an operator $\tilde{N}_h : X \to Y_h$, easier to approximate than $\bar{R}_h N$, but such that for all sufficiently smooth $q \in X$

$$(\tilde{N}_h\, q)_{i,j} - (\bar{R}_h\, Nq)_{i,j} = O(h^2)\,. \tag{2.15}$$

If we are able to construct $N_h^p : X_h \to Y_h$, $p=1,2$ such that

$$(N_h^p\, R_h\, q)_{i,j} - (\tilde{N}_h\, q)_{i,j} = O(h^p)\,, \tag{2.16}$$

then also

$$(N_h^p\, R_h\, q)_{i,j} - (\bar{R}_h\, Nq)_{i,j} = O(h^p)\,. \tag{2.17}$$

This means that after the construction of a $\tilde{N}_h$ satisfying (2.15) we may restrict ourselves to the approximation of $\tilde{N}_h$ instead of $\bar{R}_h N$. An obvious choice for the operator $\tilde{N}_h$ is

$$(\tilde{N}_h\, q)_{i,j} = \frac{1}{h} \cdot \{ (y_{\eta_{i+½,j}} f - x_{\eta_{i+½,j}} g)(\bar{q}_{i+½,j}) - (y_{\eta_{i-½,j}} f - x_{\eta_{i-½,j}} g)(\bar{q}_{i-½,j})$$

$$+(x_{\xi_{i,j+½}} g - y_{\xi_{i,j+½}} f)(\overline{q}_{i,j+½}) - (x_{\xi_{i,j-½}} g - y_{\xi_{i,j-½}} f)(\overline{q}_{i,j-½}) \} , \tag{2.18}$$

where

$$y_{\eta_{i+½,j}} = \frac{1}{h}\cdot(y_{i+½,j+½} - y_{i+½,j-½}) , \; x_{\eta_{i+½,j}} = \frac{1}{h}\cdot(x_{i+½,j+½} - x_{i+½,j-½}) ,$$

$$x_{\xi_{i,j+½}} = \frac{1}{h}\cdot(x_{i+½,j+½} - x_{i-½,j+½}) , \; y_{\xi_{i,j+½}} = \frac{1}{h}\cdot(y_{i+½,j+½} - y_{i-½,j+½}) , \tag{2.18a}$$

and

$$\overline{q}_{i+½,j} = \frac{1}{h}\cdot \int_{\Gamma_{i+½,j}} q((i + ½)\cdot h,\eta)d\eta . \tag{2.18b}$$

Thus $\overline{q}_{i+½,j}$ is the mean value of $q(\xi,\eta)$ at the cell boundary $\Gamma_{i+½,j}$. In a similar way $\overline{q}_{i-½,j}$, $\overline{q}_{i,j+½}$ and $\overline{q}_{i,j-½}$ are defined. By (2.18a) we see that the derivatives of the mapping , at the cell boundaries, are approximated by central differences.

By elementary interpolation theory it can be seen that (2.15) holds (assuming a sufficiently smooth mapping).

We wish to deal with upwind schemes therefore we have to introduce an approximate Riemann-solver or equivalently a numerical flux function. This can be easily done by using the property of rotational invariance of the Euler equations i.e.

$$\cos\phi\cdot f(q) + \sin\phi\cdot g(q) = T(\phi)^{-1} f(T(\phi)q) , \tag{2.19}$$

with

$$T(\phi) = \begin{bmatrix} 1 & 0 & 0 & 0 \\ 0 & \cos\phi & \sin\phi & 0 \\ 0 & -\sin\phi & \cos\phi & 0 \\ 0 & 0 & 0 & 1 \end{bmatrix} , \tag{2.20}$$

and q , $f(q)$ and $g(q)$ as in (2.2), $\phi\in\mathbb{R}$.

Define $l_{i+½,j}$, $l_{i,j+½}$ by

$$l_{i+½,j} = (y_{\eta_{i+½,j}}{}^2 + x_{\eta_{i+½,j}}{}^2)^{½} , \; l_{i,j+½} = (x_{\xi_{i,j+½}}{}^2 + y_{\xi_{i,j+½}}{}^2)^{½} , \tag{2.21a}$$

and $\phi_{i+½,j}$, $\phi_{i,j+½}$ by

$$l_{i+½,j}\cdot\cos\phi_{i+½,j} = y_{\eta_{i+½,j}} \quad , \; l_{i+½,j}\cdot\sin\phi_{i+½,j} = -x_{\eta_{i+½,j}} ,$$

$$l_{i,j+½}\cdot\cos\phi_{i,j+½} = -y_{\xi_{i,j+½}} \quad , \; l_{i,j+½}\cdot\sin\phi_{i,j+½} = x_{\xi_{i,j+½}} , \tag{2.21b}$$

then using (2.19)-(2.21), (2.18) yields

$$(\tilde{N}_h q)_{i,j} = \frac{1}{h}\cdot\{ l_{i+½,j}\, T^{-1}_{i+½,j}\cdot f(T_{i+½,j}\,\overline{q}_{i+½,j}) - l_{i-½,j}\, T^{-1}_{i-½,j}\cdot f(T_{i-½,j}\,\overline{q}_{i-½,j})$$

$$+ l_{i,j+½}\, T^{-1}_{i,j+½}\cdot f(T_{i,j+½}\,\overline{q}_{i,j+½}) - l_{i,j-½}\, T^{-1}_{i,j-½}\cdot f(T_{i,j-½}\,\overline{q}_{i,j-½}) \} , \tag{2.22}$$

where $T_{i+½,j} = T(\phi_{i+½,j})$ etc. This formula strongly suggest how to define the first and second order upwind schemes $N^p_h : X_h \to Y_h$, $p = 1,2$ namely

$$(N^p_h q)_{i,j} = \frac{1}{h}\cdot\{ l_{i+½,j}\, T^{-1}_{i+½,j}\, f(T_{i+½,j}\, q^-_{i+½,j} , T_{i+½,j}\, q^+_{i+½,j}) -$$

$$l_{i-½,j}\, T^{-1}_{i-½,j}\, f(T_{i-½,j}\, q^-_{i-½,j} , T_{i-½,j}\, q^+_{i-½,j}) + l_{i,j+½}\, T^{-1}_{i,j+½}\, f(T_{i,j+½}\, q^-_{i,j+½} , T_{i,j+½}\, q^+_{i,j+½}) -$$

$$l_{i,j-½}\, T^{-1}_{i,j-½}\, f(T_{i,j-½}\, q^-_{i,j-½} , T_{i,j-½}\, q^+_{i,j-½}) \} , \tag{2.23}$$

where

$$q^{-}_{i+\frac{1}{2},j} = q_{i,j} \ , \ q^{+}_{i+\frac{1}{2},j} = q_{i+1,j} \ ,$$
$$q^{-}_{i,j+\frac{1}{2}} = q_{i,j} \ , \ q^{+}_{i,j+\frac{1}{2}} = q_{i,j+1} \ , \tag{2.23a}$$

for the first order approximation ($p=1$) or

$$q^{-}_{i+\frac{1}{2},j} = q_{i,j} + \tfrac{1}{2}\cdot(q_{i,j} - q_{i-1,j}) \ , \ q^{+}_{i+\frac{1}{2},j} = q_{i+1,j} + \tfrac{1}{2}\cdot(q_{i+1,j} - q_{i+2,j}) \ ,$$
$$q^{-}_{i,j+\frac{1}{2}} = q_{i,j} + \tfrac{1}{2}\cdot(q_{i,j} - q_{i,j-1}) \ , \ q^{+}_{i,j+\frac{1}{2}} = q_{i,j+1} + \tfrac{1}{2}\cdot(q_{i,j+1} - q_{i,j+2}) \ , \tag{2.23b}$$

for the second order approximation ($p=2$) and $q \in X_h$. Furthermore $f(\ ,\ ) : \mathbb{R}^4 \times \mathbb{R}^4 \to \mathbb{R}^4$ is one of the numerical flux functions found in the literature [5],[11],[12],[13],[15]. For consistency we only need that the numerical flux $f(\ ,\ )$ is consistent with the physical flux i.e.

$$f(q\ ,\ q) = f(q) \tag{2.24}$$

To proof that $(N^p_h R_h q)_{i,j} - (\tilde{N}_h q)_{i,j} = O(h^p)$, $p=1,2$ it is easily seen that we only need to show that for all sufficiently smooth $q \in X$

$$\frac{1}{h}\cdot[\, l_{i+\frac{1}{2},j}\, T^{-1}_{i+\frac{1}{2},j}\{ f(T_{i+\frac{1}{2},j}\, q^{-}_{i+\frac{1}{2},j}\ ,\ T_{i+\frac{1}{2},j}\, q^{+}_{i+\frac{1}{2},j}) - f(T_{i+\frac{1}{2},j}\, \overline{q}_{i+\frac{1}{2},j})\} -$$
$$l_{i-\frac{1}{2},j}\, T^{-1}_{i-\frac{1}{2},j}\{ f(T_{i-\frac{1}{2},j}\, q^{-}_{i-\frac{1}{2},j}\ ,\ T_{i-\frac{1}{2},j}\, q^{+}_{i-\frac{1}{2},j}) - f(T_{i-\frac{1}{2},j}\, \overline{q}_{i-\frac{1}{2},j})\}\,] = O(h^p), \tag{2.25}$$

with the same conventions as in (2.18b) and (2.23); thus $q_{i,j} = (R_h q)_{i,j}$ in (2.23a,b). In order to prove (2.25) first define

$$\overline{q}(\xi,\eta) = \frac{1}{h^2}\cdot \int_{\xi-\frac{h}{2}}^{\xi+\frac{h}{2}} \int_{\eta-\frac{h}{2}}^{\eta+\frac{h}{2}} q(\alpha\ ,\ \beta)\, d\alpha\, d\beta\ , \tag{2.26}$$

then

$$\overline{q}(ih\ ,\ jh) = \frac{1}{h^2}\cdot \iint_{\Omega_{i,j}} q(\alpha\ ,\ \beta)\, d\alpha\, d\beta = (R_h q)_{i,j}\ , \tag{2.27}$$

and

$$\frac{\partial}{\partial \xi}\, \overline{q}(\xi,\eta)\,|_{(ih,jh)} = \frac{1}{h}\cdot(\overline{q}_{i+\frac{1}{2},j} - \overline{q}_{i-\frac{1}{2},j})\ . \tag{2.28}$$

From (2.27) and (2.23a,b) we see that

$$\frac{1}{h}\cdot(q^{\pm}_{i+\frac{1}{2},j} - q^{\pm}_{i-\frac{1}{2},j}) = \frac{\partial}{\partial \xi}\, \overline{q}(\xi,\eta)\,|_{(ih,jh)} + O(h^p)\ . \tag{2.29}$$

Thus, using (2.28), (2.29) yields

$$q^{\pm}_{i+\frac{1}{2},j} - \overline{q}_{i+\frac{1}{2},j} = q^{\pm}_{i-\frac{1}{2},j} - \overline{q}_{i-\frac{1}{2},j} + O(h^{p+1})\ . \tag{2.30}$$

Furthermore it is clear that

$$q^{\pm}_{i+\frac{1}{2},j} - \overline{q}_{i+\frac{1}{2},j} = O(h^p)\ . \tag{2.31}$$

Assuming that the numerical flux function is sufficiently smooth, it follows by a Taylor expansion that

$$f(q^{-}_{i+\frac{1}{2},j}, q^{+}_{i+\frac{1}{2},j}) = f(\overline{q}_{i+\frac{1}{2},j}, \overline{q}_{i+\frac{1}{2},j}) + f'_1(\overline{q}_{i+\frac{1}{2},j}, \overline{q}_{i+\frac{1}{2},j})(q^{-}_{i+\frac{1}{2},j} - \overline{q}_{i+\frac{1}{2},j}) +$$
$$f'_2(\overline{q}_{i+\frac{1}{2},j}, \overline{q}_{i+\frac{1}{2},j})(q^{+}_{i+\frac{1}{2},j} - \overline{q}_{i+\frac{1}{2},j}) + O(\,|q^{-}_{i+\frac{1}{2},j} - \overline{q}_{i+\frac{1}{2},j}|^2, |q^{+}_{i+\frac{1}{2},j} - \overline{q}_{i+\frac{1}{2},j}|^2). \tag{2.32}$$

Due to the consistency of the numerical flux function and using (2.31), (2.32) yields

$$f(T_{i+\frac{1}{2},j} q^{-}_{i+\frac{1}{2},j}\ ,\ T_{i+\frac{1}{2},j} q^{+}_{i+\frac{1}{2},j}) - f(T_{i+\frac{1}{2},j} \overline{q}_{i+\frac{1}{2},j}) =$$

$$f'_1(T_{i+1/2,j}\bar{q}_{i+1/2,j}\,,\,T_{i+1/2,j}\bar{q}_{i+1/2,j})\cdot T_{i+1/2,j}\cdot(q^-_{i+1/2,j}-\bar{q}_{i+1/2,j})\,+$$
$$f'_2(T_{i+1/2,j}\bar{q}_{i+1/2,j}\,,\,T_{i+1/2,j}\bar{q}_{i+1/2,j})\cdot T_{i+1/2,j}\cdot(q^+_{i+1/2,j}-\bar{q}_{i+1/2,j})+O(h^{2p}). \qquad (2.33)$$

If the mapping is sufficiently smooth then

$$l_{i+1/2,j}-l_{i-1/2,j}=O(h)\,,\;T_{i+1/2,j}-T_{i-1/2,j}=O(h)\,,\;T^{-1}_{i+1/2,j}-T^{-1}_{i-1/2,j}=O(h)\,. \qquad (2.34)$$

From (2.33),(2.30),(2.31) and (2.34) it is easily derived that (2.25) holds. Hence, N^p_h , p=1,2 are p-order accurate approximations of N.

Remarks

-The smoothness of the coordinate transformation (2.5) is necessary to obtain first and second order accurate space discretizations.

-The equation $(N^p_h q)_{i,j}=0$ is equivalent with $(h^2 N^p_h q)_{i,j}=0$. After multiplication of (2.23) by h^2 it is clear that $(h^2 N^p_h q)_{i,j}$ is determined completely by the coordinates $\{(x_{i+1/2,j+1/2}\,,\,y_{i+1/2,j+1/2})\}$ of the vertices of the mesh in the physical space (and, of course, by the numerical flux function f(,)). This means that in the physical space on an irregular mesh, a first and second order finite volume scheme can be constructed without the actual need of a coordinate transformation. The finite volume space discretization on an irregular mesh behaves in a first or second order manner only if the irregular mesh is chosen in accordance with a smooth transformation.

-Consider the steady Euler equations in the computational space $N(q)=0$ and their discretization $(N^p_h q)_{i,j}=0,\,p=1,2$ (see respectively (2.9),(2.23)). Let $q^p\in X_h,\,p=1,2$ denote the solutions of the discrete equations and $q=\tilde{q}(\xi,\eta)\in X$ be the solution of the continuous equation. If the operators $N^p_h,\,p=1,2$ are stable then

$$q^p_{i,j}=\bar{q}_{i,j}+O(h^p)\,, \qquad (2.35)$$

where $\bar{q}_{i,j}=(R_h\tilde{q})_{i,j}=\frac{1}{h^2}\cdot\iint_{\Omega_{i,j}}\tilde{q}(\xi,\eta)d\xi d\eta$. Define

$$\bar{Q}_{i,j}=\frac{1}{V_{i,j}}\cdot\iint_{\Omega_{i,j}}\tilde{q}(\xi,\eta)\cdot J(\xi,\eta)d\xi d\eta\,,$$

with

$$V_{i,j}=\iint_{\Omega_{i,j}}J(\xi,\eta)d\xi d\eta\,.$$

Then $V_{i,j}$ is the area of $\Omega^*_{i,j}$ in the physical space and $\bar{Q}_{i,j}$ is the mean value of $\tilde{q}(x,y)$ in $\Omega^*_{i,j}$. Assuming sufficient smoothness of the Jacobian $J(\xi,\eta)$ it is easily seen that

$$\frac{1}{h^2}\cdot\iint_{\Omega_{i,j}}J(\xi,\eta)\,\tilde{q}(\xi,\eta)d\xi d\eta=\frac{1}{h^2}\cdot\iint_{\Omega_{i,j}}J(\xi,\eta)d\xi d\eta\cdot\frac{1}{h^2}\cdot\iint_{\Omega_{i,j}}\tilde{q}(\xi,\eta)d\xi d\eta+O(h^2).$$

So, $\bar{Q}_{i,j}=\bar{q}_{i,j}+O(h^2)$ and (2.35) yields $q^p_{i,j}=\bar{Q}_{i,j}+O(h^p)$, $p=1,2$. Therefore, $q^p_{i,j}$ may be considered as a p-order accurate approximation of $\bar{q}_{i,j}$ as well as $\bar{Q}_{i,j}$.

-In the literature [16],[4],[1] one encounters generalizations of (2.23b) namely

$$q^-_{i+1/2,j}=q_{i,j}\;+\frac{1+\kappa}{4}\cdot(q_{i+1,j}-q_{i,j})+\frac{1-\kappa}{4}\cdot(q_{i,j}-q_{i-1,j})\,,$$
$$q^+_{i+1/2,j}=q_{i+1,j}+\frac{1+\kappa}{4}\cdot(q_{i,j}-q_{i+1,j})+\frac{1-\kappa}{4}\cdot(q_{i+1,j}-q_{i+2,j})\,. \qquad (2.36)$$

and similar $q^{\pm}_{i,j+1/2}$, $\kappa\in[-1\,,\,1]$.

It can be easily seen that (2.29) and (2.31) hold for each κ. Therefore (2.36) results in a second order accurate space discretization as well as (2.23b).

-Even when $x(\xi,\eta) = \xi$, $y(\xi,\eta) = \eta$ it can be seen that for all smooth $q \in X$

$$(\tilde{N}_h q)_{i,j} - (R_h N q)_{i,j} = O(h^2).$$

So it is plausible that $R_h N$ cannot be approximated more accurately than second order if the flux computation is based on the calculation of constant states at the cell boundaries.This is due to the fact that at a cell boundary the mean flux differs from the flux calculated in the mean state. So this result is typical for 2 and 3 dimensional problems. A formal proof of this result can be given.

3. THE DEFECT CORRECTION METHOD

In section 2 the description of first and second order upwind space discretizations of the 2D steady Euler equations has been given.

In this section the steady Euler equations and their first and second order space discretizations are denoted by

$$N q = r \; ; \; N_h^1 q_h = r_h \; ; \; N_h^2 q_h = r_h \; , \tag{3.1}$$

with h the meshsize of the (finest) grid.

Let $q_h^{1^*}$ and $q_h^{2^*}$ be the solutions of respectively $N_h^1 q_h = r_h$ and $N_h^2 q_h = r_h$. Then $q_h^{1^*}$ can be calculated efficiently by the multigrid method. We wish to use the defect correction method and the second order space discretization operator N_h^2 to improve $q_h^{1^*}$.

In a first glance the iterative defect correction (DeC) method seems applicable:

$$\begin{cases} q_h^1 := q_h^{1^*} , \\ \\ N_h^1 q_h^{n+1} = N_h^1 q_h^n + (r_h - N_h^2 q_h^n) \qquad n = 1,2.... \end{cases} \tag{3.2}$$

Unfortunately, dealing with the steady Euler equations, this iteration process is impractical because of the following reasons:

−Suppose, for the moment, that N , N_h^1 and N_h^2 are linear scalar operators with N_h^1 and N_h^2 first and second order discretizations of N . Then the symbols $N(\omega)$, $N_h^1(\omega)$ and $N_h^2(\omega)$ of respectively N , N_h^1 and N_h^2 are defined by

$$N(e^{i\omega x}) = N(\omega)\cdot e^{i\omega x} \; ; \; N_h^{1,2}(e^{i\omega x}) = N_h^{1,2}(\omega)\cdot e^{i\omega x} ,$$

and the accuracy of N_h^1 and N_h^2 can be expressed by

$$N(\omega) - N_h^1(\omega) = O(h^{p_1}) \; ; \; N(\omega) - N_h^2(\omega) = O(h^{p_2})$$

(ω fixed).

If we define the error $v_h^n = q_h^{2^*} - q_h^n$ then

$$N_h^1 v_h^{n+1} = (N_h^1 - N_h^2) v_h^n .$$

Taking $v_h^n = e^{i\omega x}$ then

$$v_h^{n+1} = \left\{ 1 - \frac{N_h^2(\omega)}{N_h^1(\omega)} \right\} \cdot v_h^n .$$

On a fixed grid with $\omega \to 0$, we see that the amplification factor goes to zero (due to the consistency of N_h^1 and N_h^2). This means that low frequency error components are damped effectively by the iterative DeC method. But for high frequency error components ($\frac{\pi}{2h} \leq \omega \leq \frac{\pi}{h}$) the difference between $N_h^1(\omega)$ and $N_h^2(\omega)$ can be quite large, causing slow or no convergence of the iteration process. Indeed, dealing with the Euler equations it is a practical experience that (3.2) converges very slowly or even not at all. Thus we may not expect to reach the fixed point $q_h^{2^*}$ of (3.2). Therefore it makes no sense to make N_h^2 TVD (by a fluxlimiter) because the TVD property only ensures that $q_h^{2^*}$ is wiggle free.

On the other hand, after one defect correction iteration q_h^2 obeys

$$N_h^{dc}\, q_h^2 := N_h^1\,(2N_h^1 - N_h^2)^{-1}\, N_h^1\, q_h^2 = r_h\,.$$

Let $N_h^{dc}(\omega)$ denote the symbol of N_h^{dc}. Then

$$N_h^{dc}(\omega) = \frac{\{N_h^1(\omega)\}^2}{2N_h^1(\omega) - N_h^2(\omega)}\,,$$

and it is easily seen that

$$\begin{aligned} N(\omega) - N_h^{dc}(\omega) &= \frac{N(\omega)\cdot\{N(\omega) - N_h^2(\omega)\} - \{N(\omega) - N_h^1(\omega)\}^2}{N(\omega) - 2\cdot\{N(\omega) - N_h^1(\omega)\} + \{N(\omega) - N_h^2(\omega)\}} \\ &= O(h^{min(2p_1,p_2)}) \end{aligned}$$

Because $p_1 = 1$ and $p_2 = 2$, this implies that N_h^{dc} is a second order space discretization of N. So q_h^2 is a second order accurate approximation. In the same way it can be seen that all q_h^2 , $n \geqslant 2$ are second order accurate approximations. Therefore, one or more DeC iteration steps are sufficient to improve the order of accuracy. This is a well known result which also holds for nonlinear problems [2].

From these arguments it is clear that we will apply the DeC method just a few times (at most four iteration steps). Then we will obtain a second order accurate approximation which suffers from wiggles in the neighbourhood of discontinuities. Because (3.2) starts with the solution of a first order scheme, which is a monotone scheme, we can prevent these wiggles by looking more closely to the correction $\delta q_h^n = q_h^n - q_h^{1^*}$. After each iteration in (3.2) we will change δq_h^n, as less as possible, but such that adding the changed correction to $q_h^{1^*}$ no new local extrema are created. In this way no wiggles can be introduced. This changing of δq_h^n is in fact a limiting process, therefore we may speak of a limited defect correction process. A simple limited defect correction process is:

step1: $\tilde{q}_h := q_h^{1^*}$.

step2:

– Calculate q_h from

$$N_h^1\, q_h = N_h^1\, \tilde{q}_h + (r_h - N_h^2\, \tilde{q}_h)\,. \tag{3.3}$$

– Calculate the correction

$$\delta q_h := q_h - q_h^{1^*}\,, \tag{3.4}$$

and set

$$q_h := q_h^{1^*}\,.$$

Then $q_h + \delta q_h$ is the solution of (3.3). Now change q_h and δq_h by scanning the grid in several directions (for instance from north-east to south-west and vice versa and from north-west to south-east and vice versa) and change the states $q_{h_{i,j}}$ and $\delta q_{h_{i,j}}$ in the visiting control volume $\Omega_{i,j}$ by the following algorithm:

$$q_{h_{i,j}} := q_{h_{i,j}} + \delta\, {q_h^{new}}_{i,j}\,,$$

$$\delta q_{h_{i,j}} := \delta q_{h_{i,j}} - \delta q_{h\ i,j}^{new} , \tag{3.5}$$

where

$$\delta q_{h\ i,j}^{new} = H(\delta q_{h_{i,j}}) \cdot \min\left(\delta q_{h_{i,j}} , \mid \delta q_{h\ i,j}^{+} \mid\right) + \\ + \{1 - H(\delta q_{h_{i,j}})\} \cdot \max\left(\delta q_{h_{i,j}} , - \mid \delta q_{h\ i,j}^{-} \mid\right), \tag{3.6}$$

and

$$\delta q_{h\ i,j}^{+} := \max\left(q_{h_{i-1,j}} , q_{h_{i+1,j}} , q_{h_{i,j-1}} , q_{h_{i,j+1}}\right) - q_{h_{i,j}} , \\ \delta q_{h\ i,j}^{-} := \min\left(q_{h_{i-1,j}} , q_{h_{i+1,j}} , q_{h_{i,j-1}} , q_{h_{i,j+1}}\right) - q_{h_{i,j}} , \tag{3.7}$$

and $H : \mathbb{R} \rightarrow \mathbb{R}$ denotes the Heaviside function

$$H(x) = \begin{cases} 1 & \text{if } x > 0 \\ 0 & \text{if } x < 0 \end{cases} .$$

– Finally set $\tilde{q}_h := q_h$.

Formula (3.5)-(3.7) must be applied to each component of $q_{h_{i,j}}$ and $\delta q_{h_{i,j}}$ separately.

It can be easily seen that $\delta q_{h\ i,j}^{new}$ is a smooth function of $\delta q_{h_{i,j}}$ and $q_{h_{i,j}}$.

Furthermore it is clear that in this way no new local extrema will be introduced in a visiting control volume . Although it is possible that after changing the state in the visiting control volume $\Omega_{i,j}$, a local extrema is created in one of the four neighbouring cells $\Omega_{i\pm1,j}$, $\Omega_{i,j\pm1}$. This limiter neglects this possibility, which is expected to occur rarely.

Step2 may be repeated a few times to steepen discontinuities effectively. In general, the limiter will not work in the smooth parts of the flow field, so in those parts the correction δq_h will be added to q_h completely. Only in the neighbourhood of discontinuities this limiter will do its job; preventing oscillations. Therefore this limiter also provides a way to detect discontinuities in the flow field.

4. NUMERICAL RESULTS

The numerical flux function used in these experiments is constructed with Osher's approximate Riemann-solver. Hence, both the first and second order upwind scheme are able to capture shocks and contact discontinuities. To see the improvement of the capturing property after a few DeC iteration steps, two model problems are considered. Problem 1 concerns an oblique shock reflected from a flat plate and problem 2 concerns an oblique contact discontinuity generated by the boundary conditions. The precise description of these two problems is:

Problem1: The oblique shock.
The domain Ω^* is $(0,4)\times(0,1)$. The exact solution has 3 subregions with uniform states as given in figure 4.1.

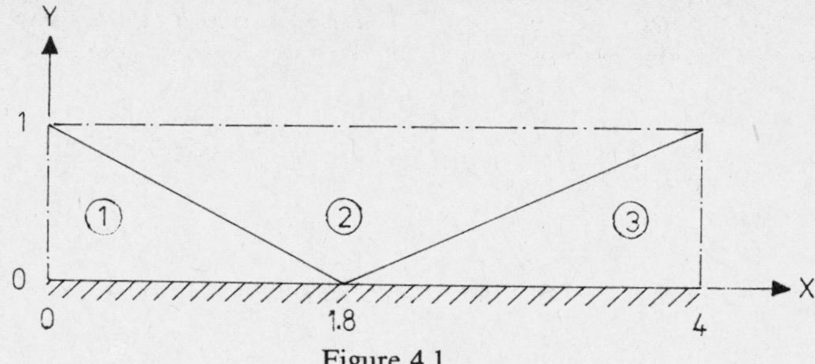

Figure 4.1.

The states are respectively:
state 1:$u=2.9, v=\ 0.0, c=1.0, p=1.0$,
state 2:$u=2.6, v=-0.5, c=1.1, p=2.1$,
state 3:$u=2.4, v=\ 0.0, c=1.2, p=4.0$.

Problem2: The contact discontinuity.
Here $\Omega^* = (0,2)\times(0,2)$. The exact solution of the problem has a discontinuity at $x+y=2$. In both parts of the domain the solution has a uniform state:
for $x+y<2$ we take $p=1.0, u=0.3, v=-0.3, c=0.6$,
for $x+y>2$ we take $p=1.0, u=0.6, v=-0.6, c=1.0$.
For the treatment of the boundary conditions see [7].

The figures 1,2,3 and 4 concern the resolution of the oblique shock and the figures 5 and 6 of the contact discontinuity.

Figure 1a,1b and 1c show the pressure contours on a 8×24 mesh, respectively obtained by the first order Osher scheme and after 1 and 4 DeC iteration steps. Figure 2a,2b and 2c show the same results but on a 16×48 mesh. In all cases the limiter, described in section 3, has been used. In figure 3a and 3b pressure distributions along the flat plate are shown (using the 16×48 mesh). In these figures results are shown, again obtained by the first order scheme and after 1 and 4 DeC iterations. Figure 3a has been obtained with, figure 3b without the limiter. Figure 4a and 4b show similar results at $y=0.5$. After 4 DeC iteration steps the quality of the shock capturing seems comparable with the results obtained by a second order TVD scheme [3].

Figure 5a,5b,5c and 6a,6b,6c show density contours on respectively a 16×16 and a 32×32 mesh. Again results of the first order scheme and after 1 and 4 DeC iteration steps are shown. For comparison see [7].

For both problems, it is clear that after a few DeC iteration steps the capturing of the discontinuities has been improved considerably.

5. CONCLUSION

This paper is concerned with the discretization of the steady Euler equations by the finite volume technique. On an irregular mesh it is shown in detail how to apply Van Leer's projection-evolution stages in the discretization. Herein, the rotational invariance of the Euler equations is effectively used. For a general numerical flux function, consistent with the physical flux, a proof is given of the order of accuracy for a first and second order upwind scheme. Hence, the results hold for all well known approximate Riemann-solvers.

Second order accurate approximations are obtained by a defect correction (DeC) method. A limiter, used in the DeC method, is constructed to maintain monotone solutions. For two typical model problems (an oblique shock and a contact discontinuity), only a few (3 or 4) DeC iteration steps are sufficient to steepen discontinuities effectively. This makes the method cheap to apply.Furthermore, the quality of the results seems comparable with results obtained by TVD schemes.

Acknowledgement. The author would like to thank P.W.Hemker, B.Koren and P.M.de Zeeuw for their cooperation and valuable suggestions.

REFERENCES

[1] Anderson, W.T., Thomas, J.L., and Van Leer, B., "A comparison of finite volume flux vector splittings for the Euler equations" AIAA Paper No. 850122.

[2] Böhmer, K., Hemker, P. & Stetter, H., "The Defect Correction Approach." Computing Suppl. 5 (1984) 1-32.

[3] Chakravarthy, S.R. and Osher, S., "High resolution applications of the Osher upwind scheme for the Euler equations." AIAA Paper 83-1943,Proc.AIAA Sixth Computational Fluid Dynamics Conf.(Danvers,Mass.July 1983),1983,pp363-372.

[4] Chakravarthy, S.R. and Osher, S., "A new class of high accuracy TVD schemes for hyperbolic conservation laws." AIAA Paper 85-0363,AIAA 23rd Aerospace Science Meeting. (Jan.14-17,1985/Reno,Nevada).

[5] Godunov, S.K., "A finite difference method for the numerical computation of discontinuous solutions of the equations of fluid dynamics." Mat.Sb.(N.S.)47(1959),271-;also Cornell Aeronautical Laboratory transl..

[6] Harten, A., Lax, P.D. & Van Leer, B., "On upstream differencing and Godunov-type schemes for hyperbolic conservation laws." SIAM Review 25 (1983) 35-61.

[7] Hemker, P.W., "Defect correction and higher order schemes for the multi grid solution of the steady Euler equations." In this volume.

[8] Hemker, P.W. & Spekreijse, S.P., "Multigrid solution of the Steady Euler Equations." In: Advances in Multi-Grid Methods (D.Braess, W.Hackbusch and U.Trottenberg eds) Proceedings Oberwolfach Meeting, Dec. 1984, Notes on Numerical Fluid Dynamics, Vol.11, Vieweg, Braunschweig, 1985.

[9] Hemker, P.W. & Spekreijse, S.P., "Multiple Grid and Osher's Scheme for the Efficient Solution of the the Steady Euler Equations." Report NM-8507, CWI, Amsterdam, 1985.

[10] Mulder, W.A. "Multigrid Relaxation for the Euler equations." To appear in: J. Comp. Phys. 1985.

[11] Osher, S & Solomon, F., "Upwind difference schemes for hyperbolic systems of conservation laws." Math. Comp. 38 (1982) 339-374.

[12] Roe, P.L., "Approximate Riemann solvers, parameter vectors and difference schemes." J. Comp. Phys. 43 (1981) 357-372.

[13] Steger, J.L. & Warming, R.F., "Flux vector splitting of the inviscid gasdynamics equations with applications to finite difference methods." J. Comp. Phys. 40 (1981) 263-293.

[14] Sweby, P.K. "High resolution schemes using flux limiters for hyperbolic conservation laws", SIAM J.Numer.Anal. 21 (1984) 995-1011.

[15] Van Leer, B., "Flux-vector splitting for the Euler equations." In: Procs. 8th Intern. Conf. on numerical methods in fluid dynamics, Aachen, June, 1982. Lecture Notes in Physics 170, Springer Verlag.

[16] Van Leer, B., "Upwind difference methods for aerodynamic problems governed by the Euler equations." Report 84-23, Dept. Math. & Inf., Delft Univ. Techn., 1984.

[17] Van Leer, B., "Towards the ultimate conservative difference scheme.2. Monotonicity and conservation combined in a second order scheme." J.Comp.Phys.14,361-370(1974).

[18] Van Leer, B. & Mulder, W.A., "Relaxation methods for hyperbolic equations." Report 84-20, Dept. Math. & Inf., Delft Univ. Techn., 1984.

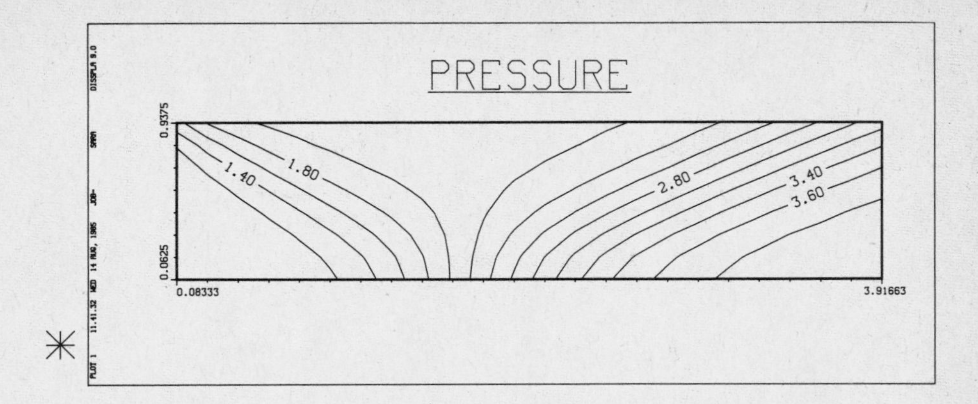

Figure 1a.

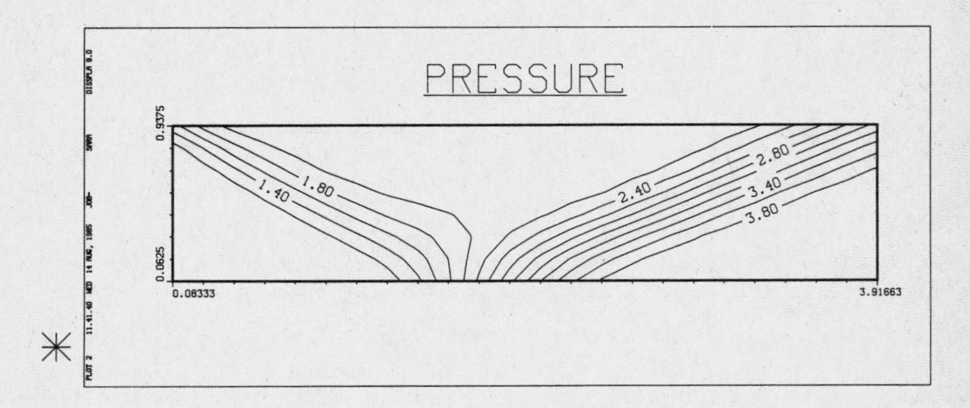

Figure 1b.

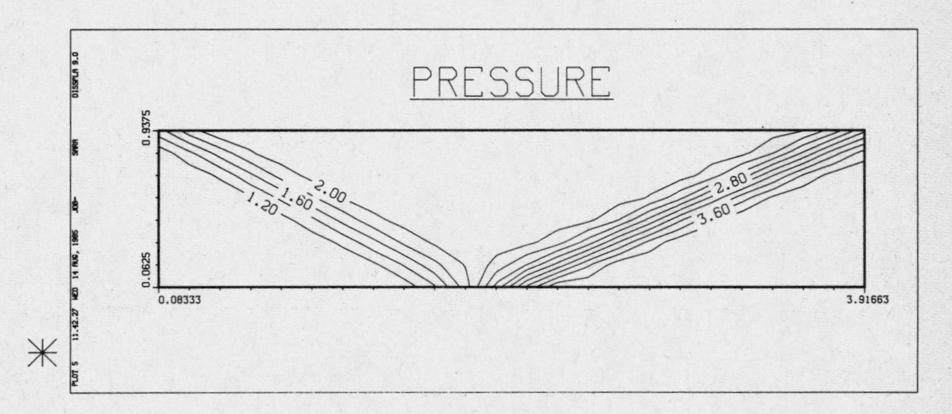

Figure 1c.

Pressure contours of an oblique shock on a 8×24 mesh, obtained by the first order upwind scheme and after 1 and 4 DeC iteration steps.

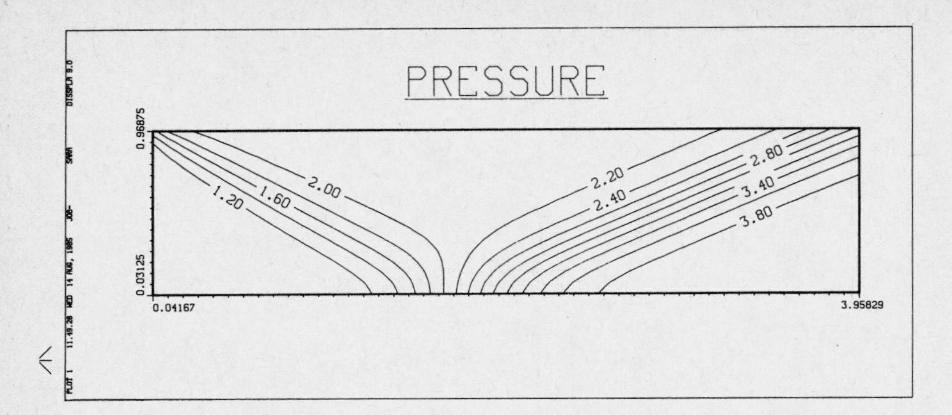

Figure 2a.

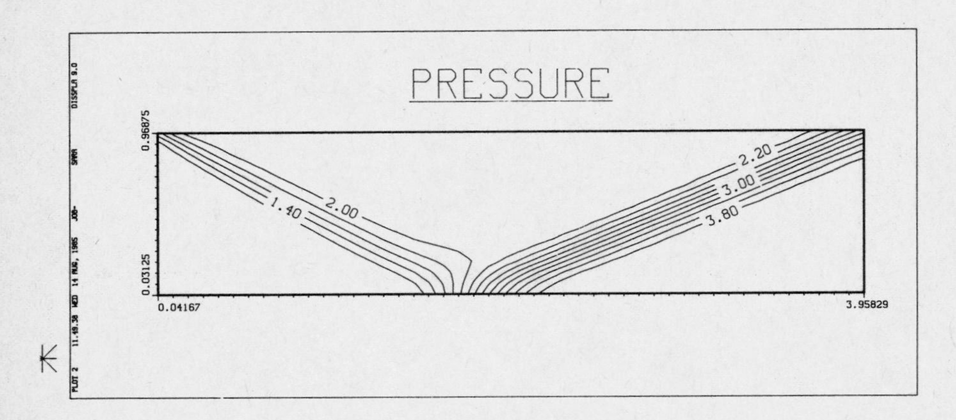

Figure 2b.

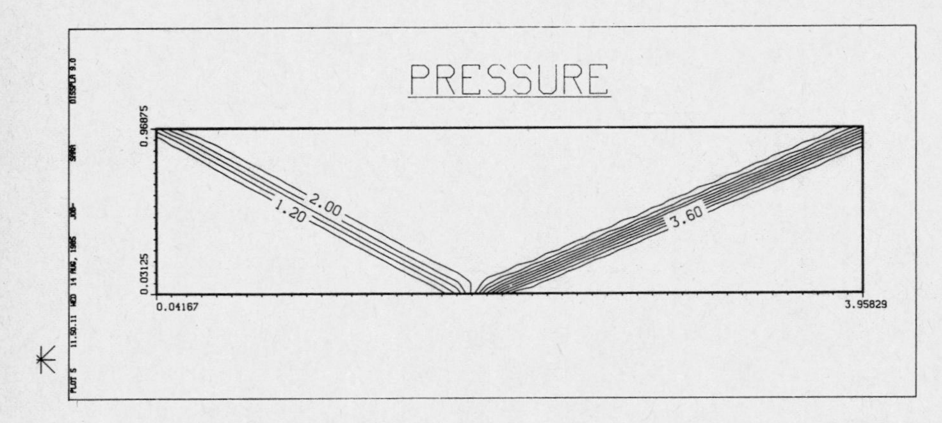

Figure 2c.

Pressure contours on a 16×48 mesh.

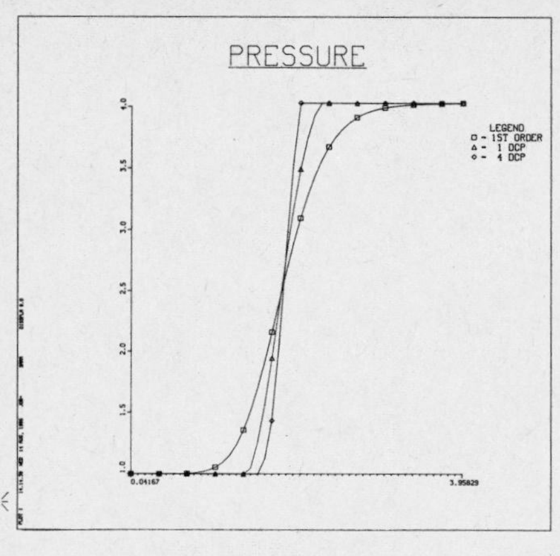

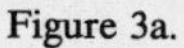

Figure 3a.

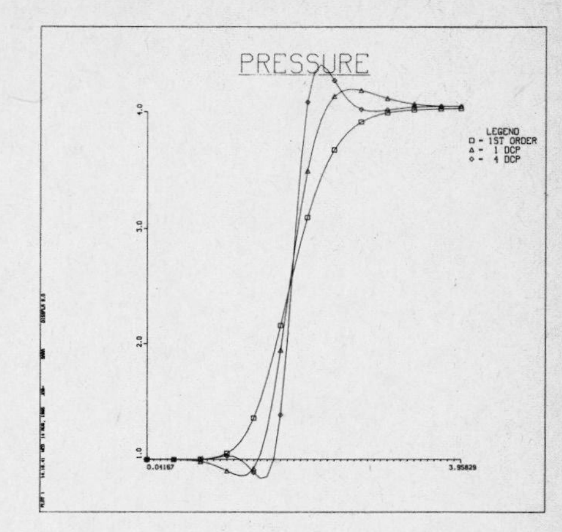

Figure 3b.

Pressure profiles computed at the surface of the flat plate, using the first order scheme and after 1 and 4 DeC iteration steps. Figure 3a has been obtained with a limiter, figure 3b without.

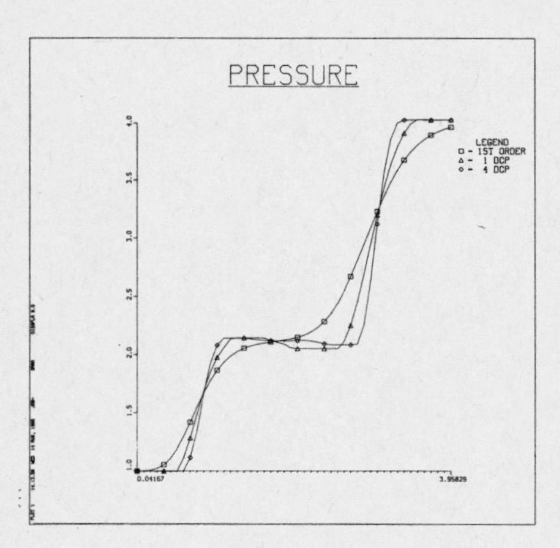

Figure 4a.

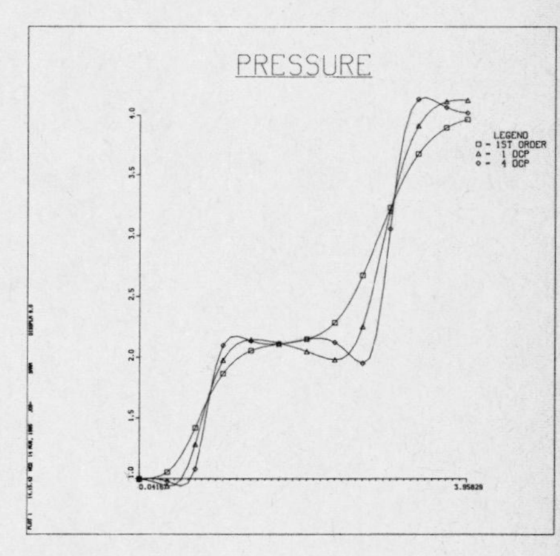

Figure 4b.

Pressure profiles at $y = 0.5$.

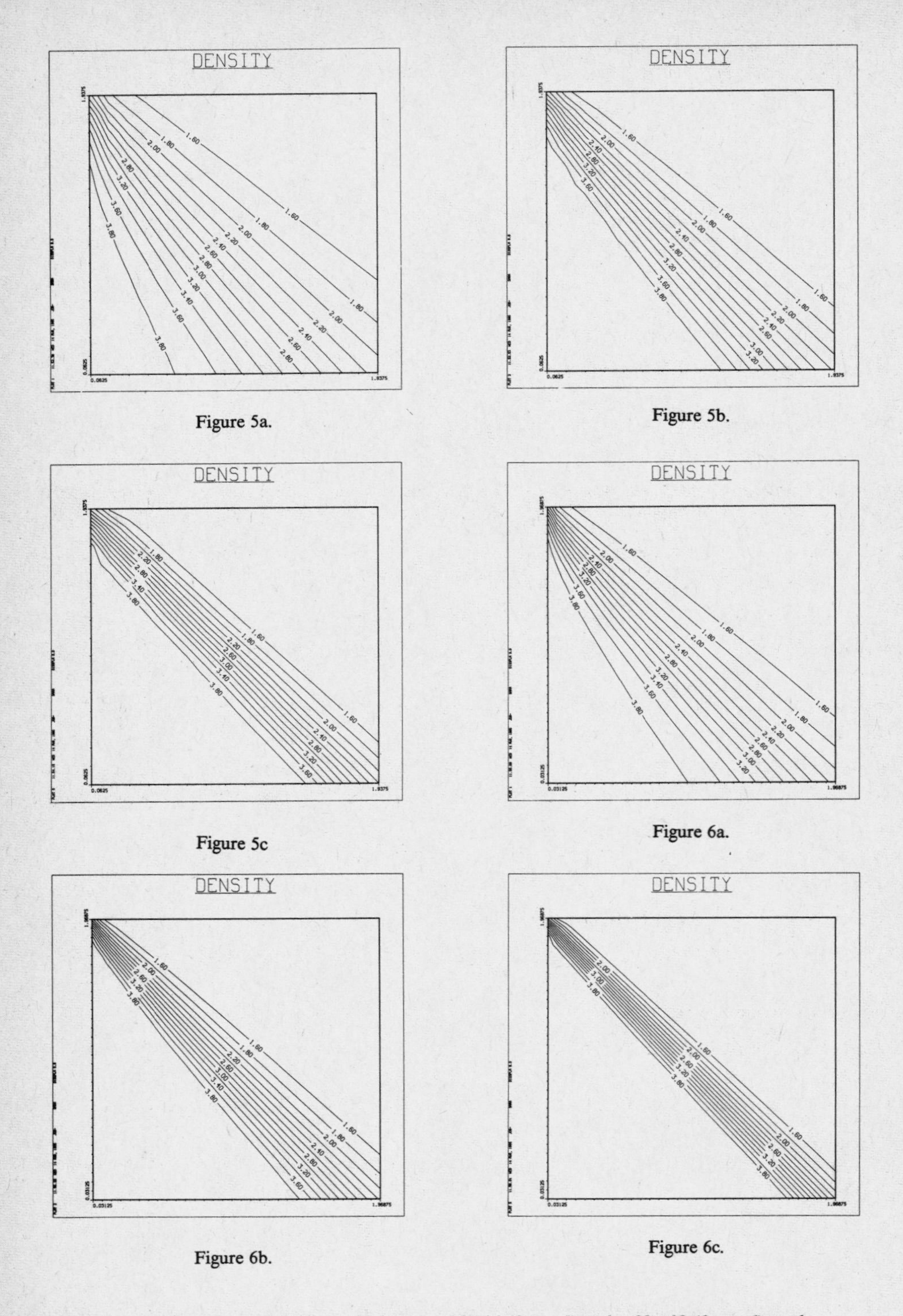

Figure 5a.

Figure 5b.

Figure 5c

Figure 6a.

Figure 6b.

Figure 6c.

Density contours of a contact discontinuity on a 16×16 (figure 5) and a 32×32 (figure 6) mesh, obtained by the first order scheme and after 1 and 4 DeC iteration steps.

MULTIGRID METHOD IN SUBSPACE AND DOMAIN PARTITIONING IN THE DISCRETE SOLUTION OF ELLIPTIC PROBLEMS

Panayot Vassilevski
Bulgarian Academy of Sciences, Institute of Mathematics
1090 Sofia, P.O.Box 373, Bulgaria

In solving elliptic boundary value problems it is very natural to partition the original region into a number of subregions where independent, say problems with zero Dirichlet data, are to be solved and a problem on the separator lines (surfaces) remains to be solved, which compensates the difference between the normal derivatives of the solutions of these subproblems. (The last problem, in its discrete variant is called sometimes capacitance matrix equation, see e.g., [6 - 7]). This strategy is very attractive when in the subregions there exist fast elliptic solvers or available software which is wellsuited to such regions and subproblems. In these cases it is not necessary (and not preferable) to compute the capacitance matrix explicitely, since it is sufficient to compute the product matrix x vector in an appropriate fast way. This strategy has been developed in [11] (see also [12]) [6 - 7], [3] and [15] . Related is also [8] . Multigrid method in domain partitioning is considered in [10] . There a Schwartz alternating algorithm is utilized in order to obtain a parallelized version of an algorithm for solving discrete problems on composite meshes from [16].

1. Model Problem Analysis

Our consideration is related to the following model problem on a T - shaped region (see fig. 1) for Poisson's equation

$$\Delta u = f \quad \text{in } D, \; f \in H^{-1}(D), \qquad u|_{\partial D} = 0 . \tag{1.1}$$

As domain partitioning technique is wellsuited to parallel computations we derive here a straightforward approximation of the capacitance equation on the separator boundary $\Gamma = \Gamma_2$ (see fig. 1) . We consider a problem and a region D, which can be split into the form $D = D_1 \cup \Gamma \cup D_2$, and in D_i the variables can be separated. Such a straightforward approximation can be achieved without having in mind any special approximations in the subregions D_i, $i = 1, 2$.

The next question is what a method can be used for solving the corresponding linear algebraic problem. One way is to look for an appropriate pre-conditioner for the Pre-conditioned CG method ([5]) in order to obtain an optimal iterative method with respect to the number of operations. Such a strategy is studied in [11 - 12], [6 - 7], [3], [15]. Another approach is to apply the general concept of the MG method with an appropriate choice of its parameters - smoothing procedure, prolongation and restriction operators. This is very useful in the case when it is not clear what a preconditioner can be chosen for solving the capacitance equation. Our recommendation is to use in MG method the procedure generated by a few steps of the CG method as a smoothing procedure (as in [2]) and naturally constructed prolongation and restriction operators. In our model problem analysis this strategy is described and studied below. The variable coefficients case is studied in §2.

In the formulation of the capacitance matrix problem we need the following useful definition

Definition 1 (Poincare - Steklov operators, [13])

Consider the following problem:

Find $u \in H_0^1(\Omega, \Gamma)$, such that

$$\Delta u = 0 \text{ in } \Omega ,$$

$$\frac{d}{dn_1} u = w \text{ on } \Gamma, \ \Gamma \subset d\Omega, \ w \in (H^{\frac{1}{2}}_{o,o}(\Gamma))^*.$$

The operator $S : (H^{\frac{1}{2}}_{o,o}(\Gamma))^* \longrightarrow H^{\frac{1}{2}}_{o,o}(\Gamma)$, defined by $Sw = u|_\Gamma$ is called Poincare - Steklov operator.

By definition

$$H_0^1(\Omega, \Gamma) = \{ u \in H^1(\Omega) \text{ with vanishing trace on } d\Omega \setminus \Gamma \} ;$$

$$H^{\frac{1}{2}}_{o,o}(\Gamma) = \{ v \in L_2(\Gamma): \ \bar{v} \in H^{\frac{1}{2}}(d\Omega), \text{ where } \bar{v} = 0 \text{ on } d\Omega \setminus \Gamma \text{ and } \bar{v} = v \text{ on } \Gamma \}$$

$$\overset{[14]}{=} \{ v \in H^{\frac{1}{2}}(\Gamma): \ \|r^{-\frac{1}{2}} v\|_{L_2(\Gamma)} < \infty , \text{ r is the distance to the boundary of } \Gamma \} .$$

Further we shall need the following spaces

Definition 2

$$H^{-\frac{1}{2}+s}(\Gamma) = (H^{\frac{1}{2}-s}(\Gamma))^*, \ 0 < s < \tfrac{1}{2} \ \text{ and } \ H^{-\frac{1}{2}}(\Gamma) = (H^{\frac{1}{2}}_{o,o}(\Gamma))^* .$$

Domain Partitioning Technique

i) Solve the auxiliary problems

$$(1.2)\qquad \Delta u^{(i)} = f \text{ in } D_i, \quad u^{(i)} = 0 \text{ on } d D_i, \; i = 1,2$$

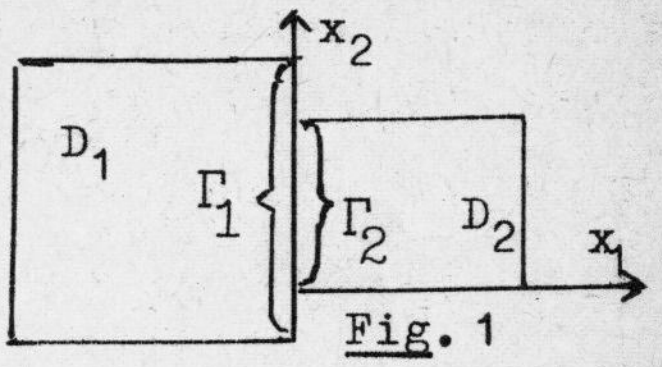

and determine

$$(1.3)\qquad v = -\sum_{i=1}^{2} \left(\frac{d}{dn_{D_i}} u^{(i)}\right)\Big|_{\Gamma}$$

Fig. 1

(n_{D_i} is the outward unit normal vector with respect to $d D_i$). It is not hard to be shown that $v \in (H^{\frac{1}{2}}_{o,o}(\Gamma))^*$ using Green's formula ;

ii) Let $\overset{o}{u} \in H^{\frac{1}{2}}_{o,o}(\Gamma)$ be the solution of the problem (the capacitance equation)

$$(1.4)\qquad (J^* S_1^{-1} J + S_2^{-1}) \overset{o}{u} = v, \; (v \text{ from } (1.3)), \; v \in (H^{\frac{1}{2}}_{o,o}(\Gamma))^*).$$

S_i are the Poincare - Steklov operators corresponding to the regions D_i and boundaries Γ_i, $i = 1, 2$, respectively; J is a prolongation operator defined by $J w = w$ on Γ_2 and $J w = 0$ on $\Gamma_1 \setminus \Gamma_2$. J is a continuous mapping from $H^{\frac{1}{2}}_{o,o}(\Gamma_2)$ on $H^{\frac{1}{2}}_{o,o}(\Gamma_1)$.

The main result is

<u>Proposition</u> 1 $u|_{\Gamma_2} = \overset{o}{u}$, where u is the solution of (1.1) and $\overset{o}{u}$ - the solution of (1.4) .

It can easily be derived using Green's formula.

<u>Regularity Property of Problem (1.4)</u>

<u>Proposition</u> 2 The following a priori estimate holds true

$$(1.5)\qquad \|\overset{o}{u}\|_{H^1_o(\Gamma_2)} \leq C \|v\|_{L_2(\Gamma_2)} .$$

(For a proof see part 2.)

<u>Properties of Poincare - Steklov Operators</u>, [13]

<u>Proposition</u> 3 Each operator S (see <u>definition</u> 1) is selfadjoint and positive definite and if Ω is the rectangle $(0, A_1) \times (0, A_2)$ and $\Gamma = \{0\} \times (0, A_2)$ the eigenvalues of S can be found analytically, i.e.

$$\lambda_k(S) = (\text{th}(k\Pi A_1/A_2))/(k\Pi/A_2), \quad k = 1,2,\ldots$$

The case of axi - symmetric Laplacian is considered in [19] .

<u>Discretization of the Problem (1.4)</u>

Let us introduce the following uniform grids Γ_{i,h_i} on Γ_i, $i = 1,2$ respectively. Then the following discrete problem can be put forward

$$(1.6)\qquad ((J^{h_1}_{h_2})^* S^{-1}_{1,h_1} J^{h_1}_{h_2} + S^{-1}_{2,h_2})\, y = v_{h_2} ,$$

where to each Poincare - Steklov operator, operating on the space of functions, defined on (0 , b) the following discrete counterpart is defined $(S_h^{-1}y)(x) = \sum_{kh<b} y_k/\lambda_k(S)\sin(\frac{k\Pi x}{b})$, $x \in \Gamma_h$ an uniform grid on (0 , b) with meshsize h ; $\lambda_k(S)$ are the eigenvalues of S ;

(1.7) $y_k = \frac{2h}{b} \sum_{x\in\Gamma_h} y(x)\sin(\frac{k\Pi x}{b})$, $kh<b$ (discrete Fourier coefficients of y) ;

Our operators S^{-1}_{i,h_i} , i = 1, 2 are constructed in the described way. We have used also

$$(J^{h_1}_{h_2}y)(x) = \int_{-1}^{1} \hat{y}^{h_2}(x+th_1)\, s(t)\, dt, \quad \hat{y}^{h_2}(x) = \sum_{x_{h_2}\in\Gamma_{h_2}} y(x_{h_2})\, s(\frac{x - x_{h_2}}{h_2}), \quad x\in R^1$$

and $s(t) = \max(0 , 1 - |t|)$, $t \in R^1$.

As usual we assume $h_1 \cong h_2$ and denote $h = \max(h_1 , h_2)$ and by X_{h_1,h_2} the matrix of the problem (1.6) - the capacitance matrix.

Properties of the Capacitance Matrix [17]

Proposition 4

i) The condition number of X_{h_1,h_2} grows linearly with h^{-1};

ii) $S^{-1}_{2,h_2} \leq X_{h_1,h_2} \leq C\, S^{-1}_{2,h_2}$;

iii) The product $X_{h_1,h_2}\, y$ can be computed using FFT (for appropriate h_1 and h_2) for an amount of ops (arithmetic operations), proportional to $h^{-1}\log h^{-1}$.

Error Estimation [17 - 18]

We compare the solution y of (1.6) with the following discrete function

$$\overset{o}{u}_{h_2}(x) = \sum_{kh_2<b_2} \overset{o}{u}_k \sin(\frac{k\Pi x}{b_2}) , \; x\in\Gamma_{h_2} \quad (b_2 \text{ is the lenght of } \Gamma_2),$$

$$\overset{o}{u}_k = \frac{2}{b_2}\int_\Gamma \overset{o}{u}(x)\sin(\frac{k\Pi x_2}{b_2})\, dx_2, \; k = 1,2,\ldots$$

The following convergence result holds true

Theorem 1

(1.8) $$\|\overset{o}{u}_{h_2} - y\|_{H^{\frac12}_{o,o,h_2}(\Gamma_{2,h_2})} \leq C\, h^s \|\overset{o}{u}\|_{H^{\frac12+s}(\Gamma_2)} , \quad 0 < s \leq \tfrac12 .$$

(In the case s = 0 the discrete problem is also convergent in the same norm). By definition $\|y\|_{H^{\frac12}_{o,o,h_2}(\Gamma_{h_2})} = (\sum_{kh_2<b_2} k\, y_k^2)^{\frac12}$. (For y_k see (1.7))

Restriction and Prolongation Operators

Let Γ_h and Γ_k be two uniform grids on the interval (0 , b) and consider the following operator

(1.9) $$(I_h^k y)(x) = \sum_{l<\min(b/h,\,b/k)} y_l \sin\left(\frac{l\Pi x}{b}\right), \quad x \in \Gamma_k$$

and $y_l = \frac{2h}{b} \sum_{x\in\Gamma_h} y(x)\sin\left(\frac{l\Pi x}{b}\right), \quad lh<b.$

If $h<k$ I_h^k is called a restriction operator while in the case $h>k$ a prolongation one.

Two-Grid Iteration Method

Let us now concentrate on the TGM for solving $\chi y = v$, i.e. let us have grids on two levels $\Gamma_{1,h_{1,s}}$ on Γ_1, $\Gamma_{2,h_{2,s}}$ on Γ_2, $s=1,2$ ($h_{i,1}>h_{i,2}$, $i=1,2$) and the corresponding discrete problems $X_s y_s = v_s$, $s=1,2$. Let us denote $I_{h_{2,2}}^{h_{2,1}}$ by I_2^1 and $I_{h_{2,1}}^{h_{2,2}}$ by I_1^2 (see (1.9)).

The following expression determines the two-grid iteration matrix (e.g. [9], [2]) $M^{TGM} = (X_2^{-1} - I_1^2 X_1^{-1} I_2^1) X_2 S_t$, where the matrix (operator) S_t is called a smoother, which performs t smoothing iterations. We consider here the following cases of S_t:

i) $S_t = S^t$ and $S = I_2 - h_{22}/C_X X_2$ (modified Jacobi-like iteration);

ii) $S_t = p_t(X_2)$ and the polynomial $p_t(x) = \prod_{i=1}^{t} (1 - r_i^{-1}x)$ is defined by t steps of the CG method (cf. [2]).

The convergence of TGM will follow from the general convergence proofs (e.g. [9], [2]) if an approximation property is established.

Approximation Property [20]

<u>Theorem</u> 2 The following approximation property for the operators X_s holds true

(1.10) $$\left\| X_2^{-1} - I_1^2 X_1^{-1} I_2^1 \right\|_{L_2 \longrightarrow B} \leq C_A h_{2,1}^{\frac{1}{2}} .$$

L_2 is the discrete l_2 space of gridfunctions, defined on $\Gamma_{h_{2,2}}$ and B is $H^{\frac{1}{2}}_{o,o,h_{2,2}}(\Gamma_{h_{2,2}})$.

The proof follows from <u>theorem</u> 1 (error estimate (1.8)), <u>proposition</u> 2 (regularity property) and definition (1.9).

Smoothing Property

A smoothing property is called the following estimate [9]

(1.11) $$\| X_2 S_t \|_{B \longrightarrow L_2} \leq C_S \eta(t) h_{22}^{-\frac{1}{2}}, \text{ where } \eta(t) \rightarrow 0, \ t \rightarrow \infty .$$

In our particular cases the estimate (1.11) holds true for $\eta(t) = t^{-\frac{1}{2}}$. The proof in the case i) - Jacobi-like iteration is an obvious modification of the proof in [9]. In the case ii) it follows from the estimate of convergence of the CG method (see e.g., <u>theorem</u>

5.1 in [4]). As a corollary the following main result can be formulated

Theorem 3 MG method is an optimal iterative method for solving the problem (1.6) (MG method in subspace), which requires $O(h^{-1}\log h^{-1})$ ops (arithmetic operations) per iteration.

Above we have considered only parallel dissections of the region. The case of perpendicular dissections is also of interest. For example let us consider a partitioning as on fig. 2 . A problem arises for computation the Poincare - Steklov operator, defined by

$$\begin{aligned}
&-\Delta u = 0 \text{ in } D = (0, A_1) \times (0, A_2),\\
(1.12)\qquad &\frac{d}{dn}u = w \in H^{-\frac{1}{2}}(\Gamma) \text{ on } \Gamma = \Gamma_1 \cup \Gamma_2,\ \Gamma_1 = (0, b_1) \times \{0\},\\
&\qquad \Gamma_2 = \{0\} \times (0, b_2),\ b_i < A_i,\ i = 1, 2,\\
&u \in H_0^1(D, \Gamma),
\end{aligned}$$

i.e. $Sw = u|_\Gamma$.

First we shall derive a splitting of S^{-1}, which is constructed by the following sequence of problems

$$\begin{aligned}
&-\Delta u^1 = 0 \text{ in } D_1 = (0, b_1) \times (0, A_2),\\
(1.13)\qquad &\frac{d}{dn}u^1 = C \text{ on } \Gamma_1,\ u^1 = v_2 \text{ on } \Gamma_2,\\
&u^1 \in H_0^1(D_1, \Gamma).
\end{aligned}$$

The task is to determine $w^1 = \frac{d}{dn}u^1$ on $\Gamma_{2,1} = \{b_1\} \times (0, A_2)$, $(w^1 \in H^{-\frac{1}{2}}(\Gamma_{2,1}))$, $v_1^1 = u^1|_{\Gamma_1}$ and $\frac{d}{dn}u^1$ on Γ_2 ;

$$\begin{aligned}
&-\Delta u^{(2)} = 0 \text{ in } D,\\
(1.14)\qquad &[u^{(2)}] = 0,\ \left[\frac{d}{dn}u^{(2)}\right] = -w^1 \text{ on } \Gamma_{2,1},\\
&u^{(2)} \in H_0^1(D).
\end{aligned}$$

Here we have to determine $w_1^{(2)} = \frac{d}{dn}u^{(2)}$ on Γ_1 and $w_2^{(2)} = \frac{d}{dn}u^{(2)}$ on Γ_2. If we denote $v_{2,1} = u^{(2)}$ on $\Gamma_{2,1}$ the problem for $v_{2,1}$ is

$$(1.14)' \qquad (\bar{S}_{1,1}^{-1} + \bar{S}_{1,2}^{-1})\, v_{2,1} = -w^1 ,$$

where $\bar{S}_{1,i}$ are the Poincare - Steklov operators for the regions D_i and boundary $\Gamma_{2,1}$ $(D_2 = (b_1, A_1) \times (0, A_2))$. Having $v_{2,1}$ problem (1.14) is formulated in the rectangle D_1

$$\begin{aligned}
&-\Delta u^{(2)} = 0 \text{ in } D_1,\\
(1.14)''\qquad &u^{(2)} = v_{2,1} \text{ on } \Gamma_{2,1},\\
&u^{(2)} \in H_0^1(D_1, \Gamma_{2,1}).
\end{aligned}$$

Up to here we have found the solution of the following problem

$$\begin{aligned}
&-\Delta \overset{1}{u} = 0 \text{ in } D,\\
&\overset{1}{u} = v_1^1 \text{ on } \Gamma_1,\ \overset{1}{u} = v_2 \text{ on } \Gamma_2,\\
&\overset{1}{u} \in H_0^1(D, \Gamma).
\end{aligned}$$

$(\overset{1}{u} = u^1 + u^{(2)}$ in D_1 and $\overset{1}{u} = u^{(2)}$ in $D_2)$. And if at last we solve the problem

$$(1.15)\qquad \begin{aligned} &-\Delta \overset{o}{u} = 0 \text{ in } D,\\ &\overset{o}{u} = v_1 - v_1^1 \text{ on } \Gamma_1, (v_1 - v_1^1 \in H^{\frac{1}{2}}_{o,o}(\Gamma_1)),\\ &\overset{o}{u} \in H^1_o(D, \Gamma_1) \end{aligned}$$

also determining $\frac{d}{dn}\overset{o}{u}\big|_{\Gamma}$ the problem of evaluation $S^{-1}v$ is solved, i.e. the following proposition is valid

<u>Proposition</u> 5 Let S be the Poincare - Steklov operator defined by (1.12) Then the value $S^{-1}v$ can be computed by solving each of the problems (1.13), (1.14)', (1.14)'' and (1.15) in a rectangle, setting $v_1 = v\big|_{\Gamma_1}$ and $v_2 = v\big|_{\Gamma_2}$. (The problem (1.14)' has **been considered in** the previous part, studying parallel dissections of regions).

Summarizing, we can conclude that it is necessary to derive procedures for solving the following basic problems in a rectangle :

$$(1.16)\qquad \begin{aligned} &-\Delta u = 0 \text{ in } D,\\ &u\big|_{\Gamma_2} \text{ given (in an appropriate space class)}\\ &\frac{d}{dn}u\big|_{\Gamma_1} = 0,\\ &u \in H^1_o(D, \Gamma). \end{aligned}$$

The task is to determine $\frac{d}{dn}u$ on Γ_2 **and** on $\Gamma_{2,1}$, and $u\big|_{\Gamma_1}$ $(\Gamma_{2,1} = \{A_1\} \times (0, A_2))$

$$(1.17)\qquad \begin{aligned} &-\Delta u = 0 \text{ in } D,\\ &u\big|_{\Gamma_2} \in H^{\frac{1}{2}}_{o,o}(\Gamma_2) \text{ given},\\ &u \in H^1_o(D, \Gamma_2), \end{aligned}$$

and to determine $\frac{d}{dn}u$ on Γ

Let us consider, for example, problem (1.16) in a discrete situation. I.e. we have an uniform grid $\Gamma_{2,h}$ on Γ_2 and a gridfunction v_h defined on it, which vanishes on the toppoint of Γ_2. Expand v_h in its discrete Fourier series

$$(1.18)\qquad v_h(x) = \sum_{k h \leq A_2} v_k \cos\left(\frac{k-\frac{1}{2}}{A_2}\Pi x\right), \quad x \in \Gamma_{2,h},$$

where $$v_k = \frac{2}{A_2} \sum_{x \in \Gamma_{2,h}} h\, v_h(x) \cos\left(\frac{k-\frac{1}{2}}{A_2}\Pi x\right), \quad k = 1, 2, \dots, A_2/h.$$

Let us also have an uniform grid on Γ_1, Γ_{1,h_1}. Then construct the following function

$$u^h(x) = \sum_{k h \leq A_2} v_k \cos\left(\frac{k-\frac{1}{2}}{A_2}\Pi x_2\right) m_k(x_1),$$

where $m_k(x_1) = \mathrm{sh}\left(\frac{k-\frac{1}{2}}{A_2}\Pi (A_1 - x_1)\right) / \mathrm{sh}\left(\frac{k-\frac{1}{2}}{A_2}\Pi A_1\right)$. We have

$$\begin{aligned} &-\Delta u^h = 0 \text{ in } D,\\ &u^h = v^h = \text{the trigonometric polynomial defined by (1.18) for } x \in \Gamma_2,\\ &\frac{d}{dn}u^h = 0 \text{ on } \Gamma_1, \end{aligned}$$

We approximate $v_1 = u\big|_{\Gamma_1}$ by the following gridfunction

(1.19) $$v_{1,h_1}(x_1) = \sum_{k h \leq A_2} v_k \, s_k(h_1)\, m_k(x_1), \quad (0, x_1) \in \Gamma_{1,h_1},$$

where $s_k(h_1)$ are some smoothing coefficients, e.g., $s_k = (\frac{\operatorname{sh} t}{t})^2$, $t = (\frac{k-\frac{1}{2}}{2A_1}\Pi h_1)$.

There arises a problem for fast evaluation of the sums (1.19) for $x_1 = i h_1$, $i = 1,2,\ldots, A_1/h_1 - 1$. For a given tolerance $\varepsilon > 0$ and a fixed $x_1 = i_o h_1$ the sum is computed approximately summing over all k between 1 and $O(\frac{\log(h_1\varepsilon)^{-1}}{i_o h_1})$. Then it is seen that the total computational work is proportional to $h_1^{-1}(\log^2 h_1 + \log h_1 \log \varepsilon)$. Computation of such sums arises also in the approximation of problem (1.17). (Such a method for computing of sums like (1.19) is used in [1]).

The approximation of $w_2 = \frac{d}{dn}u$ on Γ_2 is given by the following grid function

$$w_{2,h}(x_2) = \sum_{k h \leq A_2} v_k/\lambda_k \cos(\frac{k-\frac{1}{2}}{A_2}\Pi x_2), \quad x_2 \in \Gamma_{2,h},$$

where

$$\lambda_k = \operatorname{th}(\frac{k-\frac{1}{2}}{A_2}\Pi A_1) / (\frac{k-\frac{1}{2}}{A_2}\Pi), \quad k \geq 1.$$

We summarize

Proposition 6 The computation of the product capacitance matrix x vector (gridfunction) in the case of perpendicular dissections of regions requires $O(h^{-1}\log^2 h)$ ops.

For each reasonable prolongation and restriction operators we can modify theorem 3 as follows

Theorem 4 MG method in subspace is an optimal iterative method for solving the capacitance equation, arising in perpendicular dissections of the region, which requires $O(h^{-1}\log^2 h)$ ops per iteration.

Further details will be given in [20] (see also the next part of the paper).

2. The Case of Variable Coefficients Operator

Our results will be established for conforming, Lagrangian finite element approximations of Dirichlet problem for second order, self - adjoint elliptic problems in plane regions,

$$Lu = -\sum_{i,j} \frac{\partial}{\partial x_i} a_{ij}(x) \frac{\partial}{\partial x_j} u + a_o(x)\, u = f(x), \quad x \in D,$$

$$u = 0 \text{ on } \Gamma = dD.$$

The operator L has real, sufficiently smooth coefficients, $a_{ij}(x) = a_{ji}(x)$ and the bilinear form

$$a(u, v) = \int_D \sum_{i,j} a_{ij}(x) \frac{\partial}{\partial x_i} u \frac{\partial}{\partial x_j} v + a_o(x) uv\, dx$$

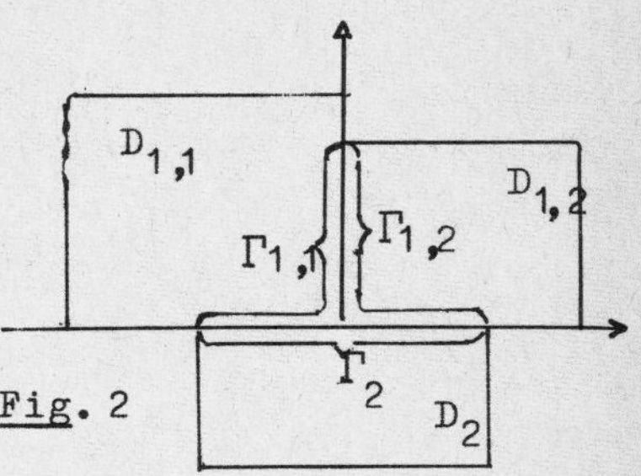

Fig. 2

satisfies

$$\frac{1}{C} \|u\|^2_{H^1(D)} \leq a(u, u)$$

and

$$a(u, v) \leq C \|u\|_{H^1(D)} \|v\|_{H^1(D)} .$$

Let us partition the region D into three parts (see fig. 2). This is not the general case of partitioning, but it contains the main difficulties which arise in such a partitioning. Thus we have (see fig.2)

$$D = D_{1,1} \cup D_{1,2} \cup D_2 \cup \Gamma_1 \cup \Gamma_2$$

and denote

$$\Gamma_{1,1} = dD_{1,1} \cap \Gamma, \quad \Gamma_{1,2} = dD_{1,2} \cap \Gamma, \quad \Gamma = \Gamma_1 \cup \Gamma_2, \quad \Gamma_1 = \Gamma_{1,1} \cap \Gamma_{1,2} .$$

Using the domain partitioning technique the following problem on the separator lines arises to be solved

$$(2.1) \qquad \begin{aligned} &L \overset{o}{u} = 0 \text{ in } D_{1,1} \cup D_{1,2} \cup D_2, \\ &\left[\overset{o}{u}\right] = 0, \left[\frac{d}{dn_L} \overset{o}{u}\right] = w \text{ on } \Gamma, \quad \overset{o}{u} \in H^1_o(D), \end{aligned}$$

where w belongs to the space $(H^{\frac{1}{2}}_{o,o}(\Gamma))^*$. The brackets denote the jump of the functions on the boundary Γ. In variational formulation it becomes

$$(2.2) \qquad a(\overset{o}{u}, v) = \int_\Gamma w\, v\, ds, \ \forall v \in H^1_o(D),$$

and we wish to find u on Γ .

This is our problem in subspace. As in general there do not exist exact representations for Poincare - Steklov operators in this case, it follows that the problem (2.2) (the capacitance equation) cannot be approximated in a straightforward fashion. That is why we use finite element method.

We describe below the domain partitioning technique for the discrete problem. First, solve the auxiliary problems

$$a_{D_{1,i}}(\overset{o}{u}{}^{(1,i)}_h, v) = \int_{D_{1,i}} f\, v\, dx, \forall v \in V_h \cap H^1_o(D_{1,i}), \ \overset{o}{u}{}^{(1,i)}_h \in V_h \cap H^1_o, \quad i = 1, 2,$$

and

$$a_{D_2}(\overset{o}{u}{}^{(2)}_h, v) = \int_{D_2} f\, v\, dx, \ \forall v \in V_h \cap H^1_o(D_2), \ \overset{o}{u}{}^{(2)}_h \in V_h \cap H^1_o(D_2) .$$

Let $w|_{\Gamma_2} = -\frac{d}{dn_L}\overset{o}{u}{}_h^{(1,i)}(x) - \frac{d}{dn_L}\overset{o}{u}{}_h^{(2)}(x)$ if $x \in \Gamma_{1,i}$, $i = 1,2$ and $w = -\left[\frac{d}{dn_L}\overset{o}{u}{}_h^{(1,i)}\right]$ on Γ_1. (n denotes the unit outward normal vector with respect to the boundary of each considered region).

Afterwards solve the problem

$$(2.3) \qquad a(\overset{o}{u}_h, v) = \int_\Gamma w\, v\, ds, \quad \forall v \in V_h \cap H_o^1(D), \quad \overset{o}{u}_h \in V_h \cap H_o^1(D).$$

This can be rewritten in the following blockmatrix form

$$K\,x = \begin{bmatrix} K^{(1,1)} & 0 & 0 & K_{13}^{(1,1)} \\ & K^{(1,2)} & 0 & K_{13}^{(1,2)} \\ & & K^{(2)} & K_{23}^{(2)} \\ \text{symmetric} & & & K^{(3)} \end{bmatrix} \begin{bmatrix} x_{1,1} \\ x_{1,2} \\ x_2 \\ x_3 \end{bmatrix} = \begin{bmatrix} 0 \\ 0 \\ 0 \\ b_3 \end{bmatrix}.$$

Here $K^{(1,1)}$, $K^{(1,2)}$, $K^{(2)}$, $K^{(3)}$ are the stiffness submatrices which represent the couplings between pairs of nodes in $D_{1,1}$, $D_{1,2}$, D_2 and Γ, respectively, while $K_{13}^{(1,1)}$, $K_{13}^{(1,2)}$, $K_{23}^{(2)}$ represent couplings between $D_{1,1}$ and Γ, $D_{1,2}$ and Γ, and D_2 and Γ, respectively.

Our attempts are to derive a method for solving the following reduced system of linear algebraic equations

$$\begin{aligned} X\,x_3 = (K^{(3)} - (K_{13}^{(1,1)})^*(K^{(1,1)})^{-1}K_{13}^{(1,1)} - (K_{13}^{(1,2)})^*(K^{(1,2)})^{-1}(K_{13}^{(1,2)}) \\ (2.4) \qquad - (K_{23}^{(2)})^*(K^{(2)})^{-1}(K_{23}^{(2)})\,)\,x_3 = b_3 . \end{aligned}$$

Computation of the product Xx_3 for a given vector x_3 can be done at the expense of solving homogeneous problems with nonzero Dirichlet data (x_3) in each subregion. By assumption, the finite element problems on the subregions are solved exactly. We therefore concentrate our study on the solution of the problem (2.4). Even if the original problem has a very smooth boundary, its partition into subregions allows us to consider regions with corners. For our further study we shall assume that the partition has some regularity properties. We assume that there exists $0 < s \le \frac{1}{2}$ such that:

i) for the problems

$$a_{D_i}(u, v) = 0, \ \forall v \in H_0^1(D_i, \Gamma_2), \ u \in H_o^1(D_i, \Gamma_2),$$
$$u|_{\Gamma_2} = z \in H_{o,o}^{\frac{1}{2}+s}(\Gamma_2), \ i = 1,2,$$

the estimates

$$\frac{1}{C}\left\|\frac{d}{dn_L}u\right\|_{H^{-\frac{1}{2}+s}(\Gamma_2)} \le \|u\|_{H^{1+s}(D_i)} \le C\,\|z\|_{H_{o,o}^{\frac{1}{2}+s}(\Gamma_2)}, \ i = 1,2$$

hold true;

ii) for the problems

$$a_{D_{1,i}}(u\,,v) = 0,\ \forall v\in H_o^1(D_{1,i},\Gamma_{1,i}),\ u\in H_o^1(D_{1,i},\Gamma_{1,i}),$$

$$u\big|_{\Gamma_1} = z\in H_{o,\cdot}^{\frac{1}{2}+s}(\Gamma_1)=\left\{v\in H^{\frac{1}{2}+s}(\Gamma_1)\ :\ v=0 \text{ on } \bar{\Gamma}_1\cap d\,D\right\}$$

the estimates

$$\frac{1}{C}\left\|\frac{d}{dn_L}u\right\|_{H^{-\frac{1}{2}+s}(\Gamma_1)}\leq \|u\|_{H^{1+s}(D_{1,i})}\leq C\,\|z\|_{H_{o,\cdot}^{\frac{1}{2}+s}(\Gamma_1\ )}\ ,\ i=1,2$$

have to be true ;

iii) for the problems

$$a_{D_i}(u\,,v)=\int_{\Gamma_2} w\ v\ ds\ ,\ \forall v\in H_o^1(D_i\,,\Gamma_2),\ u\in H_o^1(D_i\,,\Gamma_2)\ ,\ i=1,2$$

the estimates

$$\frac{1}{C}\left\|u\big|_{\Gamma_2}\right\|_{H_{o,\rho}^{\frac{1}{2}+s}(\Gamma_2)}\leq \|u\|_{H^{1+s}(D_i)}\leq C\,\|w\|_{H^{-\frac{1}{2}+s}(\Gamma_2)},\ i=1,2$$

are to be valid ;

iv) for the problems

$$a_{D_{1,i}}(u\,,v)=\int_{\Gamma_{1,i}} w\ v\ ds\ ,\ \forall v\in H_o^1(D_{1,i},\Gamma_{1,i}),\ u\in H_o^1(D_{1,i},\Gamma_{1,i}),\ i=1,2$$

the estimates

$$\frac{1}{C}\|u\|_{H_{o,\rho}^{\frac{1}{2}+s}(\Gamma_{1,i})}\leq \|u\|_{H^{1+s}(D_{1,i})}\leq C\,\|w\|_{H^{-\frac{1}{2}+s}(\Gamma_{1,i})}\ ,\ i=1,2$$

hold true .

Using the assumptions i) - iv) the following regularity property can be proven

Theorem 2.1 For the substructured problem (2.2) (or (2.4) in matrix terminology) the following a priori estimate is valid

$$\|\overset{o}{u}\|_{H_{o,\rho}^{\frac{1}{2}+s}(\Gamma)}\leq C\,\|w\|_{H^{-\frac{1}{2}+s}(\Gamma)}$$

for the given $s\in(0\,,\frac{1}{2}]$.

Proof: Solve the following auxiliary problem

$$a_{D_1}(u^1,v)=\int_{\Gamma_1} w\ v\ ds,\ \ \forall v\in H_o^1(D_1\,,\Gamma_2),\ u^1\in H_o^1(D_1\,,\Gamma_2).$$

This problem can be rewritten in the following operator form

$$(2.5)\qquad (\overset{o}{S}{}^{-1}_{1,1}+\overset{o}{S}{}^{-1}_{1,2})\,u_1^1 = w,$$

where $u_1^1=u^1$ on Γ_1 and the operators $\overset{o}{S}_{1,i}$ are defined as follows

$$\overset{o}{S}_{1,i}\,w=u_{1,i} \text{ on } \Gamma_1,$$

where $u_{1,i}$ are $H_o^1(D_{1,i},\Gamma_{1,i})$ solutions of the problems

$$a_{D_{1,i}}(u_{1,i}, v) = \int_{\Gamma_1} w\, v\, ds, \ \forall v \in H_o^1(D_{1,i}, \Gamma_{1,i}),\ i = 1, 2.$$

Using the estimates from ii) and iv) the following is obtained

$$\|\overset{o}{S}_{1,i}\bar{w}\|_{H_{o,\cdot}^{\frac{1}{2}+s}(\Gamma_1)} \le c\, \|\bar{w}\|_{H^{-\frac{1}{2}+s}(\Gamma_1)} \quad \text{and} \quad \|\overset{o}{S}{}_{1,i}^{-1} v\|_{H^{-\frac{1}{2}+s}(\Gamma_1)} \le c\, \|v\|_{H_{o,\cdot}^{\frac{1}{2}+s}(\Gamma_1)}.$$

Therefore, the following can be easily derived

$(I + \overset{o}{S}{}_{1,1}^{-1}\, \overset{o}{S}_{1,2})\, \overset{o}{S}{}_{1,2}^{-1}$ is a continuous operator from $H_{o,\cdot}^{\frac{1}{2}+s}(\Gamma_1)$ onto $H^{-\frac{1}{2}+s}(\Gamma_1)$

Then by the theorem of Banach (for the inverse operator)

(and by ii) and iv)) the estimates

$$\|u^1|_{\Gamma_{1,i}}\|_{H_{o,o}^{\frac{1}{2}+s}(\Gamma_{1,i})} \le c\, \|w\|_{H^{-\frac{1}{2}+s}(\Gamma_1)}, \quad i = 1, 2 \tag{2.6}$$

hold true.

Further, solve the problem

$$(S_1^{-1} + S_2^{-1})\, u_2^{(2)} = w_2 - S_2^{-1} u_2^1, \quad (u_2^1 = u^1 \ \text{ on } \Gamma_2 \cap \Gamma_{1,i},\ i = 1,2),$$

where S_i, $i = 1, 2$ are the Poincare - Steklov operators for the regions D_i and boundary Γ_2, respectively (in definition 1 replace Δ by L). Using the same argument as for the problem (2.5) and having in mind the estimates in i) and iii) we obtain

$$\begin{aligned} \|u_2^{(2)}\|_{H_{o,o}^{\frac{1}{2}+s}(\Gamma_2)} &\le c\, (\|w_2\|_{H^{-\frac{1}{2}+s}(\Gamma_2)} + \|u_2^1\|_{H_{o,o}^{\frac{1}{2}+s}(\Gamma_2)}) \qquad \text{(by (2.6))} \\ &\le c\, (\|w_2\|_{H^{-\frac{1}{2}+s}(\Gamma_2)} + \sum_{i=1}^{2} \|w_{1,i}\|_{H^{-\frac{1}{2}+s}(\Gamma_{1,i})}) \end{aligned} \tag{2.7}$$

($w_1 = w$ on Γ_1, $w_2 = w$ on Γ_2, $w_{1,i} = w$ on $\Gamma_{1,i}$, $i = 1, 2$).

And at last solve the problem

$$a_{D_1}(u_1, v) = 0, \ \forall v \in H_o^1(D_1, \Gamma_2),\ u_1 \in H_o^1(D_1, \Gamma_2),$$
$$u_1 = u_2^{(2)} \text{ on } \Gamma_2.$$

Using iii) we obtain

$$\|u_1\|_{H^{1+s}(D_1)} \le c\, \|u_2^{(2)}\|_{H_{o,o}^{\frac{1}{2}+s}(\Gamma_2)} \tag{2.8}$$

and by the usual trace theorem it follows

$$\|u_1|_{\Gamma_1}\|_{H^{\frac{1}{2}+s}(\Gamma_1)} \le c\, \|u_1\|_{H^{1+s}(D_1)}. \tag{2.9}$$

Form now the following function

$$u_2^1 + u_2^{(2)} \ \text{ on } \Gamma_2,$$
$$u_1^1 + u_1|_{\Gamma_1} \ \text{ on } \Gamma_1,$$

which, by construction, is equal to $\overset{o}{u}$ on Γ ($\overset{o}{u}$ the solution of (2.2)).

Using the estimates (2.6) - (2.9) the proof is completed.

For the problem

$$a_D(\overset{o}{u}_h, v) = \int_\Gamma w\, v\, ds, \quad \forall v \in V_h \cap H_o^1(D), \ \overset{o}{u}_h \in V_h \cap H_o^1(D),$$

the following error estimate holds true

$$\| P_h \overset{o}{u} - \overset{o}{u}_h \|_{H^1(D)} \leq C\, h^s\, \| \overset{o}{u} \|_{H^{1+s}(D)},$$

($P_h \overset{o}{u}$ is the interpolant of $\overset{o}{u}$ in V_h). As a consequence (and by theorem 2.1)

$$(2.10) \qquad \| P_h \overset{o}{u}|_\Gamma - \overset{o}{u}_h|_\Gamma \|_{H^{\frac{1}{2}}_{o,o}(\Gamma)} \leq C\, h^s\, \| \overset{o}{u} \|_{H^{1+s}(D)} \leq C\, h^s\, \| w \|_{H^{-\frac{1}{2}+s}(\Gamma)}$$

an error estimate in subspace is obtained.

Consider now two discrete problems on nested grids $\Gamma_h \supset \Gamma_k$

$$X_h\, x_3^{(h)} = b_3 \quad \text{and} \quad X_k\, x_3^{(k)} = b_3^{(k)}.$$

For each reasonable restriction (I_h^k) and prolongation (I_k^h) operators, $b_3^{(k)} = I_h^k b_3$, using the error estimate in subspase (2.10) the following approximation property can be established

$$(2.11) \qquad \| X_h^{-1} - I_k^h X_k^{-1} I_h^k \| \leq C_A\, h^s .$$

This is enough to prove the convergence (independent of h) of TGM applied to our substructured problem basing on the general convergence proofs in [9] . The choice of the norms in (2.11) is considered there also.

Finally we mention that a sequence of approximations on nested grids is required and therefore for the problems in the subregions such approximations are needed also. It is clear that if these subproblems are solved iteratively (in computation the product $X x_3$), e.g. by an MG method, it is not necessary to solve them almost exactly but only a few iterations are sufficient.

REFERENCES

1. Bachvalov, N.S. and M.Yu. Orechov : On fast manners for solving Poisson's equation. USSR J. Comput. Math. and Math. Physics, 22(1982) 1386 - 1392. In Russian.
2. Bank, R.E. and C.C. Douglas: Sharp estimates for multigrid rates of convergence with general smoothing and acceleration. Research Report YALEU/DCS/RR-277.
3. Bjorstad, P.E. and O.B. Widlund : Iterative methods for the solution of elliptic problems on regions partitioned into substructures. Technical Report # 136, 1984, Computer Science Department, New York University.

4. Chandra, R.: Conjugate gradient methods for PDEs. Research Report # 129, Department of Computer Science, Yale University.
5. Concus, P., G.H. Golub and D.P. O'Leary: A generalized conjugate gradient method for the numerical solution of elliptic partial differential equations, in: J.R. Bunch and D.J. Rose, eds., Sparse Matrix Computations, pp. 309 - 332, New York, Academic Press, 1976.
6. Dryja, M.: A capacitance matrix method for Dirichlet problem on polygon region. Numer. Math. 39(1982), 51-64.
7. Dryja, M.: A finite element - capacitance matrix method for elliptic problems on regions partitioned into subregions. Numer. Math. 44(1984), 153 - 168.
8. Glowinski, R., J. Periaux, and Q.V. Dinh: Domain decomposition methods for nonlinear problems in fluid dynamics. Rapports de Recherche N° 147, 1982, INRIA Centre de Rocquencourt.
9. Hackbusch, W.: Multi - grid convergence theory, in: W. Hackbusch and U. Trottenberg, eds., Multigrid methods. Lect. Notes Math. # 960, Springer-Verlag, Berlin - Heidelberg - New York, 1982, pp. 177 - 219.
10. Hackbusch, W.: Local defect correction method and domain decomposition technique. Computing, Suppl. 5(1984), 89 - 113.
11. Kuznetsov, Yu.A.: Block - relaxation methods in subspace, their optimization and applications, in: Variational - difference methods in mathematical physics, Sibirian Branch of USSR Acad. Sci., Computing center, Novosibirsk, 1978, pp. 178 - 212. In Russian.
12. Kuznetsov, Yu.A.: Computational methods in subspaces, in: Computational processes and systems, vol. 2, Nauka, Moscow, 1985, pp. 265 - 350. In Russian.
13. Lebedev, V.I., and V.I. Agoshkov: Poincare - Steklov operators and their applications in Analysis. Department of Computational Math., USSR Acad. Sci., Moscow, 1983. In Russian.
14. Lions, J.L. and E. Magenes: Nonhomogeneous boundary value problems and applications, I, Springer, NY, 1972.
15. Nepomnyashtchich, S.V.: On application of bordering method to mixed boundary value problem for elliptic equations and on network norms in $W_2^{1/2}(S)$. Preprint 106, 1984, Computing Center, Sibirian Branch of USSR Acad. Sci., Novosibirsk. In Russian.
16. Stuben, K. and U. Trottenberg: Multigrid methods: Fundamental algorithms, model problem alalysis and applications, in: W. Hackbusch and U. Trottenberg, eds., Multigrid methods. Lect. Notes Math. # 960 Springer - Verlag, Berlin - Heidelberg - New York, 1982, pp. 1-176.
17. Vassilevski, P.S.: Numerical solution of Poisson's equation on regions partitioned into substructures, I: Derivation and properties of the discrete problem. Compt.Rend.Bulg.Acad.Sci. 39(1986), in press
18. Vassilevski, P.S.: Numerical solution of Poisson's equation on regions partitioned into substructures, II: Convergence of the method Compt.Rend.Bulg.Acad.Sci. 39(1986) N°3, in press.
19. Vassilevski, P.S.: Numerical solution of axially symmetric Poisson equation on regions partitioned into substructures. Proceedings of the Third Conference on Differential equations and applications, held at Russe, June 30 - July 6, 1985, Bulgaria. In press.
20. Vassilevski, P.S.: Numerical solution of elliptic BVPs on regions partitioned into substructures. In preparation.

FAS MULTIGRID EMPLOYING ILU/SIP SMOOTHING: A ROBUST FAST SOLVER FOR 3D TRANSONIC POTENTIAL FLOW*

A.J. van der Wees
Informatics Division, National Aerospace Laboratory NLR
P.O. Box 153, 8300 AD EMMELOORD, The Netherlands

Abstract

ILU/SIP is shown to be a very efficient and robust smoothing algorithm within the multigrid method for the solution of elliptic (subsonic) and mixed elliptic/hyperbolic (transonic) potential flow problems. The algorithm is fully implicit and fairly insensitive to large grid cell aspect ratios; in the hyperbolic regions of the flow the algorithm is uniformly stable.
It will also be shown that the best multigrid performance for 3D problems is obtained by performing a priori grid optimization, for which requirements will be derived.
With an optimized grid, the method is fast for engineering applications and the physical quantities of interest are determined with great efficiency.

1. Introduction

The multigrid (MG) method has been demonstrated to be an effective tool for the construction of fast solution methods for 3D transonic potential flow. This type of flow is governed by a second order partial differential equation of mixed elliptic/hyperbolic type which allows the occurrence of shock waves, i.e. weak solutions.
The most commonly used smoothing relaxation algorithm for this type of problem is Successive Line (Over-) Relaxation (SLOR), which has been shown to be successful within the multigrid method both in two and three dimensions [1,2]. However, considerable multigrid convergence deterioration has been encountered for SLOR when used on grids with high aspect ratios; moreover, SLOR is not stable for all local flow directions in the supersonic (hyperbolic) regions of the flow.

* Part of this research has been performed under contract with the Netherlands Agency for Aerospace Programs (NIVR).

In 2D applications, apart from SLOR, also the Alternative Direction Implicit (ADI) and the Approximate Factorization (AF) algorithm have been shown to be successful within the multigrid method [3,4]. These algorithms employ (a sequence of) relaxation parameters which must be optimized for use within the multigrid method [5]. In 3D applications, so far no successes have been reported for AF or ADI within the multigrid method, but AF has been widely used in the monogrid version [6]. When using AF, special attention has to be paid to the relaxation of boundary conditions specified for auxiliary intermediate variables [7].
In the search for faster smoothing algorithms, both the Incomplete Lower Upper (ILU) decomposition and the Strongly Implicit Procedure (SIP) have been investigated for transonic flow [8,9]. In previous work of the present author [10,11] these two algorithms have been merged into one algorithm ILU/SIP. The resulting algorithm is fully implicit in all three coordinate directions and uniformly stable in the hyperbolic part of the flow. Hence, it is a very robust algorithm which has been shown to be at least twice as fast as SLOR within the multigrid method [11]. The algorithm does not employ (a sequence of) relaxation parameters; also no boundary conditions have to be specified for auxiliary variables. Nevertheless, the algorithm is costeffective, in the sense that it claims reasonable computer resources.

The MG-ILU/SIP method forms the basis of MATRICS (Multi-component Aircraft TRansonic Inviscid Computation System), that is at present being developed at NLR to calculate routinely the transonic flow around a complex aircraft configuration.
In the present paper, results will be shown for MG-ILU/SIP for the solution of transonic potential flow around realistic wings embedded in boundary conforming curvilinear grids.

2. Flow equation

Transonic potential flow is described by the mass conservation equation

$$\frac{\partial}{\partial x^i}(\rho u^i) = 0. \tag{1}$$

The velocity u^i is split in the (given) freestream velocity u^i_∞ and the perturbation velocity $\frac{\partial \varphi}{\partial x^i}$, viz.

$$u^i = u^i_\infty + \delta^{ij}\frac{\partial \varphi}{\partial x^j}. \tag{2}$$

The density ρ is given by

$$\rho = \{1 + \frac{\gamma-1}{2} M_\infty^2 (1 - u^i u_i)\}^{\frac{1}{\gamma-1}} . \tag{3}$$

The Mach number is defined as

$$M^2 = \frac{q^2}{a^2} \quad , \quad q^2 = u^i u_i, \tag{4}$$

where the speed of sound is given by

$$a^2 = \frac{1}{M_\infty^2} + \frac{\gamma-1}{2} (1 - u^i u_i). \tag{5}$$

Quasi-linearization of the flow equation (1) gives

$$(1 - M^2)\varphi_{ss} + \varphi_{mm} + \varphi_{nn} = 0, \tag{6}$$

where s is the stream direction and m and n are two mutually orthogonal directions normal to the stream direction. Equation (6) is elliptic in the subsonic part of the flow ($M < 1$), hyperbolic in the supersonic part of the flow ($M > 1$) and locally parabolic at sonic surfaces ($M = 1$).

3. Discretization by the finite volume (FV) method

The flow equation (section 2) is solved in a curvilinear computational space ξ^α, $\alpha = 1,2,3$. The grid is given by a set of volumes in the physical x^i-space, which are transformed into cubical computational cells in the ξ^α-space. Within each computational cell, the velocity disturbance potential φ and the coordinate values x^i are given in the cornerpoints of the cell. Each computational cell is considered as an isoparametric finite element, using trilinear interpolation in each cell to compute geometry and potential derivatives. The resulting scheme, presented in [12], is very compact, as it requires only one density computation per gridpoint.

In the supersonic regions of the flow, a bias has to be added to the discretization to suppress non-physical expansion shocks. This is done by upwinding the mass flux ρq against the flow direction. The scheme used is that of Osher, Hafez and Whitlow [13] and the shockpoint operator is that of Boerstoel [1].

4. Grid generation

In transonic wing calculations, curvilinear grids are used to discretize the physical space around the wing. These grids, which are boundary conforming to the wing, are often reasonably rectangular in planes perpendicular to the wing surface, but mostly skew in the spanwise direction due to the sweep of the wing (Figs. 1a, 1b). On the whole, grid skewness can be referred to as moderate. The grids have very fine, almost square, meshes near the wing surface (Fig. 1c) and are highly stretched towards the far-field boundaries (Figs. 1a, 1b). The grid generation introduces an artificial cut emanating from the trailing edge of the wing and the wing tip to infinity. At this boundary (artificial) continuity type boundary conditions will be specified.

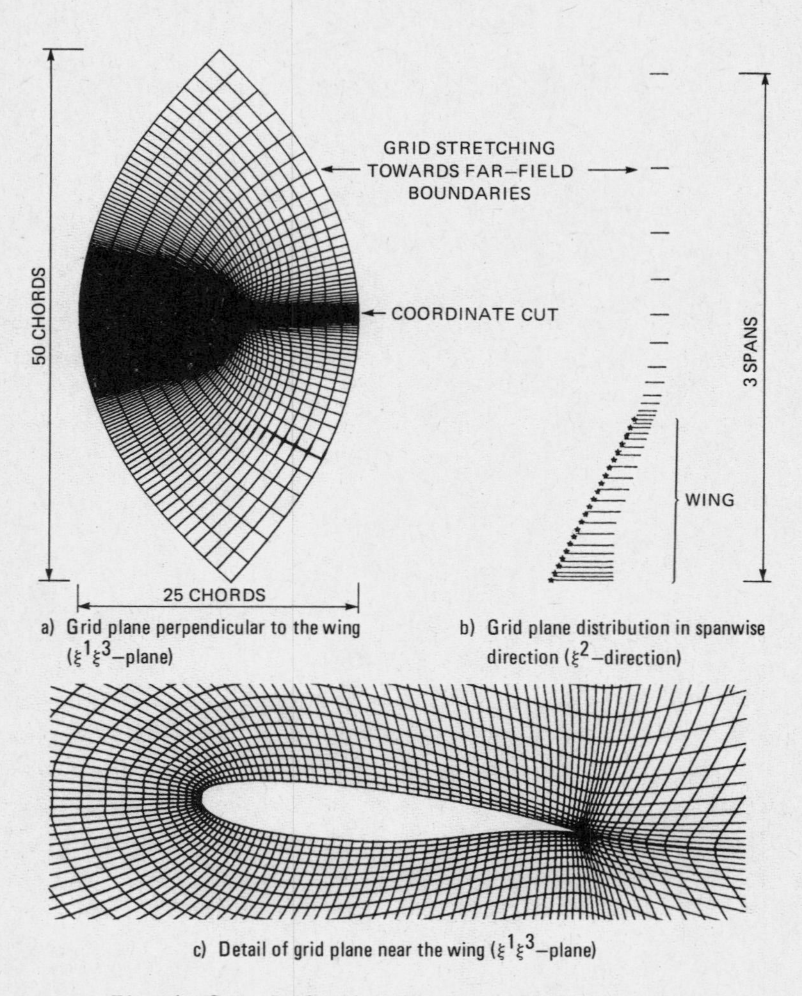

a) Grid plane perpendicular to the wing ($\xi^1\xi^3$-plane)

b) Grid plane distribution in spanwise direction (ξ^2-direction)

c) Detail of grid plane near the wing ($\xi^1\xi^3$-plane)

Fig. 1 Computational grid around DFVLR-F4-wing

5. Boundary conditions

On the wing surface and in the symmetry plane zero normal velocity is prescribed, which is a Neumann type boundary condition. On the far-field boundaries, except on the downstream boundary, the velocity perturbation potential φ is set to zero, which is a Dirichlet type boundary condition. On the downstream boundary the flow is required to follow the freestream direction, i.e. $\partial u/\partial s_\infty = 0$, which is once more a Neumann type boundary condition. Across the coordinate cut, introduced by the grid generation mapping, the continuity of the mass flux and the jump in potential are prescribed. The continuity of the mass flux can be looked upon as a special type of Neumann boundary condition.

All boundary conditions other than Dirichlet type boundary conditions are implemented using a so-called dummy gridpoint approach [10]. Adjacent to each face of the computational box a face of dummy gridpoints is defined where before each relaxation potentials are valued such that the boundary condition is satisfied at the face of the computational box.

6. Multigrid method

An excellent introduction to the multigrid method is the survey article by Stüben, Trottenberg [14], while the pioneering article by Brandt [15] has already reached the status of being an "evergreen". In the research presented in this paper the Full Approximation Scheme (FAS) of the multigrid method is chosen. The method used is described in more detail in [10,11] and solves the same equation

$$L[\varphi^k] = f^k \tag{7}$$

on a hierarchy of grids G^k, $k = 1,\ldots, N$. Here the (coarse) grid G^{k-1} is obtained from the (fine) grid G^k by deleting every other gridpoint. A restriction operator transfers variables from G^k to G^{k-1}, while a prolongation operator interpolates variables from G^{k-1} to G^k. The righthand side f^k equals f^N on the finest grid G^N, while it equals the difference between the coarse grid residual $L^{k-1}\varphi^{k-1}$ and the restricted fine grid residual $L^k\varphi^k$ on a coarse grid G^k, $k \leq N$ [10,11,14,15].

Within the multigrid method the restriction of Neumann type boundary conditions requires careful attention. As a consequence of the dummy gridpoint approach used, the Neumann boundary condition will not be satisfied exactly before each restriction. This so-called residual of the boundary condition, which can be interpreted as an inflow flux, must be restricted in the same way as the residual of the flow equation [10,11]. In doing so, no convergence problems were encountered near boundaries with prescribed Neumann boundary conditions.

7. Linearisation

The nonlinear equation $L[\varphi] = f$, section 2, is linearized in a straightforward manner by putting

$$\varphi = \phi + \Delta\varphi \quad , \quad \phi \text{ given,} \tag{8}$$

and subsequently deleting all terms of order $(\Delta\varphi)^2$ and smaller.
This leads to the equation

$$\tilde{L}[\phi]\Delta\varphi = f - L[\phi]. \tag{9}$$

The righthand side is the residual for $\varphi = \phi$ of the flow equation. The lefthand side describes an iteration scheme of a complex dense structure, which involves the grid-points $(i+p,j+q,k+r)$ where p, q and r range from -1 to 1 in the elliptic part of the flow and can be -2 or 2 in the hyperbolic part of the flow. Hence, in general there are 64 points in the scheme.
A sparser scheme can be obtained by substracting terms of order $\Delta\xi^i\Delta\xi^j$ from the lefthand side of eq. (9) by putting

$$(\Delta\varphi)_{i+p,j+q,k+r} = (\Delta\varphi)_{i+p,j,k} + (\Delta\varphi)_{i,j+q,k} + (\Delta\varphi)_{i,j,k+r} - 2(\Delta\varphi)_{i,j,k}. \tag{10}$$

This way an iteration scheme is obtained which involves terms of the type $(\Delta\varphi)_{i+p,j+q,k+r}$, where p, q and r still range from -2 to 2, but this time with the restriction that only one of them can be nonzero. Iteration schemes of such sparsity are amenable to application of the ILU/SIP relaxation scheme. The price paid is the deletion of cross-derivative terms in $\Delta\varphi$ from the iteration scheme. These terms are always present due to the linearisation of the density and in many cases also due to the skewness of the grid. Experience has shown, however, that the mathematical convergence properties of the iteration scheme are hardly damaged by this deletion.

8. ILU/SIP algorithm

An extensive description of ILU and SIP can be found in Meijerink and Van der Vorst [16] and in Stone [17] respectively. Here, only a brief description will be given.
Combination of equations (9), (10) leads to the matrix-vector equation

$$\tilde{A}[\phi]\Delta\varphi = f - A[\phi] \tag{11}$$

where the row in the lefthand side corresponding to the gridpoint (i,j,k) can be written as

$$\{\tilde{A}[\phi]\Delta\varphi\}_{i,j,k} = a^{pqr}(\Delta\varphi)_{i+p,j+q,k+r}\,, \tag{12}$$

$$p,q,r \in \{-2,\ -1,\ 0,\ 1,\ 2\}\,,\quad p \neq 0 \text{ or } q \neq 0 \text{ or } r \neq 0. \tag{13}$$

An iteration scheme to solve $A(\varphi) = f$ can be described as

$$A^{*}[\phi^{n}]\Delta\varphi^{n} = f - A[\phi^{n}]\,, \tag{14a}$$

$$\Delta\varphi^{n} = \phi^{n+1} - \phi^{n}\,, \tag{14b}$$

where the iteration matrix A^{*} is chosen easily invertible and favourably is a good approximation of the system matrix $\tilde{A}$. The error matrix B is defined by

$$A^{*} = \tilde{A} + B. \tag{15}$$

This results in the modified equation

$$\begin{aligned} B[\phi^{n}]\Delta t\, \overleftarrow{\partial}_{t}\, \phi^{n+1} &= f - \tilde{A}[\phi^{n+1}] + \tilde{A}[\phi^{n}] - A[\phi^{n}] \\ &= g[\phi^{n}] - \tilde{A}[\phi^{n+1}]. \end{aligned} \tag{16}$$

With both ILU and SIP, an incomplete lower/upper (Gauss) decomposition of the system-matrix $\tilde{A}$ is carried out. For each algorithm, this decomposition is performed using a prespecified sparse matrix pattern, here coinciding with the pattern of $\tilde{A}$. In carrying out the Gauss matrix decomposition process for the lower triangular part of $\tilde{A}$, nonzero entries will be generated outside this pattern. The treatment of these non-zero entries determines the form of the error matrix B that will be obtained. This treatment differs for ILU and SIP.
In case of ILU, the nonzero entries mentioned before are left untouched (in fact they need not even be computed) and hence $\{B\Delta\varphi\}_{i,j,k}$ has the form

$$\{B\Delta\varphi\}_{i,j,k} = b^{pqr}(\Delta\varphi)_{i+p,j+q,k+r}\,,\ p = 0 \text{ or } q = 0 \text{ or } r = 0, \tag{17}$$

corresponding to a modified equation of the form

$$\left(c + d\Delta\xi^{i}\frac{\partial}{\partial\xi^{i}} + e\Delta\xi^{i}\Delta\xi^{j}\frac{\partial}{\partial\xi^{i}}\frac{\partial}{\partial\xi^{j}} + \ldots\right)\Delta t\, \overleftarrow{\partial}_{t}\, \phi^{n+1} = g[\phi] - \tilde{A}[\phi^{n+1}]. \tag{18}$$

The structure of the matrices $\tilde{A}$ and B is sketched in figure 2 for the case that an incomplete Gauss decomposition is applied to the 7-point discretization of the Laplace operator.

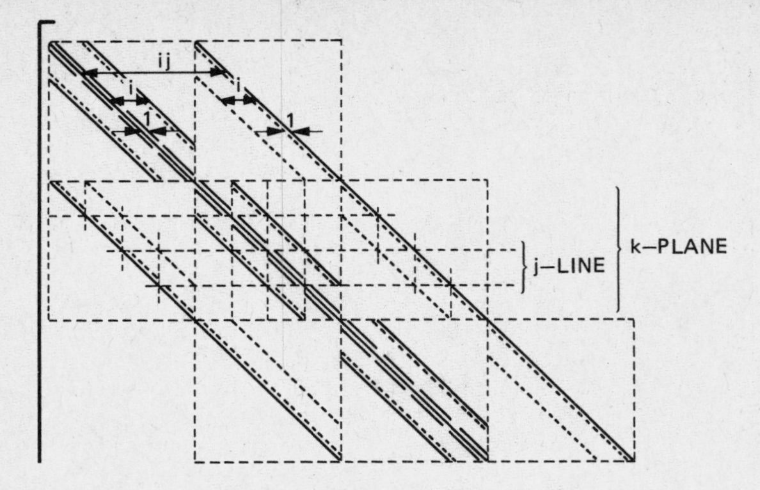

Fig. 2 Upper left corner of the patterns of the system matrix $\tilde{A}$ (drawn lines) and the error matrix B (dotted lines) for an ILU-decomposition of the 7-point discretization of the Laplace-operator on an i*j*k-grid

In case of SIP, all $\overleftarrow{\partial}_t \phi^{n+1}$ - and $\Delta\xi^i \frac{\partial}{\partial\xi^i} \overleftarrow{\partial}_t \phi^{n+1}$ - terms are annihilated in the modified equation (18) by making use of the first order Taylor expansion

$$(\Delta\varphi^n)_{i+p,j+q,k+r} \cong (\Delta\varphi^n)_{i+p,j,k} + (\Delta\varphi^n)_{i,j+q,k} + (\Delta\varphi^n)_{i,j,k+q} - 2(\Delta\varphi^n)_{i,j,k} \quad (19)$$

as follows. Each of the nonzero entries mentioned before is added to the off-diagonal entries a^{poo}, a^{oqo}, a^{oor} and twice substracted from the main diagonal entry a^{ooo} of $\tilde{A}$. All these entries are present in the pattern of $\tilde{A}$. The term $B\Delta\varphi^n$ now takes the form

$$\{B\Delta\varphi^n\}_{i,j,k} = b^{pqr} \{(\Delta\varphi^n)_{i+p,j+q,k+r} - (\Delta\varphi^n)_{i+p,j,k} - (\Delta\varphi^n)_{i,j+q,k} - (\Delta\varphi^n)_{i,j,k+r} + 2(\Delta\varphi^n)_{i,j,k}\} \quad (20)$$

and consequently the modified equation of SIP becomes (compare eq. (18)):

$$(e\ \Delta\xi^i\Delta\xi^j \frac{\partial}{\partial\xi^i}\frac{\partial}{\partial\xi^j} + \ldots)\Delta t\ \overleftarrow{\partial}_t\ \phi^{n+1} = g[\phi] - \tilde{A}[\phi^{n+1}]. \quad (21)$$

Usually, only a fraction α, $0 \leq \alpha \leq 1$ of $(\Delta\varphi^n)_{i+p,j+q,k+r}$ is approximated using equation (19). This will be denoted as SIP(α), whence SIP(0.) is identical to ILU. Therefore, whereas SIP is described here as a modification of classical ILU, ILU can also be looked upon as a special version of SIP. The family of algorithms described here will consequently be referred to as ILU/SIP(α). Usually, SIP is not described within the framework of ILU, but by a set of recursive formulas [9,17,18]. The explicit derivation of these formulas is cumbersome and is here formulated implicitly by implementing SIP as a modified form of classical ILU.

9. Properties and applications of ILU/SIP

Two important properties of the ILU/SIP algorithm are its full implicitness and the absence of a preferred sweep direction. The price paid is the necessity to store the entire upper triangular matrix U. The disk storage required is, however, well balanced against the disk storage required for the finite volume scheme. The total computation system is I/O-bound on the NLR Cyber 180-855, and hence more complex ILU algorithms are unattractive due to the extra I/O time required for the transfer of the denser upper triangular matrix U.
Multiple applications [19,20] have shown that use of ILU within the multigrid method has led to a very fast, robust and insensitive tool for the solution of a wide variety of 2D elliptic problems. An application of using SIP within the multigrid method for the solution of 2D transonic flow is given in reference 9. The applicability of MG-ILU/SIP to 3D subsonic and transonic potential flow has been shown by the present author in [10,11] for a simple non-lifting wing embedded in a rectangular grid. In the present paper the capabilities of MG-ILU/SIP will be shown for realistic wings embedded in curvilinear grids.

10. Local mode analysis

A local mode (Von Neumann) analysis, as described in [10,11], will be used in this paper to analyze the ILU/SIP smoothing properties for the Laplace equation, compare eq. (6),

$$\varphi_{\xi^1\xi^1} + \varphi_{\xi^2\xi^2} + \varphi_{\xi^3\xi^3} = 0 , \tag{22}$$

discretized using central differences on a rectangular grid with constant mesh sizes $\Delta\xi^1$, $\Delta\xi^2$ and $\Delta\xi^3$. This enables to analyze the influence of the two cell aspect ratios $\Delta\xi^i/\Delta\xi^j$, $i \neq j$.
An extensive analysis has shown that the convergence deterioration due to extreme aspect ratios is far more severe than the convergence degradation due to mesh skewness. Also the performance degradation due to severe grid stretching (variation in successive mesh ratios) is far less than the deterioration of solution accuracy due

to the same grid stretching. Hence, the analysis will be restricted here to variations in aspect ratio.

In references 10,11 expressions are derived for the computation of the smoothing factor (maximum high frequency reduction factor) $\bar{\rho}$ in case of ILU/SIP and SLOR. Here only the numerical results are given. In table 1 the smoothing factor $\bar{\rho}$ is given for a complete set of variations of $\Delta\xi^1$, $\Delta\xi^2$ and $\Delta\xi^3$. The numbers presented have been obtained for the frequency values $\pm \frac{2\pi}{K}$, K = 2, 3, 4, 8, 16, 32, 64, 128, 256, which are considered to be representative for the entire frequency domain. In table 1, SLOR-X, SLOR-Y and SLOR-Z are abbreviations for SLOR along ξ^1-, ξ^2- and ξ^3-lines respectively.

TABLE 1

Maximum reduction factors $\bar{\rho}$ of high frequency modes for various algorithms in case of the elliptic testproblem, equation (22)

$\Delta\xi^1$	$\Delta\xi^2$	$\Delta\xi^3$	SLOR-X	SLOR-Y	SLOR-Z	ILU	SIP(.7)	SIP
1.0000	1.0000	1.0000	.5000	.5000	.5000	.2195	.2132	.3798
.3162	1.0000	1.0000	.4982	.8462	.8462	.3609	.4134	.4463
1.0000	.3162	1.0000	.8462	.4982	.8462	.3609	.4134	.4463
1.0000	1.0000	.3162	.8462	.8462	.4982	.3609	.4134	.4463
3.1623	1.0000	1.0000	.9091	.8346	.8346	.7325	.7325	.7325
1.0000	3.1623	1.0000	.8346	.9091	.8346	.7325	.7325	.7325
1.0000	1.0000	3.1623	.8346	.8346	.9091	.7325	.7325	.7325
.3162	1.0000	3.1623	.8346	.9804	.9821	.8627	.8627	.8627
.3162	3.1623	1.0000	.8346	.9821	.9804	.8627	.8627	.8627
1.0000	.3162	3.1623	.9804	.8346	.9821	.8627	.8627	.8627
1.0000	3.1623	.3162	.9804	.9821	.8346	.8627	.8627	.8627
3.1623	.3162	1.0000	.9821	.8346	.9804	.8627	.8627	.8627
3.1623	1.0000	.3162	.9821	.9804	.8346	.8627	.8627	.8627

It can be verified from figure 1 that all combinations of $\Delta\xi^1$, $\Delta\xi^2$ and $\Delta\xi^3$ presented in table 1 are generally present in a realistic wing grid. The relative mesh sizes considered in table 1, i.e. $(\sqrt{10})^k$, k = -1, 0, 1, are however very moderate when compared to realistic cases. For example near the leading edge of the wing, $\Delta\xi^2$ can easily be over twenty times larger than $\Delta\xi^1$, $\Delta\xi^3$.

It can be concluded from table 1 that SLOR is very sensitive to a complete set of variations of mesh sizes. Even in the moderate case considered, an almost unacceptable value $\bar{\rho}$ = .98 is obtained if there are three mesh sizes of different order of magnitude in one computational cell. ILU/SIP is far less sensitive than SLOR. In the case of three different mesh sizes, ILU/SIP is even seven times faster than SLOR. However, the case of three different mesh sizes is obviously a very unfavourable one and should, if possible, be avoided. With the grids used in actual wing calculations, this is possible only by requiring that grids have the property $\Delta\xi^1 \cong \Delta\xi^3$. Smoothing rates $\bar{\rho}$ for that case are shown in figure 3.

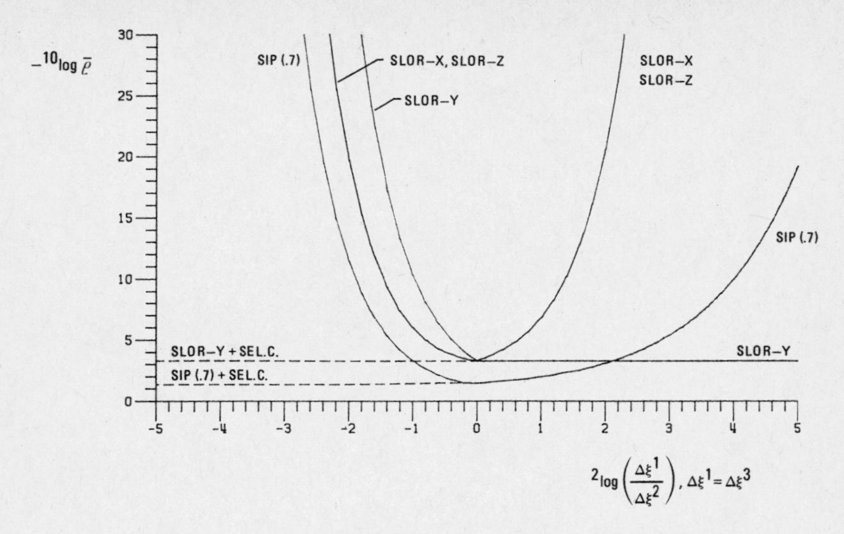

Fig. 3 Maximum reduction factors $\bar{\rho}$ of high frequency modes for various algorithms for the elliptic testproblem eq. (22), in the special case $\Delta\xi^1 = \Delta\xi^3$

It should be noted that $-{}^{10}\log \bar{\rho}$ is the number of iterations necessary to reduce the error in the high frequency modes by one order of magnitude. The curves "X + Sel.C." refer to algorithms X that are used within the multigrid method while applying selective coarsening, viz. in $\xi^1\xi^3$-planes only. A close examination of the damping of the high frequency modes reveals that the bad smoothing rate $\bar{\rho}$ for $\Delta\xi^1 << \Delta\xi^2$ is caused by a mode which has very low frequency in ξ^1- and ξ^3-direction, but very high frequency in ξ^2-direction. When there is no coarsening applied in the ξ^2-direction, this mode can equally well be smoothed on all coarser grids, and hence a "smaller spectrum" need be smoothed on the fine grid. Thus, the smoothing factor $\bar{\rho}$ improves at the cost of a less efficient coarse grid correction process, as the coarser grids will now contain more gridpoints.

The following conclusions can be drawn from figure 3:

- In general, the convergence deterioration is more severe for $\Delta\xi^1 << \Delta\xi^2$ (occurring near the wing, Fig. 1) than for $\Delta\xi^1 >> \Delta\xi^2$ (occurring in the far-field, Fig. 1).
- In general, ILU/SIP is less sensitive to the aspect ratio $\Delta\xi^1/\Delta\xi^2$, $\Delta\xi^1 = \Delta\xi^3$ than SLOR, except when $\Delta\xi^1 >> \Delta\xi^2$, where SLOR-Y does a perfect job.
- In case $\Delta\xi^1 << \Delta\xi^2$, mesh halving, i.e. doubling the number of grid points in the ξ^2-direction, can pay off in terms of total computing cost.
- Selective coarsening can lead to a convergence rate $\bar{\rho}$ which is independent of $\Delta\xi^1/\Delta\xi^2 < 1$ if $\Delta\xi^1 = \Delta\xi^3$.

It can be shown that in the general case coarsening should be applied only in the direction where $\Delta\xi^i$ is shortest. In realistic cases, where the whole variety of mesh sizes mentioned in table 1 is present, this obviously leads to a very complex multigrid method.

11. Stability analysis SIP in hyperbolic regions

The modified equation of an algorithm to solve equation (6) can be described as

$$\alpha_1 \Delta t \overleftarrow{\partial}_t \phi - \Delta t \overleftarrow{\partial}_t (\alpha_2 \Delta s \partial_s \phi + \alpha_3 \Delta m \partial_m \phi + \alpha_4 \Delta n \partial_n \phi) + \alpha_s \Delta t \overleftarrow{\partial}_t (\ldots)$$

$$+ (1 - M^2)\phi_{ss} + \phi_{mm} + \phi_{nn} = 0. \tag{23}$$

Jameson [21] has shown that a dominant ϕ_t-term (i.e. $\alpha_i = 0$, $i \neq 1$ or $\Delta m, \Delta n, \Delta s \to 0$) leads to instability in hyperbolic regions of the flow ($M^2 > 1$). Instead, "sufficient" $\alpha_2 \phi_{st}$, $\alpha_2 > 0$, must be present to obtain a stable algorithm. When applying SIP, or rather ILU/SIP(1.), all ϕ_t-, ϕ_{st}-, ϕ_{mt}- and ϕ_{nt}-terms are in fact eliminated from the modified equation and hence a so-called "temporal damping term" $\alpha_2 \phi_{st}$ must be added explicitly to make the algorithm uniformly stable in the hyperbolic region of the flow for all local flow directions. The factor α_2 is chosen of the form $C_1 + C_2(1 - M^2)$, where the C_i, $i = 1,2$, are kept as small as possible to provide a large local time step.

12. Experiments

All experiments will be performed for wings under mixed subsonic/supersonic (i.e transonic) flow conditions. Transonic flows are highly nonlinear in the vicinity of shocks, which are captured by the finite volume scheme as narrow zones of steep pressure gradients. Moreover, the positions of shocks are not known a priori and has to be found in the course of the solution process. This way, the experiments demonstrate the robustness of the algorithm for highly nonlinear mixed elliptic/hyperbolic problems. It will appear, however, that the asymptotic rate of convergence of the multigrid process is dominated by the subsonic regions of the flow. These are:

- the leading and trailing edge of the wing, where one mesh size is considerably larger than the other two ($\Delta\xi^2 \gg \Delta\xi^1, \Delta\xi^3$), especially in the far-field region outboard the wing tip (Fig. 1);
- the far-field region in grid planes perpendicular to the wing, where the mesh sizes can even have three different orders of magnitude.

In the experiments, the ILU/SIP parameter α is set to unity in hyperbolic regions of the flow, while $0 \leq \alpha \leq 1$ in elliptic regions. Either a three level or a four level multigrid method is used, employing weighted restriction of residuals, injection of potentials and bicubic prolongation [10,11]. A W-cycle multigrid strategy has been chosen, in which 12 relaxation sweeps are done on the coarsest grid, and 6 relaxation sweeps on the finer grids after each prolongation.

Firstly, a number of experiments will be presented that verify the local mode analysis presented in section 10. To this end, the transonic flow is computed about a simple swept back wing of constant chord (Fig. 4) at freestream Mach number $M_\infty = .84$ and angle of attack $\alpha = 3°$. The C-H topology grid contains 88*16*16 cells in the circumferential, wing-normal and spanwise directions. There are 56*7 cells adjacent to the wing. The multigrid method employs three gridlevels. In grid planes perpendicular to the wing, two different kinds of grids have been generated. Grid A (Fig. 4a) has the property $\Delta\xi^1 \cong \Delta\xi^3$ to a reasonable extent, while grid B (Fig. 4b) does not have this property at all and instead has computational cells with mesh sizes of three different orders of magnitude. Near the wing, both grids are approximately the same (Fig. 4d). In both cases, the grid spacing in the spanwise direction is uniform. Grid refinement in the spanwise direction will be investigated by doubling the number of cells in that direction. In such cases the grid will be referred to as AA and BB.

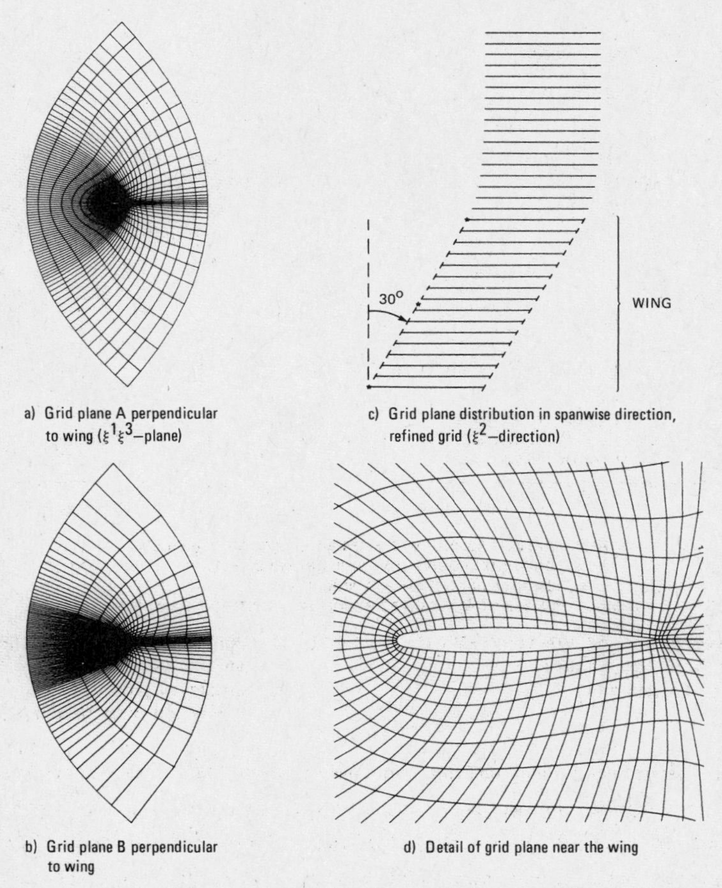

a) Grid plane A perpendicular to wing ($\xi^1\xi^3$–plane)

c) Grid plane distribution in spanwise direction, refined grid (ξ^2–direction)

b) Grid plane B perpendicular to wing

d) Detail of grid plane near the wing

Fig. 4 Computational grid used to verify the local mode analysis. Each airfoil section is an ONERA-D profile

Figure 5 shows the MG-ILU/SIP(α) convergence on grid A for several values of α in the elliptic region. Two convergence phases can be distinguished, viz. initial convergence and asymptotic convergence. The initial convergence is usually fast and corresponds to establishing the global characteristics of the flow. The asymptotic convergence, however, is in most cases much slower, because it is dominated by the cells which have the "worst" combination of mesh sizes in view of the local mode analysis presented in section 10. The convergence level obtained after the initial phase, usually about 1 - 1.5 orders of magnitude reduction in the residual, is in many cases sufficient for engineering applications. The figure also shows, that the asymptotic convergence is best for α = .70; for α = .35 and α = .0 respectively the method is asymptotically 18 % and 39 % slower. This result does not follow from the local mode analysis presented in section 10, but can possibly be explained by analyzing the convergence of a two-level multigrid cycle. The value α = 1. is generally not allowed in the elliptic region because of an insufficient relaxation of the boundary conditions.

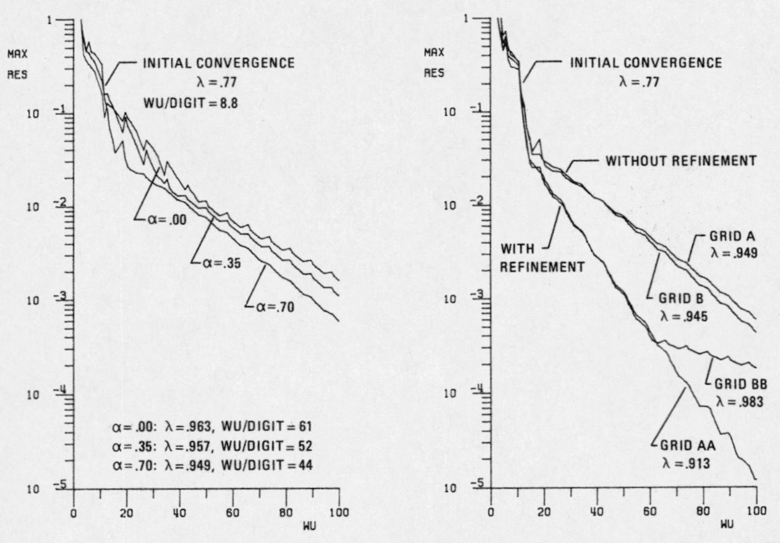

Fig. 5 Convergence of MG-ILU/SIP (α) for several values of α on grid A

Fig. 6 Convergence of MG-ILU/SIP (.7) on grids A,B and grids AA,BB (with spanwise refinement)

Figure 6 shows the effects of doubling the number of cells in the spanwise direction. The multigrid convergence on grids A and B is about the same. In both cases, the maximum residual is located at the trailing edge of the tip section, where the grids are similar. After doubling the number of cells in the spanwise direction, the convergence on grid AA becomes nearly twice as fast as on grid A. Such improved convergence was already predicted in section 10, figure 3. The result infers that doubling the resolution in the spanwise direction does not lead to an increase in computation time. The convergence on grid BB initially shows the same improvement as grid AA. However, at 60 work units there occurs a sudden slowdown of the asymptotic convergence rate. This happens as soon as the maximum residual, which was originally at the wing trailing edge, jumps to the coordinate cut in the far-field plane downstream. Here the cells have mesh sizes of three different orders of magnitude and consequently the convergence breaks down considerably, as was already predicted in section 10.

Secondly, results are shown for the transonic flow around the DFVLR-F4-wing at $M_\infty = .75$, $\alpha = .84$. The grid (Fig. 1) now has 176*32*32 (= 180224) cells, of which 112*19 are adjacent to the wing. In this case the multigrid method employs four grid levels. The computed flow solution and the MG-ILU/SIP convergence are shown in figure 7. The initial convergence is dominated in this case by the residual which is located at the extension of the leading edge outside the wing tip. Here $\Delta\xi^2 \cong 17\Delta\xi^1$, $\Delta\xi^1 \cong \Delta\xi^3$, which is indeed a very extreme aspect ratio, considering figure 3. Figure 7 also shows the convergence of the lift and the number of supersonic grid point (= size of the supersonic zone) to their final values. At about 46 work units, at only 1.4 digit reduction in the maximum residual, the lift and the supersonic zone have converged to well within 1 % of their final value, which is sufficient for engineering applications.

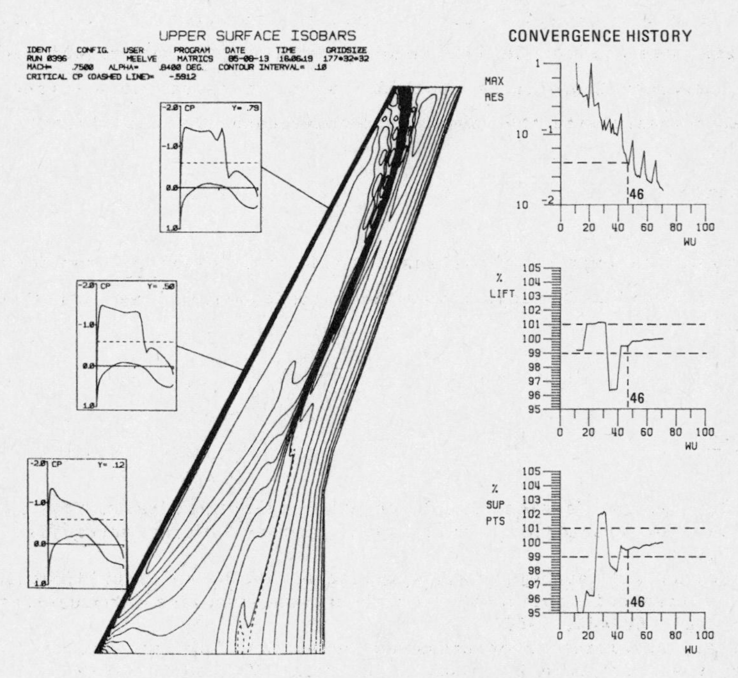

Fig. 7 Computed solution and convergence history for DFVLR-F4-wing at $M_\infty = .75$, $\alpha = .84$ on 176*32*32 C-H grid

13. Concluding remarks

The main conclusions of the research presented can be summarized as follows:

- In subsonic (elliptic) regions of the flow ILU/SIP was shown to be fairly insensitive to a wide set of variations of mesh sizes (section 10). The convergence generally improves with increasing ILU/SIP parameter α, as long as α is not chosen too close to one. A good choice is $\alpha = .70$ (section 12).
- In supersonic (hyperbolic) regions of the flow the ILU/SIP parameter α should be set to unity ($\alpha = 1.$). An explicit temporal damping term is required to obtain unconditional stability (section 11).
- The convergence of MG-ILU/SIP can be improved by performing a priori grid optimization in those regions of the grid where the flow is subsonic. It has been shown that computational cells having mesh sizes of three different orders of magnitude should be avoided. This requirement can be satisfied by using a grid which is reasonably square in grid planes perpendicular to the wing (sections 10, 12).
- In case the grid is reasonably square in grid planes perpendicular to the wing, grid refinement in the spanwise direction can improve the multigrid convergence considerably. Doubling the resolution in the spanwise direction seemingly does not have to lead to an increase in computation time (section 10, 12).
- The calculation of the transonic flow about a realistic wing indicates that MG-ILU/SIP needs only 1.4 order of magnitude reduction of the maximum residual to provide results for the lift and the size of the supersonic zone of engineering accuracy. Convergence to this engineering level of accuracy is relatively fast compared with the asymptotic convergence rate (section 12).

Acknowledgement

The author wishes to express his thanks to his colleagues J. van der Vooren and J.H. Meelker in the MATRICS project who, as a consequence, have contributed greatly to this paper.

14. References

[1] Boerstoel, J.W. and Kassies, A., Integrating multi-grid relaxation into a robust fast-solver for transonic potential flow around lifting aerofoils, AIAA Paper 83-1885, 1983.

[2] Shmilovich, A. and Caughey, D.A., Application of the multi-grid method to calculations of transonic potential flow about wing-fuselage combinations, J. Comp. Phys. 48, pp. 462-484, 1982.

[3] Jameson, A., Acceleration of transonic potential flow calculations on arbitrary meshes by the multiple grid method, AIAA Paper 79-1458 CP, 1979.

[4] Schmidt, W. and Jameson, A., Applications of multi-grid methods for transonic flow calculations, Lecture Notes in Mathematics 960, Multi-grid Methods, Proceedings Köln-Porz, 1981. Edited by W. Hackbush and U. Trottenberg, Springer-Verlag.

[5] Catherall, D., Optimum approximate factorisation schemes for 2D steady potential flows, AIAA Paper 81-1018-CP, 1981.
[6] Holst, T., Numerical solution of transonic wing flow fields, AIAA Paper 82-0105, 1982.
[7] South jr, J.C. and Hafez, M.M., Stability analysis of intermediate boundary conditions in approximate factorization schemes, AIAA Paper 83-1898-CP, 1983.
[8] Brédif, M., Finite element calculation of potential flow around wings, ONERA-TP-1984-068, 1984.
[9] Sankar, N.L., A multi-grid strongly implicit procedure for two-dimensional transonic potential flow problems, AIAA Paper 82-0931, 1982.
[10] Van der Wees, A.J., van der Vooren, J. and Meelker, J.H., Robust calculation of 3D transonic potential flow based on the non-linear FAS multi-grid method and incomplete LU decomposition, AIAA Paper 83-1950-CP, 1983.
[11] Van der Wees, A.J., Robust calculation of 3D transonic potential flow based on the non-linear FAS multi-grid method and a mixed ILU/SIP algorithm, Colloquium Topics in Numerical Analysis, J.G. Verwer (ed.), CWI Syllabus 5, 1985.
[12] Jameson, A. and Caughey, D.A., A finite volume method for transonic potential flow calculations, AIAA Paper 77-635-CP, 1977.
[13] Osher, S., Hafez, M.M. and Whithlow jr., W., Entropy condition satisfying approximations for the full potential equation of transonic flow, Math. of Comp., Vol. 44, Nr. 169, 1985.
[14] Stüben, K. and Trottenberg, U., Multigrid methods: fundamental algorithms, model problem analysis and applications, Lecture Notes in Mathematics, see [3].
[15] Brandt, A., Multi-level adaptive solutions to boundary value problems, Math. of Comp., Vol. 31, Nr. 138, 1977.
[16] Meyerink, J.A. and van der Vorst, H.A., An iterative solution method for linear problems of which the coefficient matrix is a symmetric M-matrix, Math. of Comp., Vol. 31, Nr. 137, 1977.
[17] Stone, H.L., Iterative solution of implicit approximations of multi-dimensional partial difference equations, SIAM J. Numer. Anal., Vol. 5, Nr. 3, 1968.
[18] Zedan, M. and Schneider, G.E., 3-D Modified strongly implicit procedure for finite difference heat conduction modelling, AIAA Paper 81-1136, 1981.
[19] Wesseling, P., A robust and efficient multigrid method, Lecture Notes in Mathematics, see [3].
[20] Kettler, R., Analysis and comparison of relaxation schemes in robust multigrid and preconditioned conjugate gradient methods, Lecture Notes in Mathematics, see [3].
[21] Jameson, A., Numerical solution of the three dimensional transonic flow over a yawed wing, AIAA Paper presented at the 1st AIAA Comp. Fluid, Dyn. Conf., Palm Springs, Cal., July 19-20, pp. 18-26, 1973.

Contents of GMD-Study no. 110:

Participants:

D. Albrecht	Technische Hochschule Darmstadt, West Germany
Ch. Arakawa	Universität Karlsruhe, West Germany
W. Arrenbrecht	Technische Hochschule Aachen, Inst. für Mechanik, West Germany
M.T. Arthur	Royal Aircraft Establishment, Farnborough Hants, UK
W. Auzinger	Technische Universität Wien, Austria
O. Axelsson	Catholic University of Nijmegen, The Netherlands
Herr Ballmann	Technische Hochschule Aachen, West Germany
V. Bake	Universität Münster, West Germany
R.E. Bank	University of California at San Diego, USA
S. de Barros	GMD-F1, St. Augustin, West Germany
Herr Bauer	Universität Stuttgart, West Germany
K. Becker	GMD-F1, St. Augustin, West Germany
P. Berger	Universität Stuttgart, West Germany
D. Bischoff	Universität Hannover, West Germany
P. Bjorstad	Det Norske Veritas, Oslo, Norway
H. Blum	Universität des Saarlandes, Saarbrücken, West Germany
H.G. Bock	Universität Bonn, West Germany
R. Böer	Kraftwerk Union AG, Erlangen, West Germany
R. Boyer	Universite de Provence, Marseille, France
B.J. Braams	Max-Planck-Institut für Plasmaphysik, Garching, West Germany
F. Brakhagen	GMD-F1, St. Augustin, West Germany
D. Braess	Ruhr-Universität Bochum, Math. Institut, West Germany
G. Brand	Universität Hannover, West Germany
K. Brand	GMD-F1, St. Augustin, West Germany
A. Brandt	Weizmann Institute of Science, Rehovot, Israel
U. Brockmeier	Ruhr-Universität Bochum, West Germany
O. McBryan	Courant Institute, New York, USA
Herr Burde	IABG, Ottobrunn, West Germany
N. Carmichael	Koninklijke/Shell, Rijswijk, The Netherlands
S. McCormick	University of Colorado at Denver, USA
V. Couaillier	ONERA, Chatillon, France
J. Currle	Universität Stuttgart, West Germany
K. Davstad	University of Stockholm, Sweden
B. Debus	GMD-F1, St. Augustin, West Germany
A. Delgado	Universität Essen, West Germany
O. Delgado	Universität Essen, West Germany
E. Dick	State University of Ghent, Belgium
B. Epstein	Israel Aircraft Industries, Lod, Israel
H. Finnemann	Kraftwerk Union AG, Erlangen, West Germany
F. Förtsch	Technische Hochschule Aachen, West Germany
Herr Fritsch	Universität Erlangen-Nürnberg, West Germany
L. Fuchs	Royal Institute of Technology, Stockholm, Sweden
U. Gärtel	Universität Köln, Math. Institut, West Germany
J. Genet	Universite de Pau, France
L. Geus	Universität Erlangen-Nürnberg, West Germany
N. Gluschitz	Pöring, West Germany

B. Görg	GMD-F2, St. Augustin, West Germany
W. Hackbusch	Universität Kiel, West Germany
H. Hahn	Technische Universität Braunschweig, West Germany
E. Halter	KfK/IDT, Karlsruhe, West Germany
D. Hänel	Technische Hochschule Aachen, West Germany
W. Heinrichs	Universität Düsseldorf, West Germany
P.W. Hemker	Stichting Mathematisch Centrum, Amsterdam, The Netherlands
G. Hofmann	Universität Kiel, West Germany
H. Holstein	University of Wales, Aberystwyth, UK
R.A. Hughes	University of Bristol, UK
R.K. Jain	GMD-F1, St. Augustin, West Germany
A. Jameson	Princeton University, USA
W. Joppich	GMD-F1, St. Augustin, West Germany
H. Kapitza	GKSS, Geesthacht, West Germany
H.B. Keller	CALTECH, Pasadena, USA
J. Kightley	Harwell, Oxford, UK
O. Kolp	GMD-F2, St. Augustin, West Germany
A. Kost	Technische Universität Berlin
H. Kriegel	Physikalisch Technische Bundesanstalt, Braunschweig, West Germany
N. Kroll	DFVLR, Braunschweig, West Germany
A. Kumar	DFVLR, Braunschweig, West Germany
C. Lacor	Vrije Universiteit, Brussels, Belgium
M. Lang	Bliestransbach, West Germany
E. Lewis	Bristol University, UK
H.M. Lidell	University of London, UK
J. Linden	GMD-F1, St. Augustin, West Germany
G. Lonsdale	University of Manchester, UK
R. Lorentz	GMD-F1, St. Augustin, West Germany
A. Luntz	Israel Aircraft Industries, Lod, Israel
J.F. Maitre	Ecole centrale de Lyon, France
J. Mandel	Charles University, Prague, CSSR
T. Merschen	IBM Deutschland, Böblingen, West Germany
J. Meyer	Universität Stuttgart, West Germany
H. Mierendorff	GMD-F2, St. Augustin, West Germany
H.D. Mittelmann	Arizona State University, Tempe AZ, USA
K. Morgan	University College of Swansea, UK
Z.P. Nowak	Universität Kiel, West Germany
K.-D. Oertel	GMD-F1, St. Augustin, West Germany
V. Pau	University of Bristol, UK
F. Le Piver	Commissariat a l'Energie Atomique, Villeneuve-St-Georges, France
P. Peisker	Ruhr Universität Bochum, Math. Institut, West Germany
J. Periaux	Avions Marcel Dassault, St. Cloud, France
A. Polster	Universität Erlangen-Nürnberg, West Germany
P. Puiseux	Societe Nationale ELF-Aquitaine, Pau, France
R. Rabenstein	Universität Erlangen-Nürnberg, West Germany
H.-J. Reinhardt	Battelle Institut, Frankfurt, West Germany
K. D. Reinartz	Universität Erlangen-Nürnberg, West Germany
A. Reusken	Rijksuniversiteit Utrecht, The Netherlands
U. van Rienen	Deutsches Elektronen Synchrotron, Hamburg, West Germany

R. Roche	Universite de Nancy I, France
D. Ron	Weizmann Institute of Science, Rehovot, Israel
S. Rothe	Hahn-Meitner Institut für Kernforschung, Berlin
U. Rüde	Technische Universität München, West Germany
J. Ruge	University of Colorado at Denver, USA
B. Ruttmann	GMD-F1, St. Augustin, West Germany
K.-Th. Schleicher	Technische Hochschule Darmstadt, West Germany
G.H. Schmidt	Koninklijke/Shell, Rijswijk, The Netherlands
C. Schneider	Universität Mainz, West Germany
W. Schröder	Technische Hochschule Aachen, West Germany
A. Schüller	GMD-F1, St. Augustin, West Germany
E.E. Schulman	Eta Systems Inc., Boulder CO, USA
M. Schulte	Universität Bielefeld, West Germany
H. Schütz	Technische Universität Berlin, West Germany
H. Schwichtenberg	GMD-F1, St. Augustin, West Germany
W. Seidl	Universität Erlangen-Nürnberg, West Germany
M. Smoch	Universität Münster, West Germany
K. Solchenbach	GMD-F1, St. Augustin, West Germany
S.P. Spekreijse	C.W.I, Amsterdam, The Netherlands
B. Steffen	Kernforschungsanlage Jülich, West Germany
H.J. Stetter	Technische Universität Wien, Austria
B. Stoufflet	Avion Marcel-Dassault, St Cloud, France
H. Ströll	Universität Erlangen-Nürnberg, West Germany
K. Stüben	GMD-F1, St. Augustin, West Germany
Sh. Ta'asan	ICASE, Hampton VA, USA
G. Thierauf	Universität Essen, West Germany
C.A. Thole	GMD-F1, St. Augustin, West Germany
U. Trottenberg	GMD-F1, St. Augustin, West Germany
P. Vassilevski	Bulgarian Academy of Sciences, Sofia, Bulgaria
R. Verfürth	Ruhr Universität Bochum, Math. Institut, West Germany
J. Volkert	Universität Erlangen-Nürnberg, West Germany
C. Vogt	GMD-F1, St. Augustin, West Germany
B. Wagner	Dornier GmbH, Friedrichshafen, West Germany
Herr Walter	Universität des Saarlandes, Saarbrücken, West Germany
A.J. van der Wees	National Aerospace Laboratory, Amsterdam, The Netherlands
P. Wesseling	Delft University of Technology, The Netherlands
D. Wessels	Universität Münster, West Germany
G. Winter	GMD-F1, St. Augustin, West Germany
G. Wittum	Universität Kiel, West Germany
K. Witsch	Universität Düsseldorf, West Germany
J.J. Wu	European Research Office, London, UK
H. Yserentant	Technische Hochschule Aachen, West Germany
Ch. Zenger	Technische Universität München, West Germany

Vol. 1062: J. Jost, Harmonic Maps Between Surfaces. X, 133 pages. 1984.

Vol. 1063: Orienting Polymers. Proceedings, 1983. Edited by J.L. Ericksen. VII, 166 pages. 1984.

Vol. 1064: Probability Measures on Groups VII. Proceedings, 1983. Edited by H. Heyer. X, 588 pages. 1984.

Vol. 1065: A. Cuyt, Padé Approximants for Operators: Theory and Applications. IX, 138 pages. 1984.

Vol. 1066: Numerical Analysis. Proceedings, 1983. Edited by D.F. Griffiths. XI, 275 pages. 1984.

Vol. 1067: Yasuo Okuyama, Absolute Summability of Fourier Series and Orthogonal Series. VI, 118 pages. 1984.

Vol. 1068: Number Theory, Noordwijkerhout 1983. Proceedings. Edited by H. Jager. V, 296 pages. 1984.

Vol. 1069: M. Kreck, Bordism of Diffeomorphisms and Related Topics. III, 144 pages. 1984.

Vol. 1070: Interpolation Spaces and Allied Topics in Analysis. Proceedings, 1983. Edited by M. Cwikel and J. Peetre. III, 239 pages. 1984.

Vol. 1071: Padé Approximation and its Applications, Bad Honnef 1983. Prodeedings. Edited by H. Werner and H.J. Bünger. VI, 264 pages. 1984.

Vol. 1072: F. Rothe, Global Solutions of Reaction-Diffusion Systems. V, 216 pages. 1984.

Vol. 1073: Graph Theory, Singapore 1983. Proceedings. Edited by K.M. Koh and H.P. Yap. XIII, 335 pages. 1984.

Vol. 1074: E.W. Stredulinsky, Weighted Inequalities and Degenerate Elliptic Partial Differential Equations. III, 143 pages. 1984.

Vol. 1075: H. Majima, Asymptotic Analysis for Integrable Connections with Irregular Singular Points. IX, 159 pages. 1984.

Vol. 1076: Infinite-Dimensional Systems. Proceedings, 1983. Edited by F. Kappel and W. Schappacher. VII, 278 pages. 1984.

Vol. 1077: Lie Group Representations III. Proceedings, 1982–1983. Edited by R. Herb, R. Johnson, R. Lipsman, J. Rosenberg. XI, 454 pages. 1984.

Vol. 1078: A.J.E.M. Janssen, P. van der Steen, Integration Theory. V, 224 pages. 1984.

Vol. 1079: W. Ruppert. Compact Semitopological Semigroups: An Intrinsic Theory. V, 260 pages. 1984

Vol. 1080: Probability Theory on Vector Spaces III. Proceedings, 1983. Edited by D. Szynal and A. Weron. V, 373 pages. 1984.

Vol. 1081: D. Benson, Modular Representation Theory: New Trends and Methods. XI, 231 pages. 1984.

Vol. 1082: C.-G. Schmidt, Arithmetik Abelscher Varietäten mit komplexer Multiplikation. X, 96 Seiten. 1984.

Vol. 1083: D. Bump, Automorphic Forms on GL (3,IR). XI, 184 pages. 1984.

Vol. 1084: D. Kletzing, Structure and Representations of Q-Groups. VI, 290 pages. 1984.

Vol. 1085: G.K. Immink, Asymptotics of Analytic Difference Equations. V, 134 pages. 1984.

Vol. 1086: Sensitivity of Functionals with Applications to Engineering Sciences. Proceedings, 1983. Edited by V. Komkov. V, 130 pages. 1984

Vol. 1087: W. Narkiewicz, Uniform Distribution of Sequences of Integers in Residue Classes. VIII, 125 pages. 1984.

Vol. 1088: A.V. Kakosyan, L.B. Klebanov, J.A. Melamed, Characterization of Distributions by the Method of Intensively Monotone Operators. X, 175 pages. 1984.

Vol. 1089: Measure Theory, Oberwolfach 1983. Proceedings. Edited by D. Kölzow and D. Maharam-Stone. XIII, 327 pages. 1984.

Vol. 1090: Differential Geometry of Submanifolds. Proceedings, 1984. Edited by K. Kenmotsu. VI, 132 pages. 1984.

Vol. 1091: Multifunctions and Integrands. Proceedings, 1983. Edited by G. Salinetti. V, 234 pages. 1984.

Vol. 1092: Complete Intersections. Seminar, 1983. Edited by S. Greco and R. Strano. VII, 299 pages. 1984.

Vol. 1093: A. Prestel, Lectures on Formally Real Fields. XI, 125 pages. 1984.

Vol. 1094: Analyse Complexe. Proceedings, 1983. Edité par E. Amar, R. Gay et Nguyen Thanh Van. IX, 184 pages. 1984.

Vol. 1095: Stochastic Analysis and Applications. Proceedings, 1983. Edited by A. Truman and D. Williams. V, 199 pages. 1984.

Vol. 1096: Théorie du Potentiel. Proceedings, 1983. Edité par G. Mokobodzki et D. Pinchon. IX, 601 pages. 1984.

Vol. 1097: R.M. Dudley, H. Kunita, F. Ledrappier, École d'Éte de Probabilités de Saint-Flour XII – 1982. Edité par P.L. Hennequin. X, 396 pages. 1984.

Vol. 1098: Groups – Korea 1983. Proceedings. Edited by A.C. Kim and B.H. Neumann. VII, 183 pages. 1984.

Vol. 1099: C.M. Ringel, Tame Algebras and Integral Quadratic Forms. XIII, 376 pages. 1984.

Vol. 1100: V. Ivrii, Precise Spectral Asymptotics for Elliptic Operators Acting in Fiberings over Manifolds with Boundary. V, 237 pages. 1984.

Vol. 1101: V. Cossart, J. Giraud, U. Orbanz, Resolution of Surface Singularities. Seminar. VII, 132 pages. 1984.

Vol. 1102: A. Verona, Stratified Mappings – Structure and Triangulability. IX, 160 pages. 1984.

Vol. 1103: Models and Sets. Proceedings, Logic Colloquium, 1983, Part I. Edited by G.H. Müller and M.M. Richter. VIII, 484 pages. 1984.

Vol. 1104: Computation and Proof Theory. Proceedings, Logic Colloquium, 1983, Part II. Edited by M.M. Richter, E. Börger, W. Oberschelp, B. Schinzel and W. Thomas. VIII, 475 pages. 1984.

Vol. 1105: Rational Approximation and Interpolation. Proceedings, 1983. Edited by P.R. Graves-Morris, E.B. Saff and R.S. Varga. XII, 528 pages. 1984.

Vol. 1106: C.T. Chong, Techniques of Admissible Recursion Theory. IX, 214 pages. 1984.

Vol. 1107: Nonlinear Analysis and Optimization. Proceedings, 1982. Edited by C. Vinti. V, 224 pages. 1984.

Vol. 1108: Global Analysis – Studies and Applications I. Edited by Yu.G. Borisovich and Yu.E. Gliklikh. V, 301 pages. 1984.

Vol. 1109: Stochastic Aspects of Classical and Quantum Systems. Proceedings, 1983. Edited by S. Albeverio, P. Combe and M. Sirugue-Collin. IX, 227 pages. 1985.

Vol. 1110: R. Jajte, Strong Limit Theorems in Non-Commutative Probability. VI, 152 pages. 1985.

Vol. 1111: Arbeitstagung Bonn 1984. Proceedings. Edited by F. Hirzebruch, J. Schwermer and S. Suter. V, 481 pages. 1985.

Vol. 1112: Products of Conjugacy Classes in Groups. Edited by Z. Arad and M. Herzog. V, 244 pages. 1985.

Vol. 1113: P. Antosik, C. Swartz, Matrix Methods in Analysis. IV, 114 pages. 1985.

Vol. 1114: Zahlentheoretische Analysis. Seminar. Herausgegeben von E. Hlawka. V, 157 Seiten. 1985.

Vol. 1115: J. Moulin Ollagnier, Ergodic Theory and Statistical Mechanics. VI, 147 pages. 1985.

Vol. 1116: S. Stolz, Hochzusammenhängende Mannigfaltigkeiten und ihre Ränder. XXIII, 134 Seiten. 1985.

Vol. 1117: D.J. Aldous, J.A. Ibragimov, J. Jacod, Ecole d'Été de Probabilités de Saint-Flour XIII – 1983. Édité par P.L. Hennequin. IX, 409 pages. 1985.

Vol. 1118: Grossissements de filtrations: exemples et applications. Seminaire, 1982/83. Edité par Th. Jeulin et M. Yor. V, 315 pages. 1985.

Vol. 1119: Recent Mathematical Methods in Dynamic Programming. Proceedings, 1984. Edited by I. Capuzzo Dolcetta, W.H. Fleming and T. Zolezzi. VI, 202 pages. 1985.

Vol. 1120: K. Jarosz, Perturbations of Banach Algebras. V, 118 pages. 1985.

Vol. 1121: Singularities and Constructive Methods for Their Treatment. Proceedings, 1983. Edited by P. Grisvard, W. Wendland and J.R. Whiteman. IX, 346 pages. 1985.

Vol. 1122: Number Theory. Proceedings, 1984. Edited by K. Alladi. VII, 217 pages. 1985.

Vol. 1123: Séminaire de Probabilités XIX 1983/84. Proceedings. Edité par J. Azéma et M. Yor. IV, 504 pages. 1985.

Vol. 1124: Algebraic Geometry, Sitges (Barcelona) 1983. Proceedings. Edited by E. Casas-Alvero, G.E. Welters and S. Xambó-Descamps. XI, 416 pages. 1985.

Vol. 1125: Dynamical Systems and Bifurcations. Proceedings, 1984. Edited by B.L.J. Braaksma, H.W. Broer and F. Takens. V, 129 pages. 1985.

Vol. 1126: Algebraic and Geometric Topology. Proceedings, 1983. Edited by A. Ranicki, N. Levitt and F. Quinn. V, 423 pages. 1985.

Vol. 1127: Numerical Methods in Fluid Dynamics. Seminar. Edited by F. Brezzi, VII, 333 pages. 1985.

Vol. 1128: J. Elschner, Singular Ordinary Differential Operators and Pseudodifferential Equations. 200 pages. 1985.

Vol. 1129: Numerical Analysis, Lancaster 1984. Proceedings. Edited by P.R. Turner. XIV, 179 pages. 1985.

Vol. 1130: Methods in Mathematical Logic. Proceedings, 1983. Edited by C.A. Di Prisco. VII, 407 pages. 1985.

Vol. 1131: K. Sundaresan, S. Swaminathan, Geometry and Nonlinear Analysis in Banach Spaces. III, 116 pages. 1985.

Vol. 1132: Operator Algebras and their Connections with Topology and Ergodic Theory. Proceedings, 1983. Edited by H. Araki, C.C. Moore, Ş. Strătilă and C. Voiculescu. VI, 594 pages. 1985.

Vol. 1133: K.C. Kiwiel, Methods of Descent for Nondifferentiable Optimization. VI, 362 pages. 1985.

Vol. 1134: G.P. Galdi, S. Rionero, Weighted Energy Methods in Fluid Dynamics and Elasticity. VII, 126 pages. 1985.

Vol. 1135: Number Theory, New York 1983–84. Seminar. Edited by D.V. Chudnovsky, G.V. Chudnovsky, H. Cohn and M.B. Nathanson. V, 283 pages. 1985.

Vol. 1136: Quantum Probability and Applications II. Proceedings, 1984. Edited by L. Accardi and W. von Waldenfels. VI, 534 pages. 1985.

Vol. 1137: Xiao G., Surfaces fibrées en courbes de genre deux. IX, 103 pages. 1985.

Vol. 1138: A. Ocneanu, Actions of Discrete Amenable Groups on von Neumann Algebras. V, 115 pages. 1985.

Vol. 1139: Differential Geometric Methods in Mathematical Physics. Proceedings, 1983. Edited by H. D. Doebner and J. D. Hennig. VI, 337 pages. 1985.

Vol. 1140: S. Donkin, Rational Representations of Algebraic Groups. VII, 254 pages. 1985.

Vol. 1141: Recursion Theory Week. Proceedings, 1984. Edited by H.-D. Ebbinghaus, G.H. Müller and G.E. Sacks. IX, 418 pages. 1985.

Vol. 1142: Orders and their Applications. Proceedings, 1984. Edited by I. Reiner and K. W. Roggenkamp. X, 306 pages. 1985.

Vol. 1143: A. Krieg, Modular Forms on Half-Spaces of Quaternions. XIII, 203 pages. 1985.

Vol. 1144: Knot Theory and Manifolds. Proceedings, 1983. Edited by D. Rolfsen. V, 163 pages. 1985.

Vol. 1145: G. Winkler, Choquet Order and Simplices. VI, 143 pages 1985.

Vol. 1146: Séminaire d'Algèbre Paul Dubreil et Marie-Paule Malliavin Proceedings, 1983–1984. Edité par M.-P. Malliavin. IV, 420 pages 1985.

Vol. 1147: M. Wschebor, Surfaces Aléatoires. VII, 111 pages. 1985.

Vol. 1148: Mark A. Kon, Probability Distributions in Quantum Statistical Mechanics. V, 121 pages. 1985.

Vol. 1149: Universal Algebra and Lattice Theory. Proceedings, 1984. Edited by S. D. Comer. VI, 282 pages. 1985.

Vol. 1150: B. Kawohl, Rearrangements and Convexity of Level Sets ir PDE. V, 136 pages. 1985.

Vol 1151: Ordinary and Partial Differential Equations. Proceedings 1984. Edited by B.D. Sleeman and R.J. Jarvis. XIV, 357 pages. 1985.

Vol. 1152: H. Widom, Asymptotic Expansions for Pseudodifferentia Operators on Bounded Domains. V, 150 pages. 1985.

Vol. 1153: Probability in Banach Spaces V. Proceedings, 1984. Editec by A. Beck, R. Dudley, M. Hahn, J. Kuelbs and M. Marcus. VI, 45 pages. 1985.

Vol. 1154: D.S. Naidu, A.K. Rao, Singular Pertubation Analysis o Discrete Control Systems. IX, 195 pages. 1985.

Vol. 1155: Stability Problems for Stochastic Models. Proceedings 1984. Edited by V.V. Kalashnikov and V.M. Zolotarev. VI, 44 pages. 1985.

Vol. 1156: Global Differential Geometry and Global Analysis 1984 Proceedings, 1984. Edited by D. Ferus, R.B. Gardner, S. Helgasor and U. Simon. V, 339 pages. 1985.

Vol. 1157: H. Levine, Classifying Immersions into $\mathbb{R}^4$ over Stable Maps of 3-Manifolds into $\mathbb{R}^2$. V, 163 pages. 1985.

Vol. 1158: Stochastic Processes – Mathematics and Physics. Proceedings, 1984. Edited by S. Albeverio, Ph. Blanchard and L. Streit. VI, 230 pages. 1986.

Vol. 1159: Schrödinger Operators, Como 1984. Seminar. Edited by S. Graffi. VIII, 272 pages. 1986.

Vol. 1160: J.-C. van der Meer, The Hamiltonian Hopf Bifurcation. VI, 115 pages. 1985.

Vol. 1161: Harmonic Mappings and Minimal Immersions, Montecatin 1984. Seminar. Edited by E. Giusti. VII, 285 pages. 1985.

Vol. 1162: S.J.L. van Eijndhoven, J. de Graaf, Trajectory Spaces, Generalized Functions and Unbounded Operators. IV, 272 pages. 1985.

Vol. 1163: Iteration Theory and its Functional Equations. Proceedings, 1984. Edited by R. Liedl, L. Reich and Gy. Targonski. VIII, 231 pages. 1985.

Vol. 1164: M. Meschiari, J.H. Rawnsley, S. Salamon, Geometry Seminar "Luigi Bianchi" II – 1984. Edited by E. Vesentini. VI, 224 pages. 1985.

Vol. 1165: Seminar on Deformations. Proceedings, 1982/84. Editec by J. Ławrynowicz. IX, 331 pages. 1985.

Vol. 1166: Banach Spaces. Proceedings, 1984. Edited by N. Kaltor and E. Saab. VI, 199 pages. 1985.

Vol. 1167: Geometry and Topology. Proceedings, 1983–84. Edited by J. Alexander and J. Harer. VI, 292 pages. 1985.

Vol. 1168: S.S. Agaian, Hadamard Matrices and their Applications. III, 227 pages. 1985.

Vol. 1169: W.A. Light, E.W. Cheney, Approximation Theory in Tensor Product Spaces. VII, 157 pages. 1985.

Vol. 1170: B.S. Thomson, Real Functions. VII, 229 pages. 1985.

Vol. 1171: Polynômes Orthogonaux et Applications. Proceedings, 1984. Edité par C. Brezinski, A. Draux, A.P. Magnus, P. Maroni et A. Ronveaux. XXXVII, 584 pages. 1985.

Vol. 1172: Algebraic Topology, Göttingen 1984. Proceedings. Editec by L. Smith. VI, 209 pages. 1985.